For a listing of recent titles in the *Artech House Universal Personal Communications Series,* turn to the back of this book.

WCDMA: Towards IP Mobility and Mobile Internet

WCDMA: Towards IP Mobility and Mobile Internet

Tero Ojanperä
Ramjee Prasad

Editors

Artech House
Boston • London
www.artechhouse.com

Library of Congress Cataloging-in-Publication Data
WCDMA : towards IP mobility and mobile internet / Tero Ojanperä, Ramjee Prasad,
    editors.
      p. cm. — (Artech House universal personal communications series)
    Includes bibliographical references and index.
    ISBN 1-58053-180-6 (alk. paper)
      1. Code division multiple access. 2. Wireless communication systems—Standards.
    3. Mobile communication systems—Standards. 4. Internet. I. Ojanperä, Tero.
    II. Prasad, Ramjee. III. Series.

    TK5103.452 .W395   2000   00-050812
    621.3845—dc21   CIP

British Library Cataloguing in Publication Data
WCDMA : towards IP mobility and mobile Internet.—(Artech
    House universal personal communications series)
    1. Code division multiple access 2. Wireless communication
    systems—Standards
    I. Ojanpera, Tero II. Prasad, Ramjee
    621.3'845

    ISBN 1-58053-180-6

Cover design by Igor Valdman

International Standard Book Number: 1-58053-180-6
Library of Congress Catalog Card Number: 00-050812

10 9 8 7 6 5 4 3 2 1

Books are the joy of life, and learning is a never-ending experience.

To my wife Tiina and to our sons Eerik and Elias
—Tero Ojanperä

To my wife Jyoti, to our daughter Neeli,
and to our sons Anand and Rajeev
—Ramjee Prasad

CONTENTS

CHAPTER 3   MOBILE INTERNET — 65

Tero Ojanperä

CHAPTER 4 RADIO OPERATING ENVIRONMENTS AND THEIR IMPACT ON SYSTEM DESIGN — 83

Tero Ojanperä and Ramjee Prasad

CHAPTER 6   WCDMA   171

Tero Ojanperä

CHAPTER 7   WCDMA TIME DIVISION DUPLEX   207

Tero Ojanperä, Antti Toskala, and Harri Holma

Seppo Hämäläinen, Harri Lilja, and Tero Ojanperä

Timo Eriksson and Tero Ojanperä

PREFACE

Since the first edition of this book *Wideband CDMA for Third Generation Mobile Communications* was published two years ago, enormous progress has been made in the development of third generation mobile networks. After the selection of WCDMA as a main approach in Japan and Europe in 1998, and cdma2000 and EDGE in the United States, there have been two direct sequence wideband CDMA proposals: asynchronous WCDMA and synchronous cdma2000. In 1999, however, the two proposals were harmonized, resulting in three modes: direct sequence based on WCDMA, multicarrier based on cdma2000, and time division duplex. These standards—as well as EDGE—have been now finalized and are termed release'99 standards.

Meanwhile, regulatory hurdles to third-generation implementation have been cleared. In many countries, the third-generation licenses have already been awarded using either "beauty contents" or an auction process. In WARC'2000, extension bands for third-generation networks were identified.

In addition, the standardization environment has changed dramatically. A new standardization body, the Third Generation Partnership Project (3GPP), was formed at the end of 1998 with a focus on the WCDMA air interface and GSM core network. Following the establishment of 3GPP, another partnership project, 3GPP2, was founded to develop specifications for cdma2000-based networks. In 2000, the EDGE standardization was transferred from ETSI to 3GPP. In addition, several new industrial interest groups have emerged. 3G.IP and the Mobile Wireless Internet Forum (MWIF) were initiated to drive development of Internet Protocol (IP)-based third-generation networks.

While waiting for the deployment of third-generation networks technology, second-generation networks have also developed further and the Internet has continued to develop at an unanticipated speed. Today there are more than 600 million mobile subscribers; this number may rise to one billion by 2002. Wireless Application Protocol (WAP) and i-Mode in Japan have already brought the Internet into pocket. The mobile Internet has become known to most of us—if not in practice, then from the wide media coverage.

As a result of the success of the Internet and developments in the mobile Internet, the

xvii

third-generation network has now been focused on IP based networks, termed all-IP networks or IP mobility networks. The goal is to specify a single, unified IP based core network to which different radio access networks can be connected. An all-IP approach is expected to result in increased flexibility and cost savings.

In addition to the core network development, the WCDMA, EDGE and cdma2000 air interfaces are developed further to provide higher data rates and better spectrum efficiency.

The changes in the standardization environment, the mobile Internet and all-IP networks are shaping the direction of development of third-generation networks and the whole wireless communications. This led us to the idea of revising the book and renaming it in order to better reflect the direction in which the industry is heading, *WCDMA: Towards IP Mobility and Mobile Internet.*

This book provides a comprehensive introduction to wideband CDMA and third-generation networks. It provides the technical background necessary to understand how wideband CDMA air interfaces are designed, starting from system requirements, applications, and radio environments, combined with a detailed treatment of technical solutions for spreading codes, coding, modulation, RAKE receiver, and soft handover. It also provides a review of WCDMA and cdma2000. Radio network planning is a key competence for network operators. This book gives an introduction to the art of network planning with problems and solutions specific to CDMA. The standardization and regulation environment for development of third-generation networks is very complex. This book provides insight into the structure and operations of different standardization bodies, industry interest groups, and regulatory bodies.

Recently, mobile Internet and all-IP (Internet Protocol)-based networks have emerged as cornerstones for the future growth of mobile communications. This book provides an introduction to both schemes with insight into the development of wireless communications concepts.

Figure P.1 illustrates the coverage of the book. Chapter 1 introduces basic definitions and the background of third-generation systems development. Chapter 2 explains the basic principles of CDMA. Chapter 3 introduces mobile Internet, IMT-2000 service targets and applications. Chapter 4 presents radio environments and their characteristics. Chapter 5 covers all the main aspects of wideband CDMA air interface design, including frame structures and physical channel design, spreading codes and their properties, and radio resource management aspects such as handover and power control. Chapters 6, 7, and 8 review WCDMA, WCDMA TDD, and cdma2000 air interfaces and their future evolution. In Chapter 9, radio performance of wideband CDMA, including spectrum efficiency and range, is discussed. Hierarchical cell structures (HCS) are discussed in Chapter 10. Chapter 11 describes the implementation of wideband CDMA mobiles and base stations. CDMA network planning is discussed in Chapter 12. Chapter 13 introduces the basic principles of IP. Chapter 14 presents network architectures, interfaces, and protocols for 3GPP, IS-41, and all-IP networks. Chapter 15 describes the standardization and regulation environment of third-generation networks. Finally, the future directions of the wireless multimedia communications networks are discussed in Chapter 16.

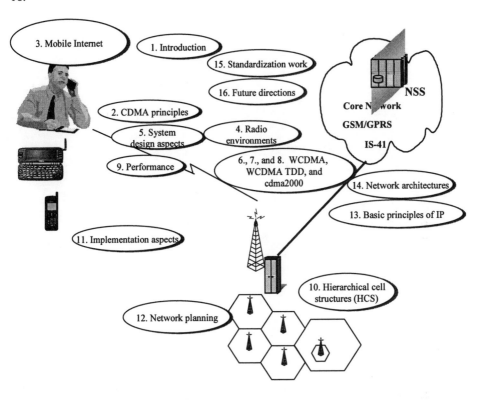

Figure P.1   Illustration of the coverage of this book.

This book is intended for everybody involved in the field of mobile radio systems. It provides different levels of material suitable for managers, researchers, system designers, and graduate students. The structure of the book facilitates two reading approaches. A general overview of wideband CDMA, third-generation systems, and IP, suitable for managers, can be obtained by reading Chapter 1; Chapter 2; Sections 3.1 to 3.6; Sections 4.1 to 4.3; Chapters 6, 7, and 8 (overview parts); Chapter 13; Chapter 14 selectively; and Chapters 15 and 16. A more thorough reader interested in technical details should read all sections. Section 3.11 provides a more mathematical view of applications covering traffic models. Sections 4.4–4.7 provide detailed propagation and mobility models. Chapter 5 contains an in-depth treatment of wideband CDMA system design aspects.

The views expressed in this book are those of the authors and do not represent the views of their employers.

ACKNOWLEDGMENTS

The material in this book originates from several projects at Nokia, Delft University of Technology (DUT), and Aalborg University (AAU) with a common goal of defining the third-generation of mobile communications. The research program to produce wideband CDMA specifications within Nokia dates back to the early 1990s when an experimental wideband CDMA testbed project was started. Tero was intensively involved with the start of CDMA system studies in 1993. Based on these studies, a project called CSS2000 was launched in 1994. Tero had the pleasure of being a project leader with an ambitious—at that time almost overwhelming—goal of producing a wideband CDMA air interface for UMTS. Later the research was expanded to a research program investigating several aspects such as 2-Mbps transmission, radio resource management, and packet aspects for UMTS. The book *CDMA for Wireless Personal Communications* (Artech House, 1996) by Ramjee proved to be a good starting point for this book. Ramjee initiated the CDMA research activities in DUT in 1989 and currently the Center for PersonKommunikation (CPK) of AAU is very active in the research and development of wireless IP-based CDMA systems.

In 1995, the FRAMES project brought together a number of talented individuals from several companies, laboratories, and universities, namely CNET, CSEM, Chalmers University of Technology, Ericsson, ETHZ, Instituto Superior Técnico, Nokia, Siemens, Roke Manor Research, The Royal Institute of Technology, DUT, the University of Kaiserslautern, and University of Oulu. The goal of FRAMES was to study and define a proposal for the UMTS air interface. FRAMES also created the seeds for this book by getting us acquainted with each other. We decided to put the results of various CDMA and third-generation related projects into a book format.

Several individuals contributed to the research and writing of this book. The efforts of the other contributors to this book—Timo Eriksson, Harri Holma, Seppo Hämäläinen, Martijn Kuipers, Harri Lilja, Carl Wijting, and Antti Toskala—are especially appreciated.

António Trindade from DUT helped to prepare the complete manuscript of the first edition. He also provided very constructive suggestions to improve the book and corrected numerous mistakes. Without his strong dedication it would not have been possible to complete the first edition of the book. Rajeev Prasad and Albena Mihovska of CPK helped to prepare the second edition manuscript, freeing us from an enormous editorial task.

Within Nokia we would like to thank the following individuals: Ari Hottinen, Chi-Zhun Honkasalo, Hannu Häkkinen, Kari Kalliojärvi, Ilkka Keskitalo, Jorma Lilleberg, Peter Muszynki, Kari Pehkonen, and Kari Rikkinen, who have been the technical forces behind several CDMA projects. We are deeply indebted to a number of individuals — Kwang-Cheng Chen, Stefano Faccin, George Fry, Steven Gray, Pertti "Bertil" Huuskonen, Markku Juntti, Matti Latva-aho, Tom Leskinen, Khiem Le, Pertti Lukander, Janne Parantainen, Riku Pirhonen, Harri Posti, Gordon Povey, Rauno Ruismäki, Tom Sexton, Alan Shu, and Markku Verkama — who helped to make this book possible by providing helpful comments and improvements. We also wish to thank Nokia management, especially Dr. Heikki Huomo, for the opportunity to write this book. In addition, Tero wishes to thank Dr. J. T. Bergqvist for the encouragement during this project. Furthermore, we appreciate the support of AAU management. The FRAMES project was one of the projects that allowed UMTS to happen. We would like to thank colleagues from the companies, laboratories, and universities participating in FRAMES.

Tero wishes to thank his dear friends Pekka Koponen and Kari Pulli for their support and constructive comments. In addition, he wants to thank his mother Maiju, brother Aki, mother-in-law Aila, and father-in-law Markku Kesti for their encouragement and support while he was writing this book.

We hope this book will help to explain the rationale behind UMTS wideband CDMA and the other wideband CDMA air interfaces, and third-generation systems in general, including all-IP-based mobile networks.

Tero Ojanperä
Ramjee Prasad
January 2001

Chapter 1

INTRODUCTION

Recently, extensive investigations have been carried out into the application of a code division multiple access (CDMA) system as an air interface multiple access scheme for IMT-2000 / UMTS (International Mobile Telecommunications System 2000 / Universal Mobile Telecommunications System). Many research and development (R&D) projects in the field of wideband CDMA have been going on in Europe, Japan, the United States, and Korea [1–4]. It has become evident that wideband CDMA is the appropriate answer to the question: "What will be the multiple access scheme for IMT-2000 / UMTS?"

1.1 MULTIPLE ACCESS

The basis for any air interface design is how the common transmission medium is shared between users, that is, the multiple access scheme. The basic multiple access schemes are illustrated in Figure 1.1. In frequency division multiple access (FDMA), the total system bandwidth is divided into frequency channels that are allocated to the users. In time division multiple access (TDMA), each frequency channel is divided into time slots and each user is allocated a time slot. In CDMA, multiple access is achieved by assigning each user a pseudo-random code (also called pseudo-noise codes due to noise-like autocorrelation properties) with good auto- and crosscorrelation properties. This code is used to transform a user's signal into a wideband spread spectrum signal. A receiver then transforms this wideband signal into the original signal bandwidth using the same pseudo-random code. The wideband signals of other users remain wideband signals. Possible narrowband interference is also suppressed in this process. TDMA and CDMA usually use FDMA to divide the frequency band into smaller frequency channels, which are then divided in a time or code division fashion.

In addition to FDMA, TDMA, and CDMA, orthogonal frequency division multiplexing (OFDM), a special form of multicarrier modulation, and space division multiple access (SDMA) are used as multiple access schemes in wireless

1

communications. In OFDM, densely spaced subcarriers with overlapping spectra are generated using fast Fourier transform (FFT) [4]. Signal waveforms are selected in such a way that the subcarriers maintain their orthogonality despite the spectral overlap. Usually, CDMA or TDMA is used with OFDM to achieve multiple access capability. SDMA separates the users spatially, typically using beam-forming techniques. It can be applied to all other multiple access schemes.

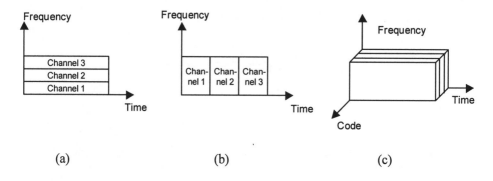

Figure 1.1   Multiple access schemes: (a) FDMA, (b) TDMA, and (c) CDMA.

1.1.1   Classification of CDMA

There are several ways to classify CDMA schemes. The most common is the division based on the modulation method used to obtain the wideband signal. This division leads to three types of CDMA: direct sequence (DS), frequency hopping (FH), and time hopping (TH) as illustrated in Figure 1.2 [1]. In DS-CDMA, spectrum is spread by multiplying the information signal with a pseudo-noise sequence, resulting in a wideband signal. In the frequency hopping spread spectrum, a pseudo-noise sequence defines the instantaneous transmission frequency. The bandwidth at each moment is small, but the total bandwidth over, for example, a symbol period is large. Frequency-hopping can either be fast (several hops over one symbol) or slow (several symbols transmitted during one hop). In the time-hopping spread spectrum, a pseudo-noise sequence defines the transmission moment. Furthermore, combinations of these techniques are possible. In this book we focus on DS-CDMA because it is the technique used for third-generation wideband CDMA proposals. Wideband CDMA is defined as a direct sequence spread spectrum multiple access scheme where the information is spread over a bandwidth of approximately 5 MHz or more. A detailed classification of CDMA is discussed in Chapter 2.

Frequency

Direct
sequence

Frequency
hopping

Time
hopping

Time

Figure 1.2   Direct sequence, frequency-hopping, and time-hopping CDMA.

1.2   CDMA: PAST, PRESENT, AND FUTURE

The origins of spread spectrum are in the military field and navigation systems. Techniques developed to counteract intentional jamming have also proved suitable for communication through dispersive channels in cellular applications. In this section we highlight the milestones in CDMA development starting from the 1950s after the invention of the Shannon theorem [5]. An extensive overview of spread spectrum history is given in [6].

In 1949 John Pierce wrote a technical memorandum where he described a multiplexing system in which a common medium carries coded signals that need not be synchronized. This system can be classified as a time-hopping spread spectrum multiple access system [6]. Claude Shannon and Robert Pierce introduced the basic ideas of CDMA in 1949 by describing the interference averaging effect and the graceful degradation of CDMA [7]. In 1950 De Rosa-Rogoff proposed a direct sequence spread spectrum system and introduced the processing gain equation and noise multiplexing idea [6]. In 1956 Price and Green filed for the antimultipath "RAKE" patent [6]. Signals arriving over different propagation paths can be resolved by a wideband spread spectrum signal and combined by the RAKE receiver. The near-far problem (i.e., a high interference overwhelming a weaker spread spectrum signal) was first mentioned in 1961 by Magnuski [6].

The cellular application spread spectrum was suggested by Cooper and Nettleton in 1978 [8]. During the 1980s, Qualcomm investigated DS-CDMA techniques, which finally led to the commercialization of cellular spread spectrum communications in the form of the narrowband CDMA IS-95 standard in July 1993. Commercial operation of IS-95 systems started in 1996. Multiuser detection (MUD) has been subject to extensive research since 1986 when Verdu formulated an optimum multiuser detection for the additive white Gaussian noise (AWGN) channel, maximum likelihood sequence estimator (MLSE) [9].

During the 1990s, wideband CDMA techniques with a bandwidth of 5 MHz or more were studied intensively throughout the world, and several trial systems were built and tested [2]. These include the FRAMES FMA2 (FRAMES Multiple Access) wideband CDMA scheme in Europe, Core-A in Japan, harmonized European/Japanese WCDMA, cdma2000 in the United States, and the TTA I and TTA II (Telecommunication Technology Association) schemes in Korea. In 1999 wideband CDMA proposals were harmonized resulting in three modes: direct sequence based on WCDMA, multicarrier, and time division duplex. Commercial opening of third-generation wireless communication systems using wideband CDMA is expected in Japan in 2001 and in many other parts of the world in 2002.

Based on the above description, the CDMA era is divided in three periods: (1) the pioneer CDMA era, (2) the narrowband CDMA era, and (3) the wideband CDMA era, as shown in Table 1.1.

Table 1.1
CDMA Era

Pioneer Era

1949	John Pierce: time-hopping spread spectrum
1949	Claude Shannon and Robert Pierce: basic ideas of CDMA
1950	De Rosa-Rogoff: direct sequence spread spectrum
1956	Price and Green: antimultipath "RAKE" patent
1961	Magnuski: near-far problem
1970s	Several developments for military field and navigation systems

Narrowband CDMA Era

1978	Cooper and Nettleton: cellular application of spread spectrum
1980s	Investigation of narrowband CDMA techniques for cellular applications
1984	DS-CDMA and hybrid CDMA/FDMA proposed as GSM multiple access
1986	Formulation of optimum multiuser detection by Verdu
1993	IS-95 standard

Wideband CDMA Era

1995	Europe : FRAMES FMA2 ⎫ Japan : Core-A ⎬ WCDMA USA : cdma2000 ⎭ Korea : TTA I, TTA II
1999	Harmonization of wideband CDMA proposals resulting in three modes: direct sequence, multicarrier, and time division duplex
2000s	Commercialization of wideband CDMA systems

1.3 MOBILE CELLULAR ERA

In 1980 the mobile cellular era started, and since then mobile communications have undergone significant changes and experienced enormous growth. First generation mobile systems using analog transmission for speech services were introduced in the 1980s. Several standards were developed: AMPS (Advanced Mobile Phone Service) in the United States, TACS (Total Access Communication System) in the United

Kingdom, NMT (Nordic Mobile Telephones) in Scandinavia, NTT (Nippon Telephone and Telegraph) in Japan, and so on.

Second-generation systems using digital transmission were introduced in the late 1980s. They offer higher spectrum efficiency, better data services, and more advanced roaming than the first-generation systems. GSM (Global System for Mobile Communications), PDC (Personal Digital Cellular), IS-136 (D-AMPS), and IS-95 (US CDMA system) belong to the second-generation systems. The services offered by these systems cover speech and low bit rate data.

The second-generation systems will further evolve towards third-generation systems and offer more advanced services such as bit rates of 100 to 200 Kbps for circuit and packet switched data. These evolved systems are commonly referred to as generation 2.5. Third-generation systems will offer even higher bit rates, more flexibility, simultaneous multiple services for one user, and services with different quality of service classes.

Figure 1.3 shows the number of subscribers for the first-, second-, and third-generation digital systems from 1995 to 2005. Today, the largest system is GSM with over 300 million subscribers and with a predicted growth of over 700 million by 2002 (i.e., over 50% of the total subscribers). From 2001 onwards, the WCDMA will grow rapidly.

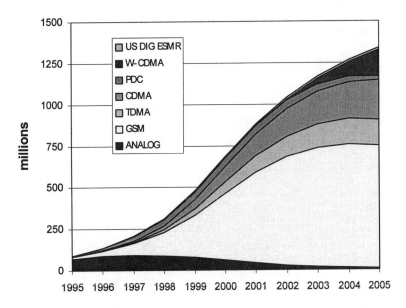

Figure 1.3   The number of subscribers for the first- and second-generation systems. Note that data from 2000 to 2005 are estimates. (*Source:* Nokia.)

Figure 1.4 depicts the penetration rates of mobile cellular radio in different countries. Since the introduction of digital cellular systems in 1992, growth has been enormously fast. The highest penetration (currently over 70%) is in Finland. High penetration rates require high-capacity networks and large amounts of frequency spectrum. This enormous growth of the second-generation systems paves the way for the introduction of third-generation networks. More and more people will become familiar with wireless access available anywhere, anytime.

Figure 1.4   Penetration of mobile radio systems. (*Source:* World Bank EMC, March 2000.)

Table 1.2 shows the technical parameters of second-generation systems. The details of the IS-95 standard are discussed in more detail in Chapter 2. All these systems are frequency division duplex (FDD) systems: they transmit and receive in different frequency bands. Time division duplex (TDD) systems are discussed in Chapter 7. In the table, the maximum data rate is the data rate the standard currently supports. The actual data rate available in commercial systems is usually much smaller. In 2000 GSM supports 43.2 Kbps, IS-95 14.4 Kbps (64 Kbps in some parts of Korea), IS-136 9.6 Kbps, and PDC 9.6 Kbps. For a more detailed discussion of these standards, refer to [10].

Table 1.2
Second-Generation Digital Systems

	GSM	IS-136	IS-95	PDC
Multiple access	TDMA	TDMA	CDMA	TDMA
Modulation	GMSK[a]	π/4-DQPSK[b] Coherent π/4-DQPSK Coherent 8-PSK	QPSK/ O-QPSK[c]	π/4-DQPSK
Carrier spacing	200 kHz	30 kHz	1.25 MHz	25 kHz
Carrier bit rate	270.833 Kbps	48.6 Kbps (π/4-PSK and π/4-DQPSK) 72.9 Kbps (8-PSK)	1.2288 Mchip/s[d]	42 Kbps
Frame length	4.615 ms	40 ms	20 ms	20 ms
Slots per frame	8/16	6	1	3/6
Frequency band (uplink/ downlink) (MHz)	880-915 /935-960 1720-1785 / 1805-1880 1930-1990 / 1850-1910	824-849 / 869-894 1930-1990 / 1850-1910	824-849 / 869-894 1930-1990 / 1850-1910	810-826 / 940-956 1429-1453 / 1477-1501
Speech codec	RPE-LTP[e] 13 Kbps Half rate 6.5 Kbps Enhanced full rate (EFR) 12.2 Kbps	VSELP[f] 8 Kbps IS-641-A: 7.4 Kbps ACELP)[g] US1: 12.2 Kbps (ACELP)	QCELP 8 Kbps CELP 8 Kbps CELP 13 Kbps	VCELP 6.7 Kbps
Maximum possible data rate	HSCSD:115.2 Kbps GPRS: 115.2 – 182.4 Kbps (depending on the coding)	IS-136+: 43.2 Kbps	IS95A: 14.4 Kbps IS95B: 115.2 Kbps	28.8 Kbps
Frequency hopping	Yes	No	N/A	No
Handover	Hard	Hard	Soft	Hard

[a] Gaussian minimum shift keying.
[b] Differential quadrature phase shift keying.
[c] Offset QPSK.
[d] A "chip" is used to denote a spread symbol in DS-CDMA systems.
[e] Regular pulse excited long term prediction.
[f] Vector sum excited linear predictive.
[g] Algebraic code excited linear predictive.

1.3.1   GSM

Table 1.3 depicts the time schedule of GSM development. The first step towards GSM was the allocation of a common frequency band in 1978, twice 25 MHz, at around 900 MHz for mobile communications in Europe [11]. The actual development of the GSM standard started in 1982 when Conference Européenne des Postes et Télécommunications (CEPT) established Groupe Special Mobile (the origin of the term GSM, which now means Global System for Mobile Communications) to develop a future Pan-European cellular mobile network. In 1984 three working parties (WP) were established to define services (WP1), radio interface (WP2), and other issues such as transmission, signaling protocols, interfaces, and network architecture (WP3). In 1986, a so-called permanent nucleus (PN), consisting of a team of full-time members, was established to continuously coordinate the GSM specification work. The European Telecommunications Standards Institute (ETSI) was established 1988, and GSM became an ETSI technical committee Special Mobile Group (TC SMG), the working parties became subtechnical committees (STCs), and the permanent nucleus became the GSM project team number 12 (PT12). In 2000, the GSM standardization was transferred into a 3GPP technical subgroup GSM/EDGE Radio Access Network (TSG GERAN). For a description of the 3GPP and TSG GERAN structure, refer to Chapter 15.

Table 1.3
GSM Development Time Schedule

1982	Groupe Special Mobile established within CEPT
1984	Several proposals for GSM multiple access: wideband TDMA, narrowband TDMA, DS-CDMA, hybrid CDMA/FDMA, narrowband FDMA
1986	Eight prototype systems tested in CNET laboratories in France Permanent nucleus is set up
1987	Basic transmission principles selected: 8-slot TDMA, 200-kHz carrier spacing, frequency hopping
1987	MoU signed
1988	GSM becomes an ETSI technical committee
1990	GSM phase 1 specifications frozen (drafted 1987–1990) GSM1800 standardization begins
1991	GSM1800 specifications are frozen
1992	GSM900 commercial operation starts
1992	GSM phase 2+ development starts
1995	GSM submitted as a PCS technology candidate to the United States
1995	PCS1900 standard adopted in the United States
1996	Enhanced full rate (EFR) speech codec standard ready
1996	14.4-Kbps standard ready GSM1900 commercial operation starts
1997	HSCSD standard ready GSM cordless system (home base station) standardization started EDGE standardization started
1998	GPRS standard ready WCDMA selected as the third-generation air interface
1999	EDGE standard ready
2000	GSM standardization moved to 3GPP and TSG GERAN established

In 1990 the GSM specifications for 900 MHz were frozen. In 1990 it was decided that GSM1800[1], the high-frequency version of GSM, was to be standardized, and the specifications were completed by 1991. The official commercial launch of GSM was in 1992. Because the GSM standardization process could not be completed for all aspects before the service launch, it was decided to divide the specifications into phases: phase 1 and phase 2 [11]. The evolution needs went beyond even these two phases, so a

---

[1] GSM1800 was originally termed DCS1800 (Digital Cellular System 1800).

workshop held in 1992 in Finland launched the idea for the phase 2+ standardization. The phase 2+ features are updated on a regular basis according to market needs and availability of specification. So, the original standard was future proof and capable of evolution. The standardization process for the U.S. equivalent, PCS1900 (Personal Communication System, now also called GSM1900), began in 1992. The GSM1900 service was launched in 1996.

A Memorandum of Understanding (MoU) was signed by telecommunication network operator organizations in 1987. The GSM MoU, renamed as the GSM Association, covers areas such as time scales for the procurement and deployment of the system, compatibility of numbering and routing plans, and concerted service introduction [11]. Today, there are over 470 GSM Association members with 372 live networks in 142 countries (August 2000). From its European origin, GSM has grown into a truly global system, and in 1997 the second largest GSM market after Italy was China.

1.3.1.1   GSM Air Interface Development

In the beginning of GSM air interface development, several multiple access schemes were proposed. The main alternatives were single channel per carrier FDMA, narrowband TDMA, wideband TDMA, and slow frequency-hopped (SFH) CDMA [12]. Comparative testing of eight prototypes (four from France and West Germany and four from Scandinavia) was carried out in Centre National d'Études des Télecommunication (CNET) laboratories in France during 1986. In 1987, based on the comparison results, the following basic principles were adopted for the GSM radio interface:

- Eight channels per carrier;
- 200-kHz carrier bandwidth;
- Slow frequency hopping.

Ten years later, in 1997, a similar evaluation process was repeated in the UMTS air interface definition process. As a result, the conclusions expressed in 1985 that "DS/CDMA systems have no chance at all to be adopted for the European system" [12] were reversed, and the WCDMA was selected for the UMTS air interface.

1.3.1.2   GSM Phase 2+

The services offered by the evolved second-generation systems will be the reference for third-generation systems. The third-generation systems have to exceed the capabilities of the second-generation systems to be commercially viable. In this section we discuss the main features of GSM phase 2+ in order to better understand what services are in the market in 2000. The most important standardized GSM phase 2+ work items from the radio access system point of view have been:

- Enhanced full rate speech codec (EFR) [13];
- Adaptive multirate codec (AMR);

- 14.4-Kbps data service [14];
- High speed circuit switched data (HSCSD);
- General Packet Radio Service (GPRS) [15];
- Enhanced data rates for global evolution (EDGE).

The EFR speech codec was first developed for the U.S. market and subsequently adopted by ETSI. It offers landline quality speech using 12.2 Kbps for speech coding and 10.6 Kbps for error protection, resulting in 22.8 Kbps [13]. AMR further improves the GSM speech service and takes advantage of changing channel conditions. Thus, radio resources can be used more efficiently while maintaining good speech quality.

The 14.4-Kbps data service provides higher throughput per time slot by puncturing (i.e., by deleting certain bits) the original rate half channel coding. HSCSD offers circuit switched data services from 9.6 Kbps with one slot up to 115.2 Kbps using eight slots. The HSCSD standard was completed in 1997, and the first infrastructure and terminal products were introduced in 1999.

General Packet Radio Service provides the mobile user with a connectionless packet access to data networks. The most important application of GPRS could be access to applications like the World Wide Web, e-mail, and Telnet. The short call setup time will facilitate transaction type and telemetric applications, like toll road systems and credit card validation. The GPRS air interface is based on the same principles as the current GSM system. In addition, new radio link control (RLC) and medium access control (MAC) layer functions allow flexible retransmission capabilities needed for packet access. GPRS offers four different bit rates per time slot (9.06, 13.4, 15.6, and 21.4 Kbps) depending on the applied channel coding. Link adaptation selects among the four possible rates the bit rate that would result in the highest throughput according to channel and interference conditions. The GPRS standard was finalized in 1998, and commercial deployment started in 2000.

The GSM evolution towards higher bit rates is further boosted by the enhanced data rates for GSM/global evolution (EDGE) concept, which has also been selected for IS-136 based third-generation systems, as discussed in Section 1.6.2. Also, more spectrum-efficient detection methods are currently under study. Joint detection detects the desired signal and dominant co-channel interferer jointly, thus improving the performance [16].

The GSM home base station concept is under standardization as well. It will offer cordless access using a standard GSM phone, thus removing the need for dual mode GSM/cordless terminals.

For more details on GSM phase 2+ and on GSM's evolution towards third-generation, refer to [17].

1.3.2   US-TDMA (IS-54/IS-136)

In September 1988, the Cellular Telecom Industry Association (CTIA) published requirements for the next generation of cellular technology in the United States. The first standard that was adopted was time division multiple access, called US-TDMA.[2] In January 1989, the TIA/CTIA (Telecommunications Industry Association/Cellular Telecom Industry Association) selected TDMA over FDMA technology. In 1995 an upbanded version of the US-TDMA was adopted for the US PCS frequency band. The US-TDMA standard with an analog control channel is termed IS-54, and the US-TDMA with digital control channel is termed IS-136, which is now the commonly used term when referring to the US-TDMA. The TIA technical committee TR45.3 is responsible for IS-136 standardization. Today, IS-136 is deployed in North and South America, and in some parts of Asia.

The IS-136 is a dual-mode standard, since it specifies both analog (AMPS) and digital (US-TDMA) modes. Due to backward compatibility to AMPS, the US-TDMA carrier spacing is 30 kHz.

The Universal Wireless Communications Consortium (UWCC) was established in 1996 to support IS-136 standards. In 1997, within the UWCC, the Global TDMA Forum set up the IS-136+ program aiming to implement higher bit rates within the 30-kHz channel. The TIA TR45.3 standardization group adopted IS-136+ in 1998. IS-136+ has two modulation schemes: coherent π/4-QPSK and 8-PSK. The control channel uses π/4-DQPSK modulation. The user information bit rates for the π/4-QPSK modulation are 9.6 Kbps (single rate), 19.2 Kbps (double rate), and 28.8 Kbps (triple rate); and for the coherent 8-PSK modulation, 14.4 Kbps (single rate), 28.8 Kbps (double rate), and 43.2 Kbps (triple rate).

1.3.3   IS-95

During the 1980s, Qualcomm carried out an in-house development and validation of a direct sequence CDMA air interface concept. In November 1989, the first IS-95 narrowband CDMA field trials were carried out using two cell sites and one mobile station in San Diego [18]. In September 1990, Qualcomm released the first version of its CDMA "Common Air Interface" Specification. On December 5, 1991, at CTIA's "Presentation of the Results of the Next Generation Cellular Field Trials," CDMA field trial results were presented. On January 6, 1992, CTIA requested that TIA should start preparing for CDMA standardization. Following that, in March 1992 a new subcommittee, TR45.5, was formed to develop the IS-95 standard. IS-95 was completed by 1993 and revised in 1995 (IS-95A). An up-banded version of IS-95 for the 1.9-GHz PCS frequency band was standardized in 1995.

IS-95 is also a dual-mode standard with an AMPS analog mode. In designing the IS-95 system, the AMPS backward compatibility was taken into account. For example, a handover from the digital to the analog mode is possible.

Further development of IS-95 towards higher bit rate services was started in 1996. This led to the completion of the IS-95B standard in 1998 [19]. While the IS-95A standard uses only one spreading code per traffic channel, IS-95B can concatenate up to eight codes for the transmission of higher bit rates. This provides a maximum data rate of 115.2 Kbps. For more details on IS-95, see Chapter 2.

---

[2] US-TDMA is also referred as D-AMPS (Digital AMPS).

CDMA Development Group (CDG) was established to promote IS-95 CDMA systems in 1994. Today, the main IS-95 markets are the United States, Japan, and Korea, the latter being the largest market with over 25 million subscribers. The success of IS-95 in Korea is based on the adoption of IS-95 as a national standard in the early 1990s. A joint effort by a number of companies under the Electronics and Telecommunications Research Institute (ETRI) led to the fast development of equipment and deployment of commercial networks.

1.3.4   Personal Digital Cellular[3]

Based on the work of a study committee organized by the Ministry of Posts and Telecommunications (MPT), the requirements for PDC were defined in 1990. In May 1990, the Japanese Research and Development Center for Radio Systems (RCR) standards committee organized subcommittees for the standardization of PDC [20]. The commercial service was started by NTT in 1993 in the 800-MHz band and in 1994 in the 1.5-GHz band. Recently, packet data has been developed for the PDC standard [21].

The PDC system, originally known as JDC (Japanese Digital Cellular) is only operated in Japan. RCR TSD27 is a common air interface specification for PDC, and it has some similarities with the American IS-54 standard. PDC operates in two frequency bands, 800 MHz and 1.5 GHz.

In 1999, NTT Docomo introduced wireless Internet access, termed i-mode, for PDC. Currently, all new terminals include i-mode, which has greatly accelerated the growth of the PDC system.

1.4   THIRD-GENERATION SYSTEMS

1.4.1   Objectives and Requirements

Emerging requirements for higher rate data services and better spectrum efficiency are the main drivers identified for the third-generation mobile radio systems. In the International Telecommunications Union (ITU), third-generation networks are called IMT-2000, and in Europe, UMTS. Since 1985, the ITU has been developing IMT-2000, previously termed Future Public Land Mobile Telephone System (FPLMTS). In ETSI, UMTS standardization started in 1990 when subtechnical committee SMG5 was established. The main objectives for the IMT-2000 air interface can be summarized as:

- Full coverage and mobility for 144 Kbps, preferably 384 Kbps;
- Limited coverage and mobility for 2 Mbps;
- High spectrum efficiency compared to existing systems;
- High flexibility to introduce new services.

---

[3] Sometimes PDC is also referred as Pacific Digital Cellular.

The bit rate targets have been specified according to the integrated services digital network (ISDN) rates. The 144-Kbps data rate provides the ISDN 2B+D channel, 384 Kbps provides the ISDN H0 channel, and 1.92 Mbps provides the ISDN H12 channel.[4] It may be, however, that the main IMT-2000 services are not ISDN-based services. It has to be noted that these figures have been subject to considerable debate. Ultimately, market demand will determine what data rates will be offered in commercial systems. Figure 1.5 describes the relation between bit rates and mobility for the second- and third-generation systems.

The targets of third-generation systems are wide, and depending on the main driver, system solutions will be different. The maturity of second-generation mobile radio systems varies, ranging from over 60% penetration in Scandinavia to a very low penetration in developing countries where cellular systems are in the beginning of their life cycle. Therefore, it is clear that the need to develop a new system varies, and the different views and needs may result in several different variants of IMT-2000. In addition, different backward compatibility requirements influence the technology applied to third-generation systems.

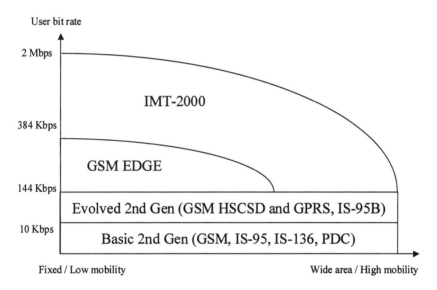

Figure 1.5   IMT-2000 user bit rate versus coverage and mobility.

1.4.2   IMT-2000 Air Interface Selections

Wideband CDMA and EDGE are the main third-generation technologies. The fast development during recent years has been due to the Japanese initiative. In the beginning of 1997, the Association for Radio Industry and Business (ARIB), a

---

[4] Even though 2 Mbps is used generally as the upper limit for IMT-2000 services, the exact service is specified to be 1.92 or 2.048 Mbps.

standardization body responsible for Japan's radio standardization, decided to proceed with detailed standardization of wideband CDMA. The technology push from Japan accelerated standardization in Europe and the United States. During 1997 joint parameters for Japanese and European wideband CDMA proposals were agreed upon. The air interface is now commonly referred as WCDMA. In January 1998, strong support behind wideband CDMA led to the selection of WCDMA as the UMTS terrestrial air interface scheme for FDD frequency bands in ETSI. The selection of wideband CDMA was also backed by Asian and American GSM operators. For TDD bands, a time division CDMA (TD-CDMA) concept was selected. In the United States in March 1998, the TIA TR45.5 committee, responsible for IS-95 standardization, adopted a framework for wideband CDMA backward compatible to IS-95, called cdma2000. Throughout 1999, the wideband CDMA schemes were further harmonized in the Operator Harmonization Group (OHG). This resulted in a three mode wideband CDMA scheme: direct sequence, multicarrier, and time division duplex. TR45.3, responsible for IS-136 standardization, adopted a TDMA-based third-generation proposal, UWC-136 (Universal Wireless Communications), based on the recommendation from the UWCC in February 1998. A key component of the UWC-136 solution is the EDGE technology, which is now used both in the GSM and IS-136 evolution.

The preferred technology for third-generation systems depends on technical, political, and business factors. Technical factors include issues such as provision of required data rates and performance. Political factors involve reaching agreement between standards bodies and taking into account the different starting points of different countries and regions. On one hand, the investments into the existing systems motivate a backward compatibility approach. On the other, new business opportunities or the possibility of changing the current situation might motivate a new approach. In the following sections we discuss the research and standardization activities that led to the selections described above.

1.4.3 Europe

1.4.3.1 European Third-Generation Radio Research Activities

Figure 1.6 illustrates European third-generation radio research activities. The development towards third-generation standards can be divided into three main phases: basic studies, development of system concepts, and comparison/consolidation. The driving force behind the European development has been the European Commission funded research programs: RACE (Research of Advanced Communication Technologies in Europe) and ACTS (Advanced Communication Technologies and Services) [22–24]. Their value has been in producing technical results for standardization, but most of all in bringing together different parties from the industry.

In Europe, the RACE I program, launched in 1988 and lasting until June 1992, started the third-generation research activities. Main studies of RACE I concentrated on individual technologies such as cellular propagation studies, handover, dynamic resource allocation, modulation, equalization, coding, channel management, and fixed network mobile functions. Many of these individual technologies were later used as a basis for the system concept development in RACE II.

Figure 1.6   European third-generation radio research activities.

Between 1992 and 1995 in the RACE II program, the CODIT (Code Division Multiple Testbed) [25] and ATDMA (advanced TDMA) [26] projects developed air interface proposals and testbeds for UMTS radio access. The ATDMA testbed provided up to a 64-Kbps user data rate with rate 1/4 coding, and the CODIT testbed provided a user data rate up to 128 Kbps with rate 1/2 coding [25]. Laboratory tests were performed for both systems. After the RACE II program, field tests for the CODIT testbed were performed [27]. A special interest group in RACE (SIG5) compared the CODIT and ATDMA air interfaces using both quantitative and qualitative criteria [28]. Depending on the selected radio environment and service scenario, each of the schemes showed advantage, but no decision to favor either one as a main candidate for UMTS was made.

ACTS was launched at the end of 1995 to support collaborative mobile research and development. Within ACTS, the FRAMES (Future Radio Wideband Multiple Access System) project investigated hybrid multiple access technologies in order to select the best combination for a basis for further detailed development of the UMTS radio access system. Based on this evaluation, a harmonized multiple access platform, FRAMES multiple access (FMA), was designed consisting of two modes: FMA1, a wideband TDMA scheme with and without spreading, and FMA2, a wideband CDMA scheme [29–34]. The FMA1 without spreading has roots in the ATDMA scheme and its subsequent developments [35]. For comparison results of FRAMES see Chapter 13.

In addition to the RACE and ACTS programs, several industrial projects have developed technologies for UMTS and IMT-2000. From 1992 to 1995, a wideband CDMA concept reported in [36, 37] was developed by Nokia. Furthermore, a wideband CDMA testbed was developed from 1992 to 1995 with transmission capabilities up to 128 Kbps for video application [38]. Field tests for the system were carried out during 1996. The test system bandwidth is configurable up to 30 MHz. Nokia's wideband CDMA was submitted to FRAMES where it was used as the basis for FMA2 development. A hybrid CDMA/TDMA concept, also called TD-CDMA, was studied in [39–42] and formed the basis of FMA1 with spreading. The UK LINK program also developed a wideband CDMA concept and testbed [43]. Telia developed the OFDM concept and testbed [44]. Telia's OFDM concept was submitted to the ETSI Gamma group.

1.4.3.2   European Standardization Activities

The tremendous success of GSM will strongly impact the standardization of third-generation systems in Europe. Europe is trying to build the third-generation systems as an evolution from the GSM system to offer a smooth transition path from the second-generation systems into the third-generation.

Within ETSI, the technical committee SMG has the standardization mandate for UMTS. Within SMG, there are several subcommittees that are responsible for the detailed technical standardization of UMTS. The subcommittee SMG2, responsible for UMTS radio access system standardization, started the definition process of UMTS terrestrial radio access (UTRA) in a UMTS workshop held in December 1996. The milestones of the process are presented below:

- M1 6/97: Definition of a limited number of UTRA concepts based on the proposed access technologies or combinations of them;
- M2 12/97: Selection of one UTRA concept;
- M3 6/98: Definition of key technical aspects of the UTRA (including carrier bandwidth, modulation, channel coding, channel types, frame structure, access protocols, channel allocation and handover mechanisms, and cell selection mechanisms).

The submitted air interface concepts were grouped into five different concept groups. FMA2 and three wideband CDMA schemes from Japan were submitted to the Alpha group. The Beta group evaluated FMA1 without spreading together with some other TDMA ideas. Two OFDM concepts were submitted to the Gamma group. The Delta group evaluated FMA1 with spreading, which has been also called the TD-CDMA scheme. In the beginning of the evaluation, another hybrid scheme, CTDMA (code time division multiple access), was also considered in the Delta group. The Epsilon group considered ODMA (opportunity driven multiple access), which is a relay technology applicable in principle to all multiple access schemes.

In the concept group evaluation process, the FMA2 proposal was considered together with the wideband CDMA schemes from Japan. At the same time, the Coordination Group (CG) in ARIB had discussions on the harmonization of different CDMA proposals. These two efforts led to the harmonization of the parameters for ETSI and ARIB wideband CDMA schemes [45]. The main parameters of the current scheme are based in the uplink on the FMA2 scheme and in the downlink on the ARIB wideband CDMA. Also, contributions from other proposals and parties were incorporated to further enhance the concept. In January 1998, ETSI SMG reached a consensus agreement on UTRA concept [46]:

- In the paired band (FDD) of UMTS the system adopts the radio access technique formerly proposed by the WCDMA group.
- In the unpaired band (TDD) the UMTS system adopts the radio access technique proposed formerly by the TD-CDMA group.
- In implementing this solution, ETSI SMG members pursue, together, the specification of UMTS with the objective of providing low cost terminals, ensuring harmonization with GSM and providing FDD/TDD dual mode operation terminals.

Consequently, technical work was initiated to achieve a fully integrated UTRA concept. In April 1998, the FDD and TDD modes were harmonized to have the same chip rate, frame length, and number of slots per frame.

1.4.4   Japan

The IMT-2000 Study Committee in ARIB was established in April 1993 to coordinate the Japanese research and development activities for IMT-2000. In October 1994, the Radio Transmission Special Group for radio transmission was formed to perform technical studies and to develop draft specifications for IMT-2000. The Special Group consisted of two ad hoc groups: CDMA and TDMA [47].

For TDMA, there were eight proposals. From these the group compiled a single carrier TDMA system, MTDMA (multimode and multimedia TDMA). For the MTDMA system, a prototype equalizer with a carrier bit rate of 1.536 Mbps and a user bit rate of 512 Kbps has been built and tested [48]. One organization proposed an OFDM scheme called BDMA (band division multiple access), and it was decided to conduct a study of this scheme. The BDMA scheme was also proposed in ETSI. However, during 1997 it became clear that the MTDMA and BDMA schemes would not be standardized for IMT-2000 within ARIB.

In Japan, several companies have developed wideband CDMA air interface proposals. Originally, 13 different wideband CDMA radio interfaces for FDD were presented to the IMT-2000 Study Committee. In early 1995, they were condensed into three FDD proposals (Core A, B, and C) and into one TDD proposal. The Core-B system is based on the wideband CDMA standardized for U.S. PCS known as IS-665. At the end of 1996, the four schemes were further combined into a single proposal where the main parameters were from Core-A, which is presented in more detail in [49–53]. Three of the original wideband CDMA proposals from Japan, two FDD and one

TDD concept, were also contributed to ETSI by their respective companies. In addition, the TDD proposal was submitted to TIA TR45.5 with modified parameters.

Both laboratory tests and field trials have been conducted for Japanese wideband CDMA proposals during 1995 and 1996. Core-A has been tested with speech and 384-Kbps video transmission in field trials [51]. Transmission of 2 Mbps has also been tested in laboratories using a bandwidth of 20 MHz. In the second phase of the Core-A trials, interference cancellation has been tested. The Core-B proposal has only been tested for speech service in the framework of PCS standardization in the United States [47]. Core-C has been tested for data rates up to 32 Kbps. Furthermore, the CDMA/TDD proposal was tested in laboratory conditions [48].

The main conclusion of the IMT-2000 Radio Transmission Special Group was that detailed studies on CDMA should be started. This meant that CDMA was the main technology choice for IMT-2000. Actually, a clear thrust for wideband CDMA came when the world's largest cellular operator, NTT DoCoMo, decided to proceed with wideband CDMA development and issued a trial system tender in 1996. All major manufacturers submitted proposals for terminal and infrastructure development. For terminal development, Matsushita, Motorola, NEC, and Nokia were selected, and for infrastructure development, Ericsson, Lucent, Matsushita, and NEC were selected. This was the first real commitment from an operator to go forward with third-generation, and it created global momentum for wideband CDMA.

1.4.5   The United States

The U.S. standardization situation is more diverse compared to Europe and Japan. Main standardization activities for wireless systems are carried out in TIA Engineering Committees TR45 and TR46, and in the T1 committee T1P1. TR45.5 is responsible for the IS-95 standardization, TR45.3 for the IS-136 standardization, and T1P1 with TR46 are jointly responsible for GSM1900 and some other technologies. In addition to standardization bodies, there are industry forums that elaborate on policy issues related to standards: UWCC for IS-136, CDMA Development Group (CDG) for IS-95, and the GSM Alliance and GSM North America for GSM.

In April 1997, CDG issued the so-called Advanced Systems initiative to develop the IS-95-based third-generation air interface proposal. During 1997 several wideband CDMA proposals from different companies, including Hughes, Lucent, Motorola, Nokia, Nortel, Qualcomm, and Samsung, were submitted to TR45.5 for the development of cdma2000 [54]. A common characteristic of all of these schemes was backward compatibility to IS-95. In March 1998, TR45.5 agreed upon the basic framework for cdma2000. The cdma2000 standard comprises cdma2000 1x and 3x options. cdma2000 1x has the same bandwidth as IS-95 (i.e., 1.25 MHz) and cdma2000 3x, also called multicarrier CDMA, three times the bandwidth of IS-95 carrier.

In the beginning of 1997, within UWCC, GTF (Global TDMA Forum) established the High Speed Data (HSD) group to evaluate air interface candidates for the IS-136 evolution towards third-generation. Several schemes, including TDMA, wideband CDMA, and two OFDM schemes, were submitted to the HSD group. Based on these proposals, UWCC developed the UWC-136 concept, which was accepted by TR45.3 in February 1998. UWC-136 consists of an enhanced IS-136 30-kHz carrier, 200-kHz high speed data carrier, and 1.6-MHz wideband TDMA carrier for an indoor radio environment. The 200-kHz HSD carrier has the same parameters as the enhanced GSM carrier (EDGE—enhanced data rates for GSM/global evolution) [55]. The wideband TDMA is based on the FRAMES FMA1 without spreading [34]. The main focus in the standardization has been on EDGE, and it seems that wideband TDMA will not be developed into a commercial system.

1.4.6   Korea

ETRI established an R&D consortium to define the Korean proposal for IMT-2000 in 1997 and 1999. A wideband CDMA proposal has been developed within ETRI [56]. This formed the basis for the TTA I scheme.

SK Telecom (previously KMT, Korea Mobile Telecom) decided in 1994 to develop a wideband CDMA air interface for IMT-2000 with a focus on a low-tier system for a pedestrian radio environment [57]. A common air interface specification was written and a test system was built. The test system provides a data rate of 36 Kbps using 5-MHz bandwidth. The final air interface specification was completed at the end of 1996 and a precommercial system was developed [57]. SK Telecom's system was merged with other industrial proposals at the beginning of 1998 and formed the TTA II scheme [58].

For the ITU radio transmission technology submission in June 1998, TTA I and TTA II were renamed Global CDMA I and II, respectively.

1.4.7   Global Harmonized Wideband CDMA

In 1999 a number of cellular operators formed the so-called Operators Harmonization Group (OHG) to drive global harmonization of third-generation air interfaces. This activity resulted in a harmonized wideband CDMA scheme consisting of three modes: direct spread (DS), multicarrier (MC), and TDD. The key parameters are outlined below:

- The harmonized standard for DS will be based on the WCDMA proposal.

- The harmonized standard for MC will be based on the cdma2000 proposal.

The modular system block diagram of the harmonized system is presented in Figure 1.7. As can be seen, each radio system can be connected either into MAP-based GSM core network or ANSI-41 core network. A third-generation operator may select one or more radio access modules together with one or more core network modules to implement a third-generation system subject to the regulatory requirements of their nation or region.

Figure 1.7 Modular 3G harmonized system.

1.4.8 Global Third-Generation Standardization

The main standardization activities are currently occuring in 3GPP. 3GPP was established in 1998 and its focus is to standardize WCDMA- and GSM-based core networks. After the 3GPP was founded, a similar standardization body (3GPP2) was formed to focus on cdma2000 standardization. In 2000, EDGE standardization was transferred into 3GPP as well.

Since the first release of third-generation standards, termed release '99, has already been finalized, the focus of standardization is further developing basic standards. In WCDMA, high-speed packet access (HSPA) will increase the maximum data rate up to 10 Mbps in the downlink. For cdma2000 1x the focus of evolution is in the improvement of cdma2000 1x, i.e., increasing bit rates within the 1.25-MHz bandwidth. The standard is termed 1xEV and features two phases; phase 1 is data only, and phase 2 includes voice and data.

1.5   FREQUENCY ALLOCATION FOR THIRD-GENERATION SYSTEMS

For IMT-2000, a total of 230 MHz was identified at the World Administrative Radio Conference (WARC'92). However, this will be allocated in different ways in different regions and countries. Figure 1.8 depicts the IMT-2000 frequency allocation in the United States, Europe, and Japan. In the United States, a part of the IMT-2000 frequency allocation is already used for PCS systems. Also, it is unlikely that the upper band of WARC'92 allocation would be used. Thus, third-generation systems have to fit into the current PCS frequency allocations. In Japan and Europe, the frequency allocation for third-generation is almost identical except that the PHS spectrum allocation partially overlaps the identified UMTS TDD spectrum.

Frequency allocations have an impact on technology choices and air interface design. Licensing policy will define the use of the IMT-2000 spectrum. If the licenses are given to the existing operators, the use of spectrum might be different than if they are given to new operators. Also, the way licenses are issued will have an impact on the market. In the United States the PCS spectrum was auctioned, which led to large investments before the actual operation started. Spectrum allocation for UMTS and spectrum licensing are further discussed in Sections 9.3.1 and 15.5, respectively.

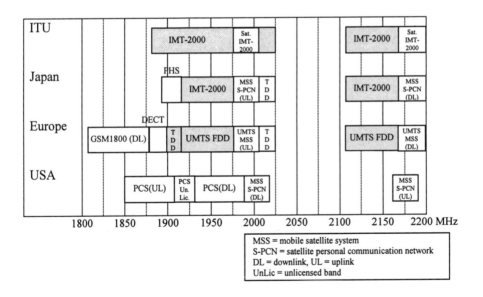

Figure 1.8   IMT-2000 frequency allocations.

Recently, work to identify UMTS extension bands has started. The target is to allocate 160 MHz more spectrum for UMTS. In the WARC'2000 the following additional bands were identified: 806-960 MHz, 1710-1885 MHz and 2500-2690 MHz. The frequency administrations are free to choose between the identified bands according their individual situation and market demand. Next, ITU-R will create a plan for the usage of the identified frequency bands, preferably taking into account the needs of the existing users and harmonization of IMT-2000.

1.6 AIR INTERFACE TECHNOLOGIES FOR THIRD-GENERATION SYSTEMS

In the search for the most appropriate multiple access technology for third-generation wireless systems, a number of new multiple access schemes have been proposed. In this section, an overview of the air interface technologies selected for the third-generation systems (wideband CDMA schemes, EDGE, and WCDMA TDD) is given. In addition, we briefly review OFDM technology, which was considered in the air interface selection of third-generation systems.

1.6.1 Wideband CDMA

Wideband CDMA has a bandwidth of 5 MHz or more. The nominal bandwidth for all third-generation proposals is 5 MHz. There are several reasons for choosing this bandwidth. First, data rates of 144 and 384 Kbps, the main targets of third-generation systems, are achievable within the 5-MHz bandwidth with a reasonable capacity. Even a 2-Mbps peak rate can be provided under limited conditions. Second, there is a lack of spectrum calls for reasonably small minimum spectrum allocation, especially if the system has to be deployed within the existing frequency bands occupied already by second-generation systems. Third, the 5-MHz bandwidth can resolve (separate) more multipaths than narrower bandwidths, increasing diversity and thus improving performance. Larger bandwidths of 10, 15, and 20 MHz have been proposed to support higher data rates more effectively. However, the main trend seems to be to evolve the standard within the original 5-MHz bandwidth.

Several wideband CDMA proposals have been made for third-generation wireless systems. They can be characterized by the following new advanced properties:

- Provision of multirate services;
- Packet data;
- Complex spreading;
- A coherent uplink using a user dedicated pilot;
- Additional pilot channel in the downlink for beamforming;
- Seamless interfrequency handover;
- Fast power control in the downlink;
- Optional multiuser detection.

The two wideband CDMA schemes for third-generation are WCDMA, which is network asynchronous, and cdma2000, which is synchronous. In network asynchronous schemes the base stations are not synchronized, while in network synchronous schemes the base stations are synchronized to each other within a few microseconds. Chapters 6 and 8 describe the WCDMA and cdma2000 schemes in more detail, respectively.

Table 1.4
Features of Wideband CDMA Schemes

	WCDMA	cdma2000
Chip rate	3.84 Mcps	1.2288 Mcps (cdma2000 1x) 3*1.2288 Mcps (downlink)/ 3.6864 Mcps (uplink), cdma2000 3x
Carrier spacing	5 MHz	1.25, 5 MHz
Frame length	10 ms	20 ms
Interbase station synchronization	Asynchronous	Synchronous

The multicarrier approach is motivated by a spectrum overlay of cdma2000 with existing IS-95 carriers. Similar to IS-95, the spreading codes of cdma2000 are generated using different phase shifts of the same M-sequence. This is possible because of the synchronous network operation.

Since WCDMA has an asynchronous network, different long codes rather than different phase shifts of the same code are used for the cell and user separation. The code structure further impacts how code synchronization, cell acquisition, and handover synchronization are performed.

1.6.2   TDMA

As already discussed, several TDMA schemes have been studied for the third-generation air interface: GSM evolution using a higher level modulation [59, 60], FMA1 without spreading (wideband TDMA) [32], ATDMA [26], GSM compatible ATDMA [61], and MTDMA [48]. The EDGE interface, whose design is based on the above-mentioned studies, will be used both for IS-136 and GSM evolution. Since the IS-136 operators have limited spectrum available for in the beginning of EDGE deployment, a requirement of an initial deployment within 1-MHz spectrum allocation was set in the UWCC. Therefore, two types of EDGE air interfaces were designed: EDGE Compact and EDGE Classic. EDGE Compact is synchronized and requires only 600-kHz spectrum allocation.

In TIA, EDGE was initially standardized for 800-MHz and 1900-MHz frequency bands and in ETSI for GSM frequency bands. In 2000, the standardization of EDGE was transferred into 3GPP. Therefore, standardization of TDMA/EDGE and GSM/EDGE will merge.

The parameters of the EDGE proposal are listed in Table 1.5. The motivation for the 200-kHz carrier of EDGE is twofold. First, the adoption of the same physical layer for IS-136 and GSM data carriers provides economics of scale and therefore cheaper equipment and faster time to market. Second, the 200-kHz carrier with higher order modulation can provide bit rates of 144 and 384 Kbps with reasonable range and capacity, fulfilling IMT-2000 requirements for pedestrian and vehicular environments.

The EDGE data frame length is 4.615 ms, and one frame consists of eight slots. The burst structure is suitable for transmission in a high delay spread environment. EDGE has two modulation schemes: GSMK and 8-PSK. The latter is used for data channels while the formed is used both for data and control channels.

Table 1.5
Basic Parameters of the EDGE Radio Standard

Duplex method	FDD	
Carrier spacing	200 kHz	
Modulation rate	270.1 symbols/s	
Frame length	4.615 ms	
Number of slots/frame	8	
Modulation	GMSK	8 PSK
Payload/burst		
• Symbols;	116	116
• Bits.	116	348
Data rate		
• One slot;	22.8 Kbps	69.6 Kbps
• One frame.	182.4 Kbps	556.8 Kbps

Source: [62].

EDGE has eight different channel coding schemes (CS): CS-1-CS-4 use GMSK modulation, and CS-5-CS-8 use 8-PSK modulation as shown in Table 1.6. Depending on the user requirements and channel conditions, suitable combination of modulation, coding, and number of data slots is selected. Asymmetrical data rates are provided by allocating a different number of time slots in the uplink and downlink. EDGE supports both real-time and non-real-time services. For optimized performance of non-real-time services incremental redundancy, a new scheme, combining FEC and retransmission, has been developed [62].

Table 1.6
Overview of EDGE Coding Schemes

Coding scheme	Code rate	Modulation	Maximum data rate/slot (Kbps)
CS-1	0.53	GSMK	8.8
CS-2	0.66	GSMK	11.2
CS-3	0.80	GSMK	14.8
CS-4	1	GSMK	17.6
CS-5	0.35	8-PSK	22.4
CS-6	0.45	8-PSK	29.6
CS-7	0.52	8-PSK	44.8
CS-8	0.70	8-PSK	59.2

1.6.3   WCDMA TDD

In the ETSI air interface selection, the TDD solution was based on the TD-CDMA principles. The role of CDMA in TD-CDMA is to multiplex the different channels within a time slot. The parameters of TD-CDMA are listed in Table 1.7.

Table 1.7
TD-CDMA Parameters

Multiple access method	TDMA/CDMA
Basic channel spacing	5 MHz
Carrier chip/bit rate	3.84 Mchip/s
Time slot structure	15 slots / TDMA frame
Frame length	10 ms
Multirate concept	Multislot and multicode
Channel coding	Convolutional codes and turbo coding
Detection	Coherent, based on training sequence
Other diversity means	Frequency/time hopping per frame or slot

1.6.4   Orthogonal Frequency Division Multiplexing

OFDM applies the principle of multicarrier modulation (MCM) by dividing the data stream into several bit streams, each of which has much lower bit rate, and using these substreams to modulate several carriers. The first applications of MCM were in military radios in the late 1950s and the early 1960s [4]. The OFDM principle was patented in 1970 [63]. The research of OFDM for Digital Audio Broadcasting (DAB) was started at the end of the 1980s, and commercial DAB systems are currently being introduced. Recently, OFDM has been applied to HIPERLAN type II and wireless asynchronous transfer mode (ATM) systems [64].

Research on OFDM for cellular applications intensified during the 1990s. Both CDMA- and TDMA-based OFDM have been studied. A multicarrier CDMA (MC-CDMA) transmitter spreads the original signal by using a given spreading code in the frequency domain. It is crucial for multicarrier transmissions to have frequency nonselective fading over each subcarrier [1]. Introduction of OFDM into the cellular world has been driven by two main benefits:

- Flexibility: each transceiver has access to all subcarriers within a cell layer;
- Easy equalization: OFDM symbols are longer than the maximum delay spread, resulting in a flat fading channel that can be easily equalized.

The main drawback of OFDM is the high peak-to-average power. This is especially severe for the mobile station and for long-range applications. Different encoding techniques have been investigated to overcome this problem. Furthermore, the possibility of accessing all the resources within the system bandwidth results in an equally complex receiver for all services, regardless of the bit rate. Of course, a partial FFT for only one OFDM block is possible for low bit rate services, but this would require an RF synthesizer for frequency hopping.

Table 1.8 shows the parameters of the OFDM air interface proposed by the Gamma group of ETSI. The Gamma concept has been developed based on contributions from different OFDM schemes described in [44, 48, 65].

Table 1.8
Main Features of OFDM Proposal

Bandslot width	100 kHz
Number of subcarriers	24
Subcarrier bandwidth	4.17 kHz
Modulation period	288.8
Block size	24 carriers / 1 symbol
Frame length	4.615 ms

Source: [65].

1.7   PREVIEW OF THE BOOK

This book consists of 16 chapters and concentrates on a study of wideband CDMA from the radio access design point of view. Propagation and application requirements for the design of wireless access systems are covered. A detailed study of wideband CDMA air interface solutions is presented. Network aspects are addressed to give an overview of a complete wireless network architecture.

Basic principles of CDMA are explained in Chapter 2. First, the CDMA concept is discussed. Basic CDMA functions such as the RAKE receiver, power control, soft handover, interfrequency handover, and multiuser detection are explained. The narrowband IS-95 CDMA standard is introduced.

In Chapter 3, the mobile Internet, third-generation applications, and user scenarios are introduced. The characteristics of UMTS teleservices such as video telephony and Internet services are discussed, and bearer service requirements are introduced. Traffic models for simulation purposes are developed.

Detailed know-how on radio propagation is a prerequisite for any radio system design. Chapter 4 presents radio environments and their characteristics. Wideband channel models are discussed in detail, as are path loss and mobility models.

Chapter 5 covers all aspects of the wideband CDMA air interface design starting from frame structures and physical channel design (including spreading codes and their properties) to radio resource management aspects (handover and power control).

Chapter 6 presents the WCDMA scheme, and its further evolution. Chapter 7 describes the WCDMA TDD scheme, and Chapter 8 the cdma2000 scheme.

Chapter 9 covers performance aspects of wideband CDMA. Simulation assumptions and the simulator used in the performance evaluation are introduced. Modeling of different CDMA functions in both the link level and the system level are discussed. The impact of power control into system performance is analyzed. Spectrum efficiency and range are two major performance measures that are considered.

Hierarchical cell structures (HCS) are discussed in Chapter 10. Because the spectrum spreading caused by nonlinear power amplifiers impacts the deployment of hierarchical cell structures, nonlinear power amplifier characteristics and power amplifier efficiency are discussed first. Two options for HCS deployment are covered: micro and macro cells at the same frequency, and micro and macro cells at different frequencies. Because the latter is more probable, its system level performance is analyzed.

Implementation issues are dealt with in Chapter 11. Optimization criteria for transceiver design are introduced. Modularity issues are discussed. Receiver and transmitter designs both in baseband and RF are explained in detail. Based on this, a typical wideband CDMA mobile terminal architecture is presented. Multimode terminals are discussed in final section.

CDMA network planning issues are discussed in Chapter 12. First, basic radio network planning procedure is explained. Next, micro cell planning in CDMA is covered. Coexistence of wideband CDMA with other air interfaces is discussed, including intermodulation aspects, deployment scenarios, guard zones and bands, and handover between systems. Finally, the transition from second-generation systems into wideband CDMA is considered from the operator's perspective.

Basic principles of Internet Protocol (IP) are explained in Chapter 13. It covers basic concepts, next generation IP protocol, IPv6, voice over IP, and Internet Engineering Task Force (IETF).

Chapter 14 presents network architectures of third-generation systems. The 3GPP specifications related to network architecture and interfaces are described, including the UMTS system architecture, protocols, and interfaces. The IS-41, IS-95 and cdma2000 network architecture, interfaces, protocols, and evolution are discussed. As an important new architecture paradigm, IP-based networks, termed "All-IP" networks, are introduced.

Chapter 15 describes the standardization environment, different standards bodies, and their role in third-generation standardization.

Chapter 16 discusses future directions of wireless systems. Evolution of different air interfaces are described, and market considerations and evolution scenarios are discussed.

REFERENCES

[1]   Prasad, R., *CDMA for Wireless Personal Communications*, Norwood, Massachusetts: Artech House, 1996.

[2]   Ojanperä, T., "Overview of Research Activities for Third-generation Mobile Communication," in *Wireless Communications TDMA vs. CDMA*, S. G. Glisic and P. A. Leppanen (eds.), Dordrecht: Kluwer Academic Publishers, 1997, pp. 415–446.

[3]   Ojanperä, T., and R. Prasad, "Overview of Air Interface Multiple Access for IMT-2000/UMTS," *IEEE Communications Magazine*, September 1998.

[4] Prasad, R., *Universal Wireless Personal Communications*, Norwood, Massachusetts: Artech House, 1998.

[5] Shannon, C. E., "A Mathematical Theory of Communication," *Bell System Technical Journal*, Vol. 27, 1948, pp. 379–423 and 623–656.

[6] Scholtz, R. A., "The Evolution of Spread-Spectrum Multiple-Access Communications," in *Code Division Multiple Access Communications*, S. G. Glisic and P. A. Leppänen (eds.), Boston: Kluwer Academic Publishers, 1995.

[7] "A Conversation with Claude Shannon," *IEEE Communications Magazine*, Vol. 22, No. 5, May 1984, pp. 123–126.

[8] Cooper, G. R., and R. W. Nettleton, "A Spread-Spectrum Technique for High-Capacity Mobile Communications," *IEEE Trans. Veh. Tech.*, Vol. 27, No. 4, November 1978, pp. 264–275.

[9] Verdu, S., "Minimum Probability of Error for Asynchronous Gaussian Multiple Access," *IEEE Trans. on IT*, Vol. IT-32, No. 1, January 1986, pp. 85–96.

[10] Goodman, D. J., *Wireless Personal Communications Systems*, Reading, Massachusetts: Addison-Wesley, 1997.

[11] Mouly, M., and M.-B. Pautet, *The GSM System for Mobile Communications*, published by the authors, 1992.

[12] Failli, R., and P. P. Giusto, "Alternative Architectures for a European Digital Mobile Radio System in the 900 MHz Band," *Proceedings of Nordic Seminar on Digital Land Mobile Radio Communications*, Espoo, Finland, February 1985.

[13] Jarvinen, K., J. Vainio, P. Kapanen, T. Honkanen, P. Haavisto, R. Salami, C. Laflamme, and J.-P. Adoul, "GSM Enhanced Full Rate Speech Codec," *Proceedings of ICASSP '97*, April 1997, pp. 771–774.

[14] Pirhonen, R., and P. Ranta, "Single Slot 14.4 Kbps Service for GSM," *Proceedings of Seventh IEEE International Symposium on Personal, Indoor and Mobile Radio Communications (PIMRC'96)*, Taipei, Taiwan, October 1–8, 1996, pp. 943–947.

[15] Hämäläinen, J., *Design of GSM High Speed Data Services*, Ph.D. Dissertation, Tampere University of Technology, Finland, 1996.

[16] Ranta, P.-A., A. Lappeteläinen, and Z.-C. Honkasalo, "Interference Cancellation by Joint Detection in Random Frequency Hopping TDMA Networks," *Proceedings of ICUPC'96 Conference*, Vol. 1, Cambridge, Massachusetts, September/October 1996, pp. 428–432.

[17] Jung, P., Z. Zvonar, and K. Kammerlander (eds.), *GSM: Evolution Towards 3rd Generation*, Boston: Kluwer Academic Publishers, 1998.

[18] Ross, A. H. M., and K. L. Gilhousen, "CDMA Technology and the IS-95 North American Standard," in *The Mobile Communications Handbook*, J. D. Gipson (ed.), Boca Raton, Florida: CRC Press, 1996, pp. 430–448.

[19] TIA/EIA IS-95B, "Mobile Station-Base Station Compatibility Standard for Dual-Mode Wideband Spread Spectrum Cellular Systems," 1998.

[20] Kinoshita, K., and M. Nagakawa, "Japanese Cellular Standard," in *The Mobile Communication Handbook*, J. D. Gibson (ed.), Boca Raton, Florida: CRC Press, 1996, pp. 449–461.

[21] Murase, A., A. Maebara, I. Okajima, and S. Hirata, "Mobile Radio Packet Data

Communications in a TDMA Digital Cellular System," *Proceedings of VTC'97*, Phoenix, Arizona, May 1997, pp. 1034–1038.

[22] "The European Path Towards UMTS," *IEEE Personal Communications*, Special Issue, Vol. 2, No. 1, February 1995.

[23] DaSilva, J. S., B. Arroyo, B. Barani, and D. Ikonomou, "European Third-Generation Mobile Systems," *IEEE Communications Magazine*, Vol. 34, No. 10, October 1996, pp. 68–83.

[24] Prasad, R., J. S. DaSilva, and B. Arroyo, "ACTS Mobile Programme in Europe," *IEEE Communications Magazine*, Vol. 36, No. 2, February 1998.

[25] Andermo, P.-G., (ed.), "UMTS Code Division Testbed (CODIT)," *CODIT Final Review Report*, September 1995.

[26] Urie, A., M. Streeton, and C. Mourot, "An Advanced TDMA Mobile Access System for UMTS," *IEEE Personal Communications*, Vol. 2, No. 1, February 1995, pp. 38–47.

[27] Ewerbring, M., J. Ferjh, and W. Granzow, "Performance Evaluation of a Wideband Testbed Based on CDMA," *Proceedings of VTC'97*, Phoenix, Arizona, May 1997, pp. 1009–1013.

[28] Pizarroso, M., and J. Jiménez (eds.), "Preliminary Evaluation of ATDMA and CODIT System Concepts," *SIG5 deliverable MPLA/TDE/SIG5/DS/P/002/b1*, September 1995.

[29] Ojanperä, T., M. Gudmundson, P. Jung, J. Sköld, R. Pirhonen, G. Kramer, and A. Toskala, "FRAMES-Hybrid Multiple Access Technology," *Proceedings of ISSSTA'96*, Vol. 1, Mainz, Germany, September 1996, pp. 320–324.

[30] Ojanperä, T., P. O. Anderson, J. Castro, L. Girard, A. Klein, and R. Prasad, "A Comparative Study of Hybrid Multiple Access Schemes for UMTS," *Proceedings of ACTS Mobile Summit Conference*, Vol. 1, Granada, Spain, October 1996, pp. 124–130.

[31] Ojanperä, T., J. Sköld, J. Castro, L. Girard, and A. Klein, "Comparison of Multiple Access Schemes for UMTS," *Proceedings of VTC'97*, Vol. 2, Phoenix, Arizona, May 1997, pp. 490–494.

[32] Klein, A., R. Pirhonen, J. Sköld, and R. Suoranta, "FRAMES Multiple Access Mode 1–Wideband TDMA with and Without Spreading," *Proceedings of PIMRC'97*, Helsinki, Finland, September 1997, pp. 37–41.

[33] Ovesjö, F., E. Dahlman, T. Ojanperä, A.Toskala, and A. Klein, "FRAMES Multiple Access Mode 2–Wideband CDMA," *Proceedings of PIMRC'97*, Helsinki, Finland, September 1997, pp. 42–46.

[34] Nikula, E., A. Toskala, E. Dahlman, L. Girard, and A. Klein, "FRAMES Multiple Access for UMTS and IMT-2000," *IEEE Personal Communications*, April 1998, pp. 16–24.

[35] Nikula, E., and E. Malkamäki, "High Bit Rate Services for UMTS Using Wideband TDMA Carriers," *Proceedings of ICUPC'96*, Vol. 2, Cambridge, Massachusetts, September/October 1996, pp. 562–566.

[36] Ojanperä, T., K. Rikkinen, H. Häkkinen, K. Pehkonen, A. Hottinen, and J. Lilleberg, "Design of a 3rd Generation Multirate CDMA System with Multiuser Detection, MUD-CDMA," *Proceedings of ISSSTA'96*, Vol. 1, Mainz, Germany,

1996, pp. 334–338.

[37] Westman, T., and H. Holma, "CDMA System for UMTS High Bit Rate Services," *Proceedings of VTC'97*, Phoenix, Arizona, May 1997, pp. 825–829.

[38] Pajukoski, K., and J. Savusalo, "Wideband CDMA Test System," *Proceedings of PIMRC'97*, Helsinki, Finland, September 1997, pp. 825–829.

[39] Klein, A., and P. W. Baier, "Simultaneous Cancellation of Cross Interference and ISI in CDMA Mobile Radio Communications," *Proceedings of IEEE International Symposium on Personal, Indoor and Mobile Radio Communications (PIMRC'92)*, Boston, Massachusetts, 1992, pp. 118–122.

[40] Klein, A., and P. W. Baier, "Linear Unbiased Data Estimation in Mobile Radio Systems Applying CDMA," *IEEE Journal on Selected Areas in Communications*, Vol. SAC-11, 1993, pp. 1058–1066.

[41] Naßhan, M. M., P. Jung, A. Steil, and P. W. Baier, "On the Effects of Quantization, Nonlinear Amplification and Band Limitation in CDMA Mobile Radio Systems Using Joint Detection," *Proceedings of the Fifth Annual International Conference on Wireless Communications WIRELESS'93*, Calgary, Canada, 1993, pp. 173–186.

[42] Jung, P., J. J. Blanz, M. M. Naßhan, and P. W. Baier, "Simulation of the Uplink of JD-CDMA Mobile Radio Systems with Coherent Receiver Antenna Diversity," *Wireless Personal Communications*, Vol. 1, 1994, pp. 61–89.

[43] Wales, S. C., "The U.K. LINK Personal Communications Programme: A DS-CDMA Air Interface for UMTS," *Proceedings of RACE Mobile Telecommunications Summit*, Cascais, Portugal, November 1995.

[44] ETSI SMG2, "Description of Telia's OFDM Based Proposal," *TD 180/97 ETSI SMG2*, May 1997.

[45] Ishida, Y., "Recent Study on Candidate Radio Transmission Technology for IMT-2000," *First Annual CDMA European Congress*, London, England, October 1997.

[46] ETSI Press Release, "Agreement Reached on Radio Interface for Third-Generation Mobile System, UMTS," January 29, 1998.

[47] Sasaki, A., "A Perspective of Third-Generation Mobile Systems in Japan," *IIR Conference Third-Generation Mobile Systems: The Route Towards UMTS*, London, England, February 1997.

[48] ARIB FPLMTS Study Committee, "Report on FPLMTS Radio Transmission Technology SPECIAL GROUP, (Round 2 Activity Report)," Draft v.E1.1, January 1997.

[49] Adachi, F., K. Ohno, M. Sawahashi, and A. Higashi, "Multimedia Mobile Radio Access Based on Coherent DS-CDMA," *Proceedings of 2nd International Workshop on Mobile Multimedia Commuications.*, Vol. A2.3, Bristol University, England, April 1995.

[50] Ohno, K., M. Sawahashi, and F. Adachi, "Wideband Coherent DS-CDMA," *Proceedings of IEEE VTC'95*, Chicago, Illinois, July 1995, pp. 779–783.

[51] Dohi, T., Y. Okumura, A. Higashi, K. Ohno, and F. Adachi, "Experiments on Coherent Multicode DS-CDMA," *Proceedings of IEEE VTC'96*, Atlanta, Georgia, pp. 889–893.

[52] Adachi, F., M. Sawahashi, and K. Ohno, "Coherent DS-CDMA: Promising

Multiple Access for Wireless Multimedia Mobile Communications," *Proceedings of IEEE ISSSTA '96*, Mainz, Germany, September 1996, pp. 351–358.

[53] Onoe, S., K. Ohno, K. Yamagata, and T. Nakamura, "Wideband-CDMA Radio Control Techniques for Third-Generation Mobile Communication Systems," *Proceedings of VTC '97*, Vol. 2, Phoenix, Arizona, May 1997, pp. 835–839.

[54] Chia, S., "Will cdmaOne Be the Third Choice?" *CDMA Spectrum*, September 1997, pp. 30–34.

[55] ETSI SMG2, "Improved Data Rates through Optimized Modulation, EDGE Feasibility Study," *Work Item 184*, Version 1.0, 1997.

[56] Hang, Y., S. C. Bang, H.-R. Park, and B.-J. Kang, "Performance of Wideband CDMA System for IMT-2000," *Proceedings of the 2nd CDMA International Conference*, Seoul, Korea, October 1997, pp. 583–587.

[57] Koo, J. M., E. K. Hong, and J. I. Lee, "Wideband CDMA Technology for FPLMTS," *Proceedings of the 1st CDMA International Conference*, Seoul, Korea, November 1996.

[58] Hong, E. K., "Intercell Asynchronous W-CDMA System for IMT-2000," *ITU-R TG 8/1 Workshop*, Toronto, Canada, September 1997.

[59] Sköld, J., P. Schramm, P.-O. Anderson, and M. Gudmundson, "Cellular Evolution into Wideband Services," *Proceedings of VTC '97*, Vol. 2, Phoenix, Arizona, May 1997, pp. 485–489.

[60] Honkasalo, H., "The Technical Evolution of GSM," *Proceedings of Telecom '95*, Geneva, October 1995, pp. 15–19.

[61] Urie, A., "Advanced GSM: A Long Term Future Scenario for GSM," *Proceedings of Telecom '95*, Vol. 2, Geneva, Switzerland, October 1995, pp. 33–37.

[62] Pirhonen, R., T. Rautava, and J. Penttinen, "TDMA Covergence for Packet Data Services," *IEEE Personal Communications*, June 1999.

[63] Chang, R. W., "Orthogonal Frequency Division Multiplexing," *U.S. Patent 3,488,455*, filed 1966, issued January 6, 1970.

[64] Mikkonen, J. K., J. P. Aldis, G. A. Awater, A. S. Lunn, and D. Hutchinson, "The Magic WAND-Functional Overview," *IEEE Journal on Selected Areas in Communications*, Vol. 16, No. 6, August 1998.

[65] Alikhani, H., R. Böhnke, and M. Suzuki, "BDMA Band Division Multiple Access — A New Air-Interface for 3rd Generation Mobile System in Europe," *Proceedings of ACTS Summit*, Aalborg, Denmark, October 1997, pp. 482–488.

Chapter 2

BASIC PRINCIPLES OF CDMA

2.1 INTRODUCTION

This chapter illustrates the basic principles of CDMA. The scope of the chapter is to give a generic understanding of CDMA without overwhelming mathematical details. Readers should refer to Chapters 5, 6, and 7 for a more extensive discussion on direct sequence CDMA air interface design aspects and wideband CDMA air interface proposals, respectively. Furthermore, [1–12] provide details of spread-spectrum (SS) and CDMA technologies.

This chapter is divided into three sections. Section 2.1 introduces the CDMA concept in general. Which criteria the transmitted signal has to fulfill in order to constitute a spread-spectrum modulation are explained. Processing gain is defined. The fundamental properties of CDMA signals, namely multiple access capability, protection against multipath interference, privacy, interference rejection, anti-jamming capability, and low probability of interception, are introduced. Different modulation methods for CDMA are treated in detail. These are direct sequence spread-spectrum, frequency-hopping spread-spectrum, time-hopping spread-spectrum, and hybrid modulation. Each modulation scheme is described with the help of block diagrams for the transmitter and the receiver. In addition, how each spread-spectrum modulation scheme achieves the above listed six properties of CDMA signals is discussed.

In Section 2.2 we review the fundamental elements of direct sequence CDMA and its application into third-generation systems, namely RAKE receiver, power control, soft handover, interfrequency handover, and multiuser detection.

In Section 2.3 the IS-95 standard is used to illustrate the downlink and uplink channel structures, as well as the power control principles of direct sequence CDMA.

2.2 CDMA CONCEPTS

In CDMA each user is assigned a unique code sequence (spreading code), which it uses

33

to encode its information-bearing signal. The receiver, knowing the code sequences of the user, decodes a received signal after reception and recovers the original data. This is possible because the crosscorrelations between the code of the desired user and the codes of the other users are small. Because the bandwidth of the code signal is chosen to be much larger than the bandwidth of the information-bearing signal, the encoding process enlarges (spreads) the spectrum of the signal and is therefore also known as spread-spectrum modulation. The resulting signal is also called a spread-spectrum signal, and CDMA is often denoted as spread-spectrum multiple access (SSMA).

The spectral spreading of the transmitted signal gives to CDMA its multiple access capability. It is therefore important to know the techniques necessary to generate spread-spectrum signals and the properties of these signals. A spread-spectrum modulation technique must fulfill two criteria:

1. The transmission bandwidth must be much larger than the information bandwidth.
2. The resulting radio-frequency bandwidth is determined by a function other than the information being sent (so the bandwidth is statistically independent of the information signal). This excludes modulation techniques like frequency modulation (FM) and phase modulation (PM).

The ratio of transmitted bandwidth to information bandwidth is called the *processing gain*, G_p, of the spread-spectrum system,

$$G_p = \frac{B_t}{B_i} \tag{2.1}$$

where B_t is the transmission bandwidth and B_i is the bandwidth of the information-bearing signal.

The receiver correlates the received signal with a synchronously generated replica of the spreading code to recover the original information-bearing signal. This implies that the receiver must know the code used to modulate the data.

Because of the coding and the resulting enlarged bandwidth, SS signals have a number of properties that differ from the properties of narrowband signals. The most interesting from the communication systems point of view are discussed below. To have a clear understanding, each property has been briefly explained with the help of illustrations, if necessary, by applying direct sequence spread-spectrum techniques.

1. *Multiple access capability.* If multiple users transmit a spread-spectrum signal at the same time, the receiver will still be able to distinguish between the users provided each user has a unique code that has a sufficiently low crosscorrelation with the other codes. Correlating the received signal with a code signal from a certain user will then only despread the signal of this user, while the other spread-spectrum signals will remain spread over a large bandwidth. Thus, within the information bandwidth the power of the desired user will be larger than the

interfering power provided there are not too many interferers, and the desired signal can be extracted. The multiple access capability is illustrated in Figure 2.1. In Figure 2.1(a), two users generate a spread-spectrum signal from their narrowband data signals. In Figure 2.1(b) both users transmit their spread-spectrum signals at the same time. At the receiver 1 only the signal of user 1 is coherently summed by the user 1 despreader and the user 1 data recovered.

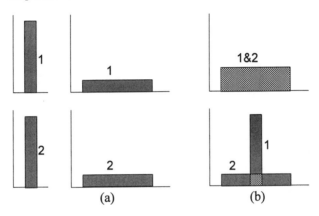

Figure 2.1 Principle of spread-spectrum multiple access.

2. *Protection against multipath interference.* In a radio channel there is not just one path between a transmitter and receiver. Due to reflections (and refractions), a signal will be received from a number of different paths. The signals of the different paths are all copies of the same transmitted signal but with different amplitudes, phases, delays, and arrival angles. Adding these signals at the receiver will be constructive at some of the frequencies and destructive at others. In the time domain, this results in a dispersed signal. Spread-spectrum modulation can combat this multipath interference; however, the way in which this is achieved depends very much on the type of modulation used. In the next section, where CDMA schemes based on different modulation methods are discussed, we show for each scheme how multipath interference rejection is obtained.

3. *Privacy.* The transmitted signal can only be despread and the data recovered if the code is known to the receiver.

4. *Interference rejection.* Crosscorrelating the code signal with a narrowband signal will spread the power of the narrowband signal thereby reducing the interfering power in the information bandwidth. This is illustrated in Figure 2.2. The receiver observes spread-spectrum signal (s) summed with a narrowband interference (i). At the receiver, the SS signal is "despread" while the interference signal is spread, making it appear as background noise compared to the despread signal.

Demodulation will be successful if the resulting background is of sufficiently weak energy in the despread information bandwidth.

5. *Anti-jamming capability, especially narrowband jamming.* This is more or less the same as interference rejection except the interference is now willfully inflicted on the system. It is this property, together with the next one, that makes spread-spectrum modulation attractive for military applications.

6. *Low probability of interception (LPI).* Because of its low power density, the spread-spectrum signal is difficult to detect and intercept by a hostile listener.

Figure 2.2   Interference rejection.

A general classification of CDMA is given in Figure 2.3. There are a number of modulation techniques that generate spread-spectrum signals. We briefly discuss the most important ones:

- *Direct sequence spread-spectrum.* The information-bearing signal is multiplied directly by a high chip rate spreading code.

- *Frequency-hopping spread-spectrum.* The carrier frequency at which the information-bearing signal is transmitted is rapidly changed according to the spreading code.

- *Time-hopping spread-spectrum.* The information-bearing signal is not transmitted continuously. Instead, the signal is transmitted in short bursts where the times of the bursts are decided by the spreading code.

- *Hybrid modulation.* Two or more of the above-mentioned SS modulation techniques can be used together to combine the advantages and, it is hoped, to combat their disadvantages. Furthermore, it is possible to combine CDMA with other multiple access methods: TDMA, multicarrier (MC), or multitone (MT) modulation. In the case of MC-CDMA, spreading is done along the frequency axis, while for MT-CDMA, spreading is done along the time axis. Note that MC-CDMA and MT-CDMA are based on orthogonal frequency division multiplexing (OFDM).

In the next section the above-mentioned pure CDMA modulation techniques are used to show the multiple access capability of CDMA. The remaining chapters will mainly concentrate on direct sequence (DS)-CDMA and its related subjects.

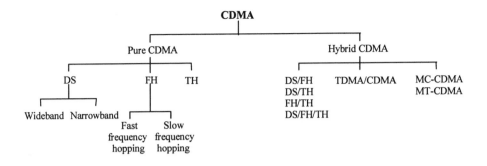

Figure 2.3   Classification of CDMA.

2.2.1   Spread-Spectrum Multiple Access

2.2.1.1   Direct Sequence

In DS-CDMA the modulated information-bearing signal (the data signal) is directly modulated by a digital, discrete time, discrete valued code signal. The data signal can be either an analog signal or a digital one. In most cases it is a digital signal. In the case of a digital signal, the data modulation is often omitted and the data signal is directly multiplied by the code signal and the resulting signal modulates the wideband carrier. It is from this direct multiplication that the direct sequence CDMA gets its name.

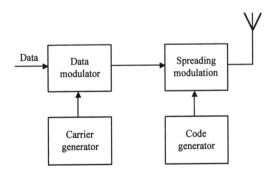

Figure 2.4   Block diagram of a DS-SS transmitter.

In Figure 2.4 a block diagram of a DS-CDMA transmitter is given. The binary data signal modulates a RF carrier. The modulated carrier is then modulated by the code signal. This code signal consists of a number of code bits called "chips" that can be either +1 or −1. To obtain the desired spreading of the signal, the chip rate of the code signal must be much higher than the chip rate of the information signal. For the spreading modulation, various modulation techniques can be used, but usually some form of phase shift keying (PSK) like binary phase shift keying (BPSK), differential binary phase shift keying (D-BPSK), quadrature phase shift keying (QPSK), or minimum shift keying (MSK) is employed.

If we omit the data modulation and use BPSK for the code modulation, we get the block diagram given in Figure 2.5. The DS-SS signal resulting from this transmitter is shown in Figure 2.6. The rate of the code signal is called the *chip rate*; one chip denotes one symbol when referring to spreading code signals. In this figure, 10 code chips per information symbol are transmitted (the code chip rate is 10 times the data rate) so the processing gain is equal to 10.

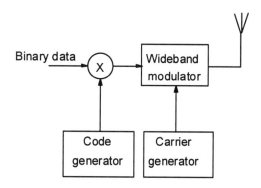

Figure 2.5   Modified block diagram of a DS-SS transmitter.

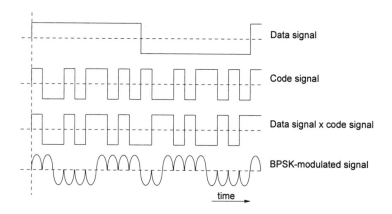

Figure 2.6   Generation of a BPSK-modulated SS signal.

After transmission of the signal, the receiver (shown in Figure 2.7) despreads the

SS signal using a locally generated code sequence. To be able to perform the *despreading* operation, the receiver must not only know the code sequence used to spread the signal, but the codes of the received signal and the locally generated code must also be synchronized. This synchronization must be accomplished at the beginning of the reception and maintained until the whole signal has been received. The code synchronization/tracking block performs this operation. After despreading, a data modulated signal results, and after demodulation, the original data can be recovered.

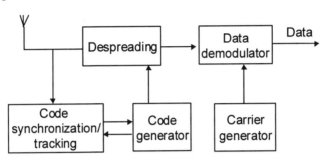

Figure 2.7   Receiver of a DS-SS signal.

In the previous section a number of advantageous properties of spread-spectrum signals were mentioned. The most important of those properties from the viewpoint of CDMA is the multiple access capability, the multipath interference rejection, the narrowband interference rejection, and with respect to secure/private communication, the LPI. We explain these four properties for the case of DS-CDMA.

- *Multiple access.* If multiple users use the channel at the same time, there will be multiple DS signals overlapping in time and frequency. At the receiver, despreading is used to remove the spreading code. This operation concentrates the power of the desired user in the information bandwidth. If the crosscorrelations between the code of the desired user and the codes of the interfering users are small, coherent detection will only put a small part of the power of the interfering signals into the information bandwidth.
- *Multipath interference.* If the code sequence has an ideal autocorrelation function, then the correlation function is zero outside the interval $[-T_c, T_c]$, where T_c is the chip duration. This means that if the desired signal and a version that is delayed for more than $2T_c$ are received, despreading will treat the delayed version as an interfering signal, putting only a small part of the power in the information bandwidth.
- *Narrowband interference.* The coherent detection at the receiver involves a multiplication of the received signal by a locally generated code sequence. As we saw at the transmitter, however, multiplying a narrowband signal with a wideband code sequence spreads the spectrum of the narrowband signal so that its power in the information bandwidth decreases by a factor equal to the

processing gain.

- *LPI.* Because the direct sequence signal uses the whole signal spectrum all the time, it will have a very low transmitted power per hertz. This makes it very difficult to detect a DS signal.

Apart from the above-mentioned properties, DS-CDMA has a number of other specific properties that we can divide into advantageous (+) and disadvantageous (−) behavior:

+ The generation of the coded signal is easy. It can be performed by a simple multiplication.

+ Because only one carrier frequency has to be generated, the frequency synthesizer (carrier generator) is simple.

+ Coherent demodulation of the DS signal is possible.

+ No synchronization among the users is necessary.

− It is difficult to acquire and maintain the synchronization of the locally generated code signal and the received signal. Synchronization has to be kept within a fraction of the chip time.

− For correct reception, the synchronization error of locally generated code sequence and the received code sequence must be very small, a fraction of the chip time. This, combined with the nonavailability of large contiguous frequency bands, practically limits the bandwidth to 10 to 20 MHz.

− The power received from users close to the base station is much higher than that received from users further away. Because a user continuously transmits over the whole bandwidth, a user close to the base will constantly create a lot of interference for users far from the base station, making their reception impossible. This near-far effect can be solved by applying a power control algorithm so that all users are received by the base station with the same average power. This control, however, proves to be quite difficult due to feedback delays, imperfect power estimates, errors in the feedback channel, and traffic conditions.

2.2.1.2   Frequency Hopping

In frequency-hopping CDMA, the carrier frequency of the modulated information signal is not constant but changes periodically. During time intervals T, the carrier frequency remains the same, but after each time interval the carrier hops to another (or possibly the same) frequency. The hopping pattern is decided by the spreading code. The set of available frequencies the carrier can attain is called the *hop-set*.

The frequency occupation of an FH-SS system differs considerably from a DS-SS system. A DS system occupies the whole frequency band when it transmits, whereas an FH system uses only a small part of the bandwidth when it transmits, but the location of this part differs in time.

The difference between the FH-SS and the DS-SS frequency usage is illustrated in Figure 2.8. Suppose an FH system is transmitting in frequency band 2 during the first time period. A DS system transmitting in the same time period spreads its signal power over the whole frequency band so the power transmitted in frequency band 2 will be much less than that of the FH system. The DS system, however, transmits in frequency band 2 during all time periods while the FH system only uses this band part of the time. On average, both systems will transmit the same power in the frequency band.

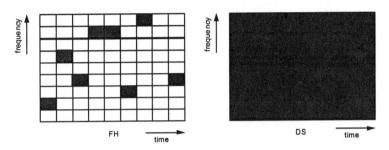

Figure 2.8   Time/frequency occupancy of FH and DS signals.

The block diagram for an FH-CDMA system is given in Figure 2.9. The data signal is baseband modulated. Using a fast frequency synthesizer that is controlled by the code signal, the carrier frequency is converted up to the transmission frequency.

The inverse process takes place at the receiver. Using a locally generated code sequence, the received signal is converted down to the baseband. The data is recovered after (baseband) demodulation. The synchronization/tracking circuit ensures that the hopping of the locally generated carrier synchronizes to the hopping pattern of the received carrier so that correct despreading of the signal is possible.

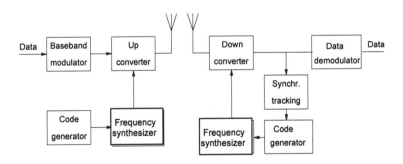

Figure 2.9   Block diagram of an FH-CDMA transmitter and receiver.

Within frequency-hopping CDMA a distinction is made that is based on the hopping rate of the carrier. If the hopping rate is (much) greater than the symbol rate, the modulation is considered to be fast frequency hopping (F-FH). In this case the carrier frequency changes a number of times during the transmission of one symbol, so that one bit is transmitted in different frequencies. If the hopping rate is (much) smaller than the symbol rate, one speaks of slow frequency hopping (S-FH). In this case multiple symbols are transmitted at the same frequency.

The occupied bandwidth of the signal on one of the hopping frequencies depends not only on the bandwidth of the information signal but also on the shape of the hopping signal and the hopping frequency. If the hopping frequency is much smaller than the information bandwidth (which is the case in slow frequency hopping), then the information bandwidth is the main factor that decides the occupied bandwidth. If, however, the hopping frequency is much greater than the information bandwidth, the pulse shape of the hopping signal will decide the occupied bandwidth at one hopping frequency. If this pulse shape is very abrupt (resulting in very abrupt frequency changes), the frequency band will be very broad, limiting the number of hop frequencies. If we make sure that the frequency changes are smooth, the frequency band at each hopping frequency will be about $1/T_h$ times the frequency bandwidth, where T_h is equal to the hopping frequency. We can make the frequency changes smooth by decreasing the transmitted power before a frequency hop and increasing it again when the hopping frequency has changed.

As has been done for DS-CDMA, we discuss the properties of FH-CDMA with respect to multiple access capability, multipath interference rejection, narrowband interference rejection, and probability of interception.

- *Multiple access.* It is easy to visualize how the F-FH and S-FH CDMA obtain their multiple access capability. In the F-FH, one symbol is transmitted in different frequency bands. If the desired user is the only one to transmit in most of the frequency bands, the received power of the desired signal will be much higher than the interfering power and the signal will be received correctly.

  In the S-FH, multiple symbols are transmitted at one frequency. If the probability of other users transmitting in the same frequency band is low enough, the desired user will be received correctly most of the time. For those times that interfering users transmit in the same frequency band, error-correcting codes are used to recover the data transmitted during that period.

- *Multipath interference.* In the F-FH CDMA the carrier frequency changes a number of times during the transmission of one symbol. Thus, a particular signal frequency will be modulated and transmitted on a number of carrier frequencies. The multipath effect is different at the different carrier frequencies. As a result, signal frequencies that are amplified at one carrier frequency will be attenuated at another carrier frequency and vice versa. At the receiver the responses at the different hopping frequencies are averaged, thus reducing the multipath interference. Because noncoherent combining is commonly used, this is not as effective as the multipath interference rejection in a DS-CDMA system, but it still gives quite an improvement.

- *Narrowband interference.* Suppose a narrowband signal is interfering on one of the hopping frequencies. If there are G_p hopping frequencies (where G_p is the processing gain), the desired user will (on the average) use the hopping frequency where the interferer is located $1/G_p$ percent of the time. The interference is therefore reduced by a factor G_p.

- *LPI.* The difficulty in intercepting an FH signal lies not in its low transmission power. During a transmission, it uses as much power per hertz as a continuous transmission. But the frequency at which the signal is going to be transmitted is unknown, and the duration of the transmission at a particular frequency is quite small. Therefore, although the signal is more readily intercepted than a DS signal, it is still a difficult task to perform.

Apart from the above-mentioned properties, the FH-CDMA has a number of other specific properties that we can divide into advantageous (+) and disadvantageous (−) behavior:

- + Synchronization is much easier with FH-CDMA than with DS-CDMA. FH-CDMA synchronization has to be within a fraction of the hop time. Because spectral spreading is not obtained by using a very high hopping frequency but by using a large hop-set, the hop time will be much longer than the chip time of a DS-CDMA system. Thus, an FH-CDMA system allows a larger synchronization error.

- + The different frequency bands that an FH signal can occupy do not have to be contiguous because we can make the frequency synthesizer easily skip over certain parts of the spectrum. Combined with the easier synchronization, this allows much higher spread-spectrum bandwidths.

- + The probability of multiple users transmitting in the same frequency band at the same time is small. A user transmitting far from the base station will be received by it even if users close to the base station are transmitting, since those users will probably be transmitting at different frequencies. Thus, the near-far performance is much better than that of DS.

- + Because of the larger possible bandwidth a FH system can employ, it offers a higher possible reduction of narrowband interference than a DS system.

- − A highly sophisticated frequency synthesizer is necessary.

- − An abrupt change of the signal when changing frequency bands will lead to an increase in the frequency band occupied. To avoid this, the signal has to be turned off and on when changing frequency.

- − Coherent demodulation is difficult because of the problems in maintaining phase relationships during hopping.

2.2.1.3   Time Hopping

In time-hopping CDMA the data signal is transmitted in rapid bursts at time intervals determined by the code assigned to the user. The time axis is divided into frames, and each frame is divided into M time slots. During each frame the user will transmit in one of the M time slots. Which of the M time slots is transmitted depends on the code signal assigned to the user. Since a user transmits all of its data in one, instead of *M* time slots, the frequency it needs for its transmission has increased by a factor *M*. A block diagram of a TH-CDMA system is given in Figure 2.10.

Figure 2.11 shows the time-frequency plot of the TH-CDMA systems. Comparing Figure 2.11 with Figure 2.8, we see that the TH-CDMA uses the whole wideband spectrum for short periods instead of parts of the spectrum all of the time.

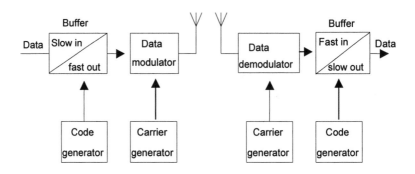

Figure 2.10   Block diagram of a TH-CDMA transmitter and receiver.

Figure 2.11   Time-frequency plot of the TH-CDMA.

Following the same procedure as for the previous CDMA schemes, we discuss the properties of TH-CDMA with respect to multiple access capability, multipath interference rejection, narrowband interference rejection, and probability of interception.

- *Multiple access.* The multiple access capability of TH-SS signals is acquired in the same manner as that of the FH-SS signals, namely, by making the probability of users' transmissions in the same frequency band at the same time small. In the case of time hopping, all transmissions are in the same frequency band, so the probability of more than one transmission at the same time must be small. This is again achieved by assigning different codes to different users. If multiple transmissions do occur, error-correcting codes ensure that the desired signal can still be recovered. If there is synchronization among the users, and the assigned codes are such that no more than one user transmits at a particular slot, then the TH-CDMA reduces to a TDMA scheme where the slot in which a user transmits is not fixed but changes from frame to frame.

- *Multipath interference.* In time-hopping CDMA, a signal is transmitted in reduced time. The signaling rate, therefore, increases, and dispersion of the signal will now lead to overlap of adjacent bits. Therefore, no advantage is to be gained with respect to multipath interference rejection.

- *Narrowband interference.* A TH-CDMA signal is transmitted in reduced time. This reduction is equal to $1/G_p$, where G_p is the processing gain. At the receiver we will only receive an interfering signal during the reception of the desired signal. Thus, we only receive the interfering signal $1/G_p$ percent of the time, reducing the interfering power by a factor G_p.

- *LPI.* With TH-CDMA, the frequency at which a user transmits is constant but the times at which a user transmits are unknown, and the durations of the transmissions are very short. This makes it difficult for an intercepting receiver to distinguish the beginning and end of a transmission and to decide which transmissions belong to which user, particularly when multiple users are transmitting.

Apart from the above-mentioned properties, the TH-CDMA has a number of other specific properties that we can divide into advantageous (+) and disadvantageous (−) behavior:

- + Implementation is simpler than that of FH-CDMA.
- + It is a very useful method when the transmitter is average-power limited but not peak-power limited since the data are transmitted in short bursts at high power.
- + As with the FH-CDMA, the near-far problem is much less of a problem since most of the time a terminal far from the base station transmits alone, and is not hindered by transmissions from stations close by.
- − It takes a long time before the code is synchronized, and the time in which the receiver has to perform the synchronization is short.
- − If multiple transmissions occur, a large number of data bits are lost, so a good error-correcting code and data interleaving are necessary.

2.2.1.4 Hybrid Systems

The hybrid CDMA systems include all CDMA systems that employ a combination of two or more of the above-mentioned spread-spectrum modulation techniques or a combination of CDMA with some other multiple access technique. By combining the basic spread-spectrum modulation techniques, we have four possible hybrid systems: DS/FH, DS/TH, FH/TH, and DS/FH/TH; and by combining CDMA with TDMA or multicarrier modulation we get two more: CDMA/TDMA and MC-CDMA.

The idea of the hybrid system is to combine the specific advantages of each of the modulation techniques. If we take, for example, the combined DS/FH system, we have the advantage of the anti-multipath property of the DS system combined with the favorable near-far operation of the FH system. Of course, the disadvantage lies in the increased complexity of the transmitter and receiver. For illustration purposes, we give a block diagram of a combined DS/FH CDMA transmitter in Figure 2.12.

The data signal is first spread using a DS code signal. The spread signal is then modulated on a carrier whose frequency hops according to another code sequence. A code clock ensures a fixed relation between the two codes.

2.3   BASIC DS-CDMA ELEMENTS

In this section, we review the fundamental elements for understanding direct sequence CDMA and its application into third-generation systems, namely, RAKE receiver, power control, soft handover, interfrequency handover, and multiuser detection.

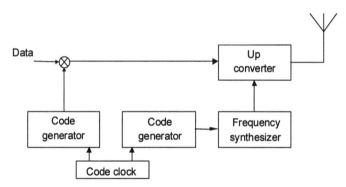

Figure 2.12   Hybrid DS-FH transmitter.

2.3.1   RAKE Receiver

A DS spread-spectrum signal waveform is well matched to the multipath channel. In a multipath channel, the original transmitted signal reflects from obstacles such as buildings and mountains, and the receiver receives several copies of the signal with different delays. If the signals arrive more than one chip apart from each other, the receiver can resolve them. Actually, from each multipath signal's point of view, other multipath signals can be regarded as interference and they are suppressed by the processing gain. A further benefit is obtained if the resolved multipath signals are combined using *RAKE receiver*. Thus, the signal waveform of CDMA signals facilitates utilization of multipath diversity. Expressing the same phenomenon in the frequency domain means that the bandwidth of the transmitted signal is larger than the coherence bandwidth of the channel and the channel is frequency selective (i.e., only part of the signal is affected by the fading).

RAKE receiver consists of correlators, each receiving a multipath signal. After despreading by correlators, the signals are combined using, for example, maximal ratio combining. Because the received multipath signals are fading independently, diversity order and thus performance are improved. Figure 2.13 illustrates the principle of RAKE receiver. After spreading and modulation, the signal is transmitted and it passes through a multipath channel, which can be modeled by a tapped delay line (i.e., the reflected signals are delayed and attenuated in the channel). In Figure 2.13 we have three multipath components with different delays (τ_1, τ_2, and τ_3) and attenuation factors (a_1, a_2, and a_3), each corresponding to a different propagation path. The RAKE receiver has a receiver finger for each multipath component. In each finger, the received signal is correlated by a spreading code, which is time-aligned with the delay of the multipath signal. After despreading, the signals are weighted and combined. In Figure 2.13, maximal ratio combining is used. That is, each signal is weighted by the path gain (attenuation factor). Due to the mobile movement, the scattering environment will change, and thus, the delays and attenuation factors will change as well. Therefore, it is necessary to measure the tapped delay line profile and to reallocate RAKE fingers whenever the delays have changed by a significant amount. Small-scale changes, less than one chip, are taken care of by a code tracking loop, which tracks the time delay of each multipath signal.

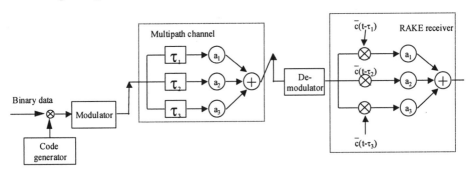

Figure 2.13 Principle of RAKE receiver.

2.3.2 Power Control

In the uplink of a DS-CDMA system, the requirement for power control is the most serious negative point. The power control problem arises because of the multiple access interference. All users in a DS-CDMA system transmit messages by using the same bandwidth at the same time, and therefore, users interfere with one another. Because of the propagation mechanism, the signal received by the base station from a user terminal close to the base station will be stronger than the signal received from another terminal located at the cell boundary. Hence, the distant users will be dominated by the close user. This is called the *near-far effect*. To achieve a considerable capacity, all signals, irrespective of distance, should arrive at the base station with the same mean power. A solution to this problem is power control, which attempts to achieve a constant received mean power for each user. Therefore, the performance of the transmitter power control (TPC) is one of the several dependent factors when deciding on the capacity of a DS-CDMA system.

In contrast to the uplink, in the downlink all signals propagate through the same channel and thus are received by a mobile station with equal power. Therefore, no power control is required to eliminate the near-far problem. The power control is, however, required to minimize the interference to other cells and to compensate for against the interference from other cells. The worst-case situation for a mobile station occurs when the mobile station is at the cell edge, equidistant from three base stations. The interference from other cells, however, does not vary very abruptly.

In addition to being useful against interfering users, power control improves the performance of DS-CDMA against fading channel by compensating the fading dips. If it followed the channel fading perfectly, power control would turn a fading channel into an AWGN channel by eliminating the fading dips completely.

There exist two types of power control principles: open loop and closed loop. The open loop power control measures the interference conditions from the channel and adjusts the transmission power accordingly to meet the desired frame error rate (FER) target. Because the fast fading does not correlate between uplink and downlink, however, open loop power control will achieve the right power target only on average. Therefore, closed loop power control is required. The closed loop power control measures the signal-to-interference ratio (SIR) and sends commands to the transmitter on the other end to adjust the transmission power.

Section 2.4.3 illustrates the different power control principles by explaining the IS-95 power control scheme. Power control is discussed in more detail in Chapters 5 and 9.

2.3.3   Soft Handover

In soft handover a mobile station is connected to more than one base station simultaneously. Soft handover is used in CDMA to reduce the interference into other cells and to improve performance through macro diversity. Softer handover is a soft handover between two sectors of a cell.

Neighboring cells of a cellular system using either FDMA or TDMA do not use the frequencies used by the given cell (i.e., there is spatial separation between cells using the same frequencies). This is called the frequency reuse concept. Because of the processing gain, such spatial separation is not needed in CDMA, and a frequency reuse factor of one can be used. Usually, a mobile station performs a handover when the signal strength of a neighboring cell exceeds the signal strength of the current cell with a given threshold. This is called *hard handover*. Since in a CDMA system the neighboring cell frequencies are the same as in the given cell, this type of approach would cause excessive interference into the neighboring cells and thus a capacity degradation. In order to avoid this interference, an instantaneous handover from the current cell to the new cell would be required when the signal strength of the new cell exceeds the signal strength of the current cell. This is not, however, feasible in practice. The handover mechanism should always allow the mobile station to connect into a cell, which it receives with the highest power (i.e., with the lowest path loss). Since in soft handover the mobile station is connected to either two or more base stations, its transmission power can be controlled according to the cell, which the mobile station receives with the highest signal strength. A mobile station enters the soft handover state when the signal strength of a neighboring cell exceeds a certain threshold but is still below the current base station's signal strength.

Fortunately, the signal structure of CDMA is well suited for the implementation of soft handover. This is because in the uplink, two or more base stations can receive the same signal because of the reuse factor of one; and in the downlink the mobile station can coherently combine the signals from different base stations since it sees them as just additional multipath components. This provides an additional benefit called macro diversity (i.e., the diversity gain provided by the reception of one or more additional signals). A separate pilot channel is usually used for the signal strength measurements for handover purposes (see Section 2.4.1).

In the downlink, however, soft handover creates more interference to the system since the new base station now transmits an additional signal for the mobile station. It is possible that the mobile station cannot collect all of the energy that the base station transmits due to a limited number of RAKE fingers. Thus, the gain of soft handover in the downlink depends on the gain of macro diversity and the loss of performance due to increased interference.

Figure 2.14 illustrates the soft handover principle with two base stations involved. In the uplink the mobile station signal is received by the two base stations, which, after demodulation and combining, pass the signal forward to the combining point, typically to the base station controller (BSC). In the downlink the same information is transmitted via both base stations, and the mobile station receives the information from two base stations as separate multipath signals and can therefore combine them.

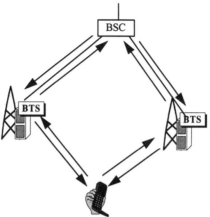

Figure 2.14   Principle of soft handover with two base station transceivers (BTS).

2.3.4   Interfrequency Handover

The third-generation CDMA networks will have multiple frequency carriers in each cell, and a hot-spot cell could have a larger number of frequencies than neigboring cells. Furthermore, in hierarchical cell structures, micro cells will have a different frequency than the macro cell overlaying the micro cells. Therefore, an efficient procedure is needed for a handover between different frequencies. A blind handover used by second-generation CDMA does not result in an adequate call quality. Instead, the mobile station has to be able to measure the signal strength and quality of an another carrier frequency, while still maintaining the connection in the current carrier frequency. Because a CDMA transmission is continuous, there are no idle slots for the interfrequency measurements as in the TDMA-based systems. Therefore, compressed mode and dual receiver have been proposed as a solution to interfrequency handover [13] (see Section 6.3.7). In the compressed mode, measurement slots are created by transmitting the data of a frame (for example, with a lower spreading ratio during a shorter period), and the rest of the frame is utilized for the measurements on other carriers. The dual receiver can measure other frequencies without affecting the reception of the current frequency.

2.3.5   Multiuser Detection

The current CDMA receivers are based on the RAKE receiver principle, which considers other users' signals as interference. In an optimum receiver, however, all signals would be detected jointly or interference from other signals would be removed by subtracting them from the desired signal. This is possible because the correlation properties between signals are known (i.e., the interference is deterministic not random).

The capacity of a direct sequence CDMA system using RAKE receiver is interference limited. In practice this means that when a new user, or interferer, enters the network, other users' service quality will degrade. The more the network can resist interference, the more users can be served. Multiple access interference that disturbs a base or mobile station is a sum of both intra- and intercell interference.

Multiuser detection (MUD), also called joint detection and interference cancellation (IC), provides means of reducing the effect of multiple access interference, and hence increases the system capacity. In the first place, MUD is considered to cancel only the intracell interference, meaning that in a practical system the capacity will be limited by the efficiency of the algorithm and the intercell interference.

In addition to capacity improvement, MUD alleviates the near-far problem typical to DS-CDMA systems. A mobile station close to a base station may block the whole cell traffic by using too high a transmission power. If this user is detected first and subtracted from the input signal, the other users do not see the interference.

Because optimal multiuser detection is very complex and in practice impossible to implement for any reasonable number of users, a number of suboptimum multiuser and interference cancellation receivers have been developed. The suboptimum receivers can be divided into two main categories: linear detectors and interference cancellation. Linear detectors apply a linear transform into the outputs of the matched filters that are trying to remove the multiple access interference (i.e., the interference due to correlations between user codes). Examples of linear detectors are decorrelator and linear minimum mean square error (LMMSE) detectors. In interference cancellation, multiple access interference is first estimated and then subtracted from the received signal. Parallel interference cancellation (PIC) and successive (serial) interference cancellation (SIC) are examples of interference cancellation. For a more detailed treatment of multiuser detection and interference cancellation, refer to Chapters 5 and 10 as well as [14–17].

2.4   IS-95 CDMA

In this section, we describe the features of the IS-95 air interface according to the IS-95B standard, with a focus on the new downlink and uplink channel structure [18].[1] Main air interface parameters, downlink and uplink channel structures, power control principles, and speech coding are discussed. For a more detailed treatment of the IS-95A standard, refer to [19], and for a theoretical analysis of IS-95 air interface solutions, refer to [8].

The IS-95 air interface standard, after the first revision in 1995, was termed IS-95A [20]; it specifies the air interface for cellular, 800-MHz frequency band. ANSI J-STD-008 specifies the PCS version (i.e., the air interface for 1900 MHz). It differs from IS-95A primarily in the frequency plan and in call processing related to subscriber station identity, such as paging and call origination. TSB74 specifies the Rate Set 2 (14.4 Kbps) standard. IS-95B merges the IS-95A, ANSI J-STD-008 [21], and TSB74 standards, and, in addition, it specifies the high speed data operation using up to eight parallel codes, resulting in a maximum bit rate of 115.2 Kbps. In addition to these air interface specifications, the IS-97 [22] and IS-98 [23] standards specify the minimum

---

[1] In this book, the terms uplink and downlink are used instead of the forward and reverse link, which are used in the IS-95 standard.

performance specifications for the mobile and base station, respectively.

Table 2.1 lists the main parameters of the IS-95 air interface. Carrier spacing of the system is 1.25 MHz. Practical deployment has shown that three CDMA carriers can be fitted into 5-MHz bandwidth due to required guard bands. The network is synchronous within a few microseconds. This facilitates use of the same long code sequence with different phase offsets as pilot sequences. An external reference signal, however, such as Global Positioning System (GPS) is needed.

2.4.1   Downlink Channel Structure

Figure 2.15 shows the downlink physical channel structure. The pilot channel, the paging channel, and the synchronization channel[2] are common control channels, and traffic channels are dedicated channels. A common channel is a shared channel, and a dedicated channel is solely allocated for the use of a single user. Data to be transmitted on synchronization, paging, and traffic channels are first grouped into 20-ms frames, convolutionally encoded, repeated to adjust the data rate, and interleaved. Then the signal is spread with an orthogonal Walsh code at a rate of 1.2288 Mcps, split into the I and Q channels, and, prior to baseband filtering, spread with long PN sequences at a rate of 1.2288 Mcps.

Table 2.1
IS-95 Air Interface Parameters

Bandwidth	1.25 MHz
Chip rate	1.2288 Mcps
Frequency band uplink	869 – 894 MHz 1930 – 1980 MHz
Frequency band downlink	824 – 849 MHz 1850 – 1910 MHz
Frame length	20 ms
Bit rates	Rate set 1: 9.6 Kbps Rate set 2: 14.4 Kbps IS-95B: 115.2 Kbps
Speech codec	QCELP 8 Kbps EVRC 8 Kbps ACELP 13 Kbps
Soft handover	Yes
Power control	Uplink: Open loop + fast closed loop Downlink: Slow quality loop
Number of RAKE fingers	4
Spreading codes	Walsh+Long M-sequence

A mobile station uses the pilot channel for coherent demodulation, acquisition, time delay tracking, power control measurements, and as an aid for the handover. In order to obtain a reliable phase reference for coherent demodulation, the pilot channel is transmitted with higher power than the traffic channels. Typically about 20% of the

---

[2] In the IS-95 standard the synchronization channel is actually termed "Sync Channel."

radiated power on the downlink is dedicated to the pilot signal. After obtaining phase and code synchronization, the mobile station acquires synchronization information (data rate of the paging channel, time of the base station's pilot PN sequence with respect to the system time) from the synchronization channel. Because the synchronization channel frame has the same length as the pilot sequence, acquisition of the synchronization channel takes place easily. The synchronization channel operates at a fixed rate of 1.2 Kbps. The paging channel is used to page a mobile station. The paging channel has a fixed data rate of 9.6 or 4.8 Kbps.

Each downlink traffic channel contains one fundamental code channel and may contain one to seven supplemental code channels. The traffic channel has two different rate sets. The rate set 1 supports data rates of 9.6, 4.8, 2.4, and 1.2 Kbps (Figure 2.15(a)) and the rate set 2 supports 14.4, 7.2, 3.6, and 1.8 Kbps (Figure 2.15(b)). Only the full rate (9.6 or 14.4 Kbps) may be utilized on the supplemental code channels. The mobile station always supports the rate set 1 and it may support the rate set 2. To achieve equal power levels at the base station receiver, the base station measures the received signal and adjusts each mobile station's power levels accordingly. The 20-ms frame is divided into 16 power control groups with a duration of 1.25 ms. One power control bit is multiplexed in for the fundamental code channel for each power control group.

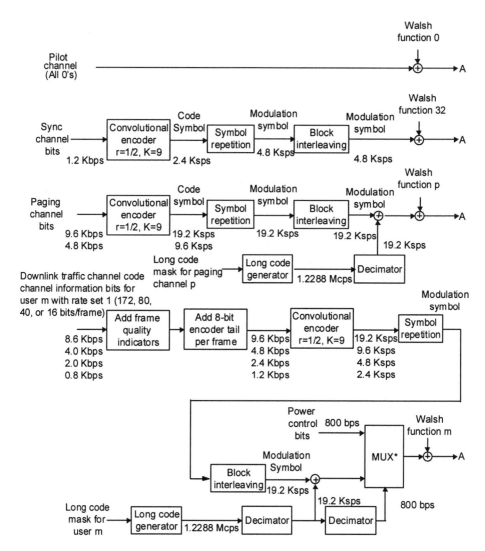

Figure 2.15(a) Rate set 1 downlink CDMA channel structure. (*Source*: [18], reproduced under written permission of the copyright holder Telecommunications Industry Association.)

Downlink traffic channel code
channel information bits for
user m with rate set 2 (267, 125,
55, or 21 bits/frame)

13.35 Kbps
6.25 Kbps
2.75 Kbps
1.05 Kbps

Add one reserved/flag bit

Add frame quality indicators

Add 8-bit encoder tail per frame

14.4 Kbps
7.2 Kbps
3.6 Kbps
1.8 Kbps

Convolutional encoder r=1/2, K=9

Code symbol

Symbol repetition

Repeated symbol

Puncture (delete) 2 of every 6 inputs

Modulation symbol

19.2 Ksps

28.8 Ksps
14.4 Ksps
7.2 Ksps
3.6 Ksps

28.8 Ksps

Power control bits 800 bps

Walsh function m

MUX*

→ A

Block interleaving

Modulation symbol

19.2 Ksps

19.2 Ksps

800 bps

Long code mask for user m

Long code generator 1.2288 Mcps

Decimator

Decimator

I-channel pilot PN sequence
1.2288 Mcps

A

I

Baseband filter

I(t)

$\cos(2\pi f_c t)$

Q

Baseband filter

Q(t)

$\sin(2\pi f_c t)$

Σ → s(t)

Q-channel pilot PN
sequence

* Power control bits are not multiplexed in for supplemental
code channels of the downlink traffic channels

Figure 2.15(b) Rate set 2 downlink CDMA channel structure. (*Source*: [18], reproduced under written permission of the copyright holder Telecommunications Industry Association.)

The transmitted data is encoded by a convolutional code with a constraint length of 9. The generator functions for this code are 753 (octal) and 561 (octal). For the synchronization channel, the paging channels, and rate set 1 on the traffic channel, a convolutional code with a rate of 1/2 is used. For the rate set 2, an effective code rate of 3/4 is achieved by puncturing two out of every six symbols after the symbol repetition.

Because the data rate on different channels varies, symbol repetition is used to achieve a fixed data rate prior to interleaving. For the synchronization channel, each convolutionally encoded symbol shall be repeated once (i.e., each symbol occurs two consecutive times). For the paging channel, each code symbol at the 4800-bps rate shall be repeated once. The code symbol repetition rate on the forward traffic channels varies with data rate. Code symbols are not repeated for the 14.4- and 9.6-Kbps data rates. Each code symbol at the 7.2- and 4.8-Kbps data rates is repeated once, at the 3.6- and 2.4-Kbps data rates three times, and at the 1.8- and 1.2-Kbps data rates seven times.

In the downlink, three types of spreading codes are used. Walsh codes of length 64 at a fixed chip rate of 1.2288 Mcps separate the physical channels. The Walsh function consisting of all zeros (W0, Walsh code number 0) is used for the pilot channel, W1-W7 are used for paging channels (unused paging channel codes can be used for traffic channels). The synchronization channel is W32, and traffic channels are W8 to W31 and W33 to W63. A pair of long M-sequences of length 16,767 ($2^{15}-1$) is used for quadrature spreading, one for the I channel and one for the Q channel. Quaternary spreading is used to obtain better interference averaging. Since the pilot channel Walsh function is all zeros, this pair of M-sequences also forms the pilot code. Different cells and sectors are distinguished with the different phase offsets of this code.

A long pseudo-random sequence with a period of $2^{42}-1$ is used for base band data scrambling (i.e., to encrypt the signal on the paging and traffic channels). It is decimated from a 1.2288-Mcps rate down to 19.2 Kbps. The long pseudo-noise sequence is the same used in the uplink for user separation, and it is generated by a modulo-2 inner product of a 42-bit mask and the 42-bit state vector of the sequence generator.

2.4.2 Uplink Channel Structure

As depicted in Figures 2.16 to 2.19, the uplink has two physical channels: a traffic channel, which is a dedicated channel, and a common access channel. A traffic channel consists of a single fundamental code channel and zero through seven supplemental code channels. Similar to the downlink, traffic channels always support the rate set 1 data rates and may support the rate set 2 data rates. The supplemental code channel can only use the full rates (9.6 or 14.4 Kbps). Data transmitted on the uplink channels are grouped into 20-ms frames, convolutionally encoded, block interleaved, and modulated by 64-ary orthogonal modulation. Then, prior to baseband filtering, the signal is spread with a long PN sequence at a rate of 1.2288 Mcps, split into the I and Q channels, and spread with in-phase and quadrature spreading sequences.

The access channel supports fixed data rate operation at 4.8 Kbps. The access channel is used by a mobile station to initiate a call, to respond to a paging channel message from the base station, and for a location update. Each access channel is associated with a downlink paging channel, and consequently there can be up to seven access channels.

The transmitted information is encoded using a convolutional code with constraint length 9 and the same generator polynomials as in the downlink. For the access channel and rate set 1 on the traffic channels, the convolutional code rate is 1/3. For rate set 2 on the traffic channels, a code rate of 1/2 is used. Similar to the downlink, code symbols output from the convolutional encoder are repeated before being interleaved when the data rate is lower than 9.6 Kbps for rate set 1 and 14.4 Kbps for rate set 2. The repeated symbols, however, are not actually transmitted. They are masked out according to a masking pattern generated by the data burst randomizer to save transmission power. For the access channel, which has a fixed data rate of 4.8 Kbps, each code symbol is repeated once. In contrast to the traffic channel, the repeated code symbols are transmitted.

The coded symbols are grouped into six-symbol groups. These groups are then used to select one of 64 possible Walsh symbols (i.e., a 64-ary orthogonal modulation is carried out to obtain good performance for noncoherent modulation). After the orthogonal modulation, the transmission rate is 307.2 Kbps. The reason to use the non-coherent modulation is the difficulty in obtaining good phase reference for coherent

Figure 2.16 Uplink CDMA channel structure for the access channel. (*Source*: [18], reproduced under written permission of the copyright holder Telecommunications Industry Association.)

demodulation in the uplink. It should be noted how the Walsh codes are used differently in the uplink and downlink. In the downlink, they are used for channelization, while in the uplink they are used for orthogonal modulation.

Each code channel in a traffic channel and each access channel are identified by a different phase of a pseudo-random M-sequence with a length of 2^{42} (see Section 5.6.2). The in-phase and quadrature spreading is performed by the same pair of M-sequences (length 2^{15}) as in the downlink (now augmented by one chip).

Figure 2.17 Uplink CDMA channel structure for fundamental code channels with rate set 1. (*Source*: [17], reproduced under written permission of the copyright holder Telecommunications Industry Association.)

Figure 2.18 Uplink CDMA channel structure for supplemental code channels with rate set 1. (*Source*: [18], reproduced under written permission of the copyright holder Telecommunications Industry Association.)

2.4.3 Power Control

IS-95 has three different power control mechanisms. In the uplink, both open loop and fast closed loop power control are employed. In the downlink, a relatively slow power control loop controls the transmission power.

2.4.3.1 Open Loop Power Control

The open loop power control has two main functions: it adjusts the initial access channel transmission power of the mobile station and compensates large abrupt variations in the pathloss attenuation. The mobile station determines an estimate of the pathloss between the base station and mobile station by measuring the received signal strength at the mobile using an automatic gain control (AGC) circuitry, which gives a rough estimate of the propagation loss for each user. The smaller the received power, the larger the propagation loss, and vice versa. The transmit power of the mobile station is determined from the equation:

mean output power (dBm) = −mean input power (dBm) + offset power + parameters   (2.2)

The offset power for the 800-MHz band mobiles (band class 0) is −73 dB and for the 1900-MHz band mobiles (band class 1) −76 dB [18]. The parameters are used to adjust the open loop power control for different cell sizes and different cell effective radiated powers (ERP) and receiver sensitivities [24]. These parameters are initially transmitted on the synchronization channel.

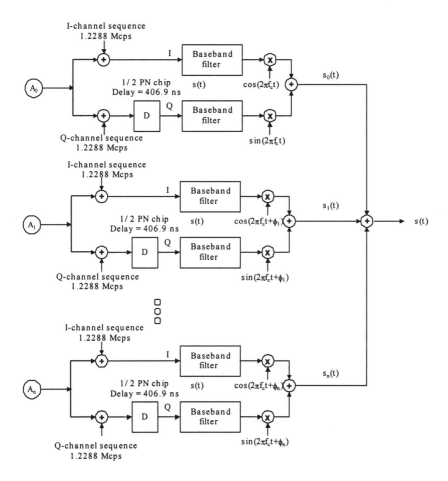

Figure 2.19 Uplink traffic channel structure including fundamental code channel and multiple supplemental code channels with rate set 1. (*Source*: [18], reproduced under written permission of the copyright holder Telecommunications Industry Association.)

The open loop power control principle is described in Figure 2.20. Because the distance (d1) of mobile station 1 to the base station (BTS) is shorter than the distance of mobile station 2 (d2) to the BTS, the signal received by the mobile station 1 has a smaller propagation loss. Assume that the mean input power of the mobile station 1 is −70 dBm (100 pW)[3] and the mean input power of the mobile station 2 is −90 dBm (1 pW). For band class 0 mobiles with no correction parameters, the mobile station transmission power to achieve equal received powers at the base station can be calculated from (2.2) to be 17 dBm (50 mW) and −7 dBm (200 µW), respectively.

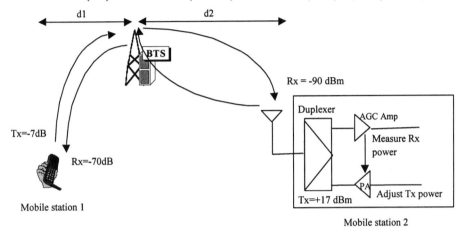

Figure 2.20 Uplink open loop power control principle.

2.4.3.2   Closed Loop Power Control

Because the IS-95 uplink and downlink have a frequency separation of 20 MHz, their fading processes are not strongly correlated. Even though the average power is approximately the same, the short term power is different, and therefore, the open loop power control cannot compensate for the uplink fading. To account for the independence of the Rayleigh fading in the uplink and downlink, the base station also controls the mobile station transmission power. Figure 2.21 illustrates the closed loop power control. The base station measures the received SIR[4] over a 1.25-ms period, equivalent to six modulation symbols, compares that to the target SIR, and decides whether the mobile station transmission power needs to be increased or decreased. The power control bits are transmitted on the downlink fundamental code channel every 1.25 ms (i.e., with a transmission rate of 800 Hz) by puncturing the data symbols. The placement of a power control bit is randomized within the 1.25-ms power control group.

---

[3] 1 dBm means 1dB over 1 mW. For example, −70 dBm is 70 dB (10 million times) less than 1 mW (i.e., 1E-12 W = 1 picoWatt).
[4] The IS-95 standard suggests that the received signal strength should be measured. In practice, however, usually the SIR or the received bit energy to noise density (E_b/I_0) is used, since they have direct impact on the bit error rate (BER).

The transmission occurs in the second power control group following the corresponding uplink traffic channel power control group in which the SIR was estimated.[5]

Because the power control commands are transmitted uncoded, their error ratio is fairly high, on the order of 5%. Because the loop is of delta modulation type (i.e., power is adjusted continuously up or down), this is tolerable. The mobile station extracts the power control bits commands and adjusts its transmission power accordingly. The adjustment step is a system parameter and can be 0.25, 0.5, or 1.0 dB. The dynamic range for the closed loop power control is ±24 dB. The composite dynamic range for open and closed loop power control is ±32 dB for mobile stations operating in band class 0, and ±40 dB for mobile stations operating in band class 1 [18]. The typical standard deviation of the power control error due to the closed loop is on the order of 1.1 to 1.5 dB [8]

The SIR required to produce a certain bit error rate varies according to radio environment and depends on the amount and type of multipath. Therefore, IS-95 employs an outer loop that adjusts the target SIR. The base station measures the signal quality (bit error rate), and based on that determines the target SIR. This outer loop, however, will increase the power control error, resulting in a total standard deviation of 1.5 to 2.5 dB [8].

Figure 2.21   Closed loop power control principle.

2.4.3.3   Downlink Slow Power Control

The base station controls its transmission power to a given mobile station according to

---

[5] For instance, if the signal is received on the reverse traffic channel in power control group number 5, and the corresponding power control bit is transmitted on the forward traffic channel during power control group number 5 + 2 = 7 [18].

the path loss and interference situation. The main purpose of the slow downlink power control is to improve the performance of mobile stations at a cell edge where the signal is weak and the interfering base station signals are strong. The downlink power control mechanism is as follows. The base station periodically reduces the transmitted power to the mobile station. The mobile station measures the frame error ratio (FER). When the FER exceeds a predefined limit, typically 1%, the mobile station requests additional power from the base station. This adjustment occurs every 15 to 20 ms. The dynamic range of the downlink power control is only ±6 dB.

Both periodic and threshold reporting may be enabled simultaneously, either one of them may be enabled, or both forms of reporting may be disabled at any given time.

2.4.4   Speech Codecs and Discontinuous Transmission

IS-95 has three speech codecs, 8-Kbps QCELP, 8-Kbps EVRC, and 13 Kbps. The higher rate codec was developed to provide better voice quality, but due to its higher bit rate it reduces the system capacity. Therefore, the enhanced variable rate codec (EVRC) operating at 8 Kbps was developed. Speech codecs are four-rate code excited linear prediction (CELP) codecs. The vocoder rates of the 8-Kbps codecs are 1, 2, 4, and 8 Kbps corresponding to channel rates of 1.2, 2.4, 4.8, and 9.6 Kbps (rate set 1). The 13-Kbps codec uses a 14.4 Kbps channel rate (rate set 2). Because the system capacity is directly proportional to interference, reduction of the transmitted data rate results in better capacity. In IS-95, data rate reduction is implemented with discontinuous transmission (DTX), which is realized by gating the transmitter in pseudo-random fashion on and off. The drawback of this approach is that it creates pulsed interference in the uplink.

REFERENCES

[1]   Prasad, R., *CDMA for Wireless Personal Communications*, Norwood, Massachusetts: Artech House, 1996.

[2]   Special issue on spread-spectrum communication, *IEEE Transactions on Communications*, Vol. COM-30, May 1982.

[3]   Simon, M. K., J. K. Omura, R. A. Scholtz, and B. K. Levitt, *Spread-Spectrum Communications*, Vol. I, II, III, Comp. Sci., 1985.

[4]   Scholtz, R. A., "The Spread-Spectrum Concept," *IEEE Transactions on Communications*, Vol. COM-25, August 1977, pp. 74–55.

[5]   Torrieri, D. J., *Principles of Secure Communication Systems*, Norwood, Massachusetts: Artech House, 1985.

[6]   Cooper, G. R., and C. D. McGillem, *Modern Communication and Spread-Spectrum*, New York: McGraw-Hill, 1986.

[7]   Glisic, S. G., and P. A. Leppanen (eds.), *Code Divison Multiple Access Communications*, Boston: Kluwer Academic Publishers, 1995.

[8]   Viterbi, A. J., *CDMA Principles of Spread-Spectrum Communications*, Reading,

Massachusetts: Addison-Wesley Publishing Company, 1995.

[9] Dixon, R. C., *Spread-Spectrum Systems,* New York: John Wiley & Sons, 1984.

[10] Glisic, S., and B. Vucetic, *Spread Spectrum CDMA Systems for Wireless Communications*, Norwood, Massachusetts: Artech House, 1997.

[11] Glisic, S., and P. Leppänen (eds.), *Wireless Communications: TDMA versus CDMA*, Boston: Kluwer Academic Publishers, 1997.

[12] Holma, H., and A. Toskala (eds.), *WCDMA for UMTS*, Chichester, England: John Wiley & Sons, 2000.

[13] Gustafsson, M., K. Jamal, and E. Dahlman, "Compressed Mode Techniques for Interfrequency Measurements in a Wideband DS-CDMA System," *Proceedings of PIMRC'97*, Helsinki, Finland, September 1997, pp. 23–35.

[14] Verdú, S., "Adaptive Multiuser Detection," in *Code Division Multiple Access Communications*, S. G. Glisic and P. A. Leppänen (eds.), Boston: Kluwer Academic Publishers, 1994, pp. 97–116.

[15] Moshavi, S., "Multiuser Detection for DS-CDMA Communications," *IEEE Communications Magazine*, October 1996, pp. 12–36.

[16] Duel-Hallen, A., J. Holtzman, and Z. Zvonar, "Multiuser Detection for CDMA Systems," *IEEE Personal Communications,* April 1995, pp. 4–8.

[17] Juntti, M., and S. Glisic, "Advanced CDMA for Wireless Communications," in *Wireless Communications TDMA versus CDMA*, S. Glisic and P. Leppanen (eds.), Boston: Kluwer Academic Publishers, 1997, pp. 447–490.

[18] TIA/EIA/SP-3693, "Mobile Station-Base Station Compatibility Standard for Dual-Mode Wideband Spread Spectrum Cellular Systems," published as ANSI/TIA/EIA-95B.[6]

[19] Garg, V. K., K. Smolik, and J. E. Wilkes, *Applications of CDMA in Wireless/Personal Communications*, Upper Saddle River, New Jersey: Prentice-Hall, 1997.

[20] TIA/EIA-95A, "Mobile Station-Base Station Compatibility Standard for Dual-Mode Wideband Spread Spectrum Cellular Systems," 1995.

[21] ANSI J-STD-008, "Personal Station-Base Station Compatibility Requirements for 1.8 to 2.0 GHz Code Division Multiple Access (CDMA) Personal Communications Systems," 1995.

[22] TIA/EIA/IS-97, "Recommended Minimum Performance Standards for Base Stations Supporting Dual-Mode Wideband Spread Spectrum Cellular Mobile Stations," December 1994.

[23] TIA/EIA/IS-98, "Recommended Minimum Performance Standards for Dual-Mode Wideband Spread Spectrum Cellular Mobile Stations," December 1994.

[24] Ross, A. H. M., and K. L. Gilhousen, "CDMA Technology and the IS-95 North American Standard," in *The Mobile Communications Handbook*, J. D. Gibson (ed.), Boca Raton, Florida: CRC Press, 1996, pp. 43–48.

---

[6] To purchase complete version of any TIA document, call Global Engineering Documents at 800-854-7179 or send a facsimile to 303-397-2740.

Chapter 3

MOBILE INTERNET

3.1   INTRODUCTION

The number of cellular subscribers has exceeded the number of personal computers as shown in Figure 3.1. In addition to the mobile revolution that has happened during the last 10 years, the Internet has been the second big revolution in the information technology industry. In the most advanced countries the Internet penetration exceeds already 40% and the growth continues in very fast pace. Figure 3.2 shows the increase in the number of Internet hosts. One reason for the increased Internet usage is the emergence of the World Wide Web (WWW), which can be accessed through a Web browser supported by virtually every personal computer. The WWW provides a common platform for application development, thus stimulating further increases in Internet traffic. Today, the Web provides access to services such as on-line shopping, location finding, banking, on-line brokering, and electronic newspapers.

Combining the Internet and mobility creates the mobile Internet. Merging the Internet with mobile radio systems will facilitate access to the Internet anywhere irrespective of the user's location. For example, a user who has lost his way could access one of the location services on the Web through his wireless terminal to get driving instructions. Mobile Internet, however, is not just a replica of the Internet delivered to wireless terminals. It is much more. This is because mobility and freedom of physical location as well as new capabilities of the wireless terminals such as locationing capability enable completely new type of applications, which are not possible on the fixed Internet. For example, your mobile terminal can locate where you are and then deliver information that is relevant to your current location such as local map or restaurants from the Internet.

The applications of a new system play a major role in capturing the attention of users who are already satisfied with existing systems. Third-generation cannot only offer totally new services, but it can also improve the existing services with its new capabilities such as the provision of simultaneous services, dynamic variable rate services and smaller delay. IMT-2000 service targets are the provision of at least 144

Kbps (preferably 384 Kbps) for wide area coverage with high mobility and 2 Mbps for local coverage with low mobility.

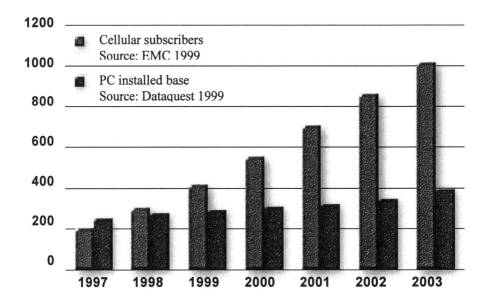

Figure 3.1 Number of cellular subscribers and personal computers (in millions).

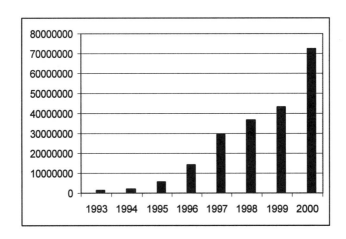

Figure 3.2 Growth of Internet hosts. (*Source*: Network Wizards, http//:www.nw.com.)

Different applications generate different traffic characteristics and thus have a different impact on the system design and on the network capacity. Service parameters include granularity, minimum transmission delay, discontinuous transmission, packet sizes, and packet distributions. The wide variety of services expected for third-generation systems will make the network planning process much more challenging compared to the second-generation systems. A video or speech call requires guaranteed quality of service, while non-real-time packet data can be offered on a best effort basis (i.e., when there is bandwidth available, the user receives it).

Section 3.2 discusses the market outlook of the mobile Internet, and Section 3.3 discusses wireless portals. Section 3.4 assesses the impact of mobile Internet on business models, and Section 3.5 reviews the technologies to implement mobile Internet: Wireless Application Protocol (WAP) and i-Mode in Japan.

Section 3.6 describes third-generation application scenarios to motivate the development of third-generation wireless systems with high data rates. In Sections 3.7 and 3.8, the Internet and wireless video traffic and their characteristics are discussed, respectively. Section 3.9 discusses multimedia services, and Section 3.10 describes generic IMT-2000 bearer service requirements. Example traffic models for the analysis and simulation of a wireless system are presented in Section 3.11.

3.2   MARKET OUTLOOK

The market outlook for mobile Internet is extremely promising. It is predicted that by 2002 there will be more mobile Internet enabled terminals than personal computers connected to the Internet, as shown by Figure 3.3. This will mean that a very large number of people will be able to access the Internet using their most personal device, a mobile phone. This opens enormous market opportunities. Because people are carrying their mobile phones everywhere they can access the Internet regardless of the place, anytime. Thus, the Internet is now accessible to a real mass market. The most bullish market predictions are saying that all wireless devices sold after 2003 are going to be Web-enabled. Therefore, in the most advanced markets mobile Internet penetration will well exceed 50% of the entire population in the near future. This shows the enormous potential of the mobile Internet.

Applications of the mobile Internet include e-mail, corporate data, news, sports, information services, travel, and banking services. It is already possible to buy a can of Coke by dialing a certain number or sending a short message from your mobile into a given number. You can also pay for a car wash using your mobile phone. These are simple but very tangible examples of mobile commerce (m-commerce), a new way of purchasing.

It has been stated that by 2005, mobile commerce could account for 15% of all business-to-consumers e-commerce revenue (Linda Mutschler, Merrill Lynch) and that all e-commerce transactions may be done wirelessly 10 years from now (Jeff Bezos, Amazon.com).

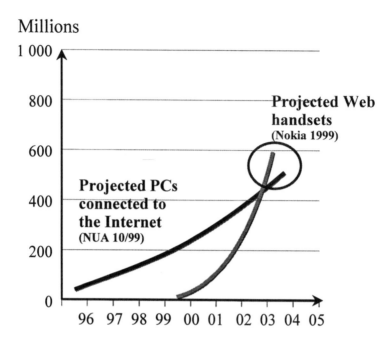

Figure 3.3 Mobile Internet-enabled handsets.

3.3 WIRELESS PORTALS

The battle for end user souls is raging hard. Everybody is trying to catch the attention of the customers. The idea of having one single access point to the Internet is called portal. It is a Web site that appears first in the terminal screen when the terminal is switched on. Examples of very well-known Internet portals are American Online (AOL) and Yahoo. Currently, operators are setting up wireless portals to capture the attention of end users and to generate new sources of revenue.

In the mobile Internet, similar to the fixed Internet today, persons interested in the same issues will form communities: mobile Internet communities. The potential number of persons participating in these communities will be the same as users having mobile handsets (i.e., more than one billion in the near future). This is far more than the number of people connected to the Internet using normal personal computers.

3.4 BUSINESS MODEL OF THE MOBILE INTERNET

A key issue in the mobile Internet is who will pay for the access. Is it going to be the users or some other parties? For example, user access could be paid for by revenues from advertising if some part of the screen could be sold to advertisers. This would enable users to access the services without cost or with reduced cost and thus drive the usage up. If users have to pay for the access by themselves, more advanced services are required to add value. These could be, for example, alerting the user when earlier ordered goods have arrived, alert about weather changes, and other services that give information that directly impacts the behavior of the user.

For the existing Internet players, the question is how to leverage the brand name of the current portal in the mobile Internet. Do they want to charge extra or use the mobile Internet to generate more traffic in the fixed Internet Web site?

3.5 MOBILE INTERNET TECHNOLOGIES

Wireless Application Protocol (WAP) and i-Mode are briefly discussed in this section.

3.5.1 WAP

WAP is a standard to provide access to Internet over wireless networks. It is a uniform technology platform with consistent content formats for delivering Internet and intranet to mobile phones, personal digital assistants, and other wireless terminals.

WAP was developed by the WAP Forum (www.wapforum.org). The idea of WAP originates from the fact that typical wireless devices have small screens and wireless networks have low bandwidth and high latency conditions. Thus, typical Internet content designed for large PC screens and demanding high bandwidth from the delivery channel is not suitable for wireless networks and devices. Therefore, WAP was designed to provide several enhancements to the standard Internet protocols to make them better suited for wireless devices. Some examples of the improvements are:

- Translation of the plain text headers of hyper text transfer protocol (HTTP) protocol to into binary code, which reduces the amount of data that must be transmitted over the air interface;
- A lightweight session establishment protocol to suspend or initiate sessions without the overhead of initial established. This allows the session to be suspended while idle to free up network resources or save battery power;
- Wireless Transmission Protocol (WTP) to provide reliable datagram service without the traditional TCP overhead. The simple WTP stack replaces the complex TCP stack in the phone resulting in significant savings in processing and memory cost in the handset.

Current implementations of WAP use circuit-switched bearers. Therefore, the connection establishment time is rather long. Sometimes, this can be rather frustrating to the user. The situation will improve when mobile packet data networks such as GPRS are taken into use. Packet-switched networks enable always connected terminals.

Figure 3.4 shows the basic WAP architecture. The center of the architecture is the WAP gateway, which acts as an intermediary by connecting the mobile network and

Internet. WAP gateway is a protocol translator, which converts the HTTP protocol into the Wireless Session Protocol (WSP). For more information on WAP, refer to [1].

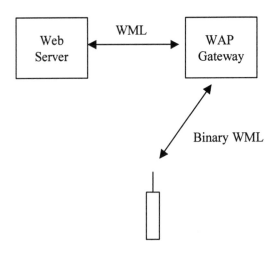

Figure 3.4   WAP basic architecture.

3.5.2   i-Mode

Another interesting development in the mobile Internet is the success of i-Mode in Japan. i-Mode was launched by NTT Docomo in 1999 and had reached over 7 million customers by May 2000. i-Mode uses a proprietary architecture that is not used outside Japan. The main difference to WAP is that i-Mode directly uses HTML language, which is also used in the normal Internet. It is not optimized for wireless devices but offers a larger availability of content. Services offered by i-Mode include e-mail, games, and access to different information services.

3.6   THIRD-GENERATION APPLICATION SCENARIO

As an example of third-generation services, we can imagine an IMT-2000 user, Mike, in an outdoor scenario. Mike would like to go to the movies and needs to find out which movies are being shown in town. Voice recognition software translates his query into a format his wireless device can understand. An intelligent search agent initiates a search on the World Wide Web for appropriate information about movies. When the information is found, it is downloaded into Mike's terminal and shown on the terminal screen. After looking at movie titles, he decides to see a preview of a film, which is downloaded to his terminal. Next, he would like to know how to get to the movie theater and downloads a digital map and a picture of the movie theater. It is clear that he could get the idea to access this kind of service anywhere and does not understand the service availability limitations imposed by the system.

In the above example, several features of IMT-2000 were used. To set up the connection, the application negotiated bearer service attributes such as bearer type, bit rate, delay, bit error rate (BER), up/down link symmetry, and error protection including no protection or unequal protection. During the session, parallel bearer services (service mix) were active (i.e., IMT-2000 is able to associate several bearers with a call, and bearers can be added and released independently during a call). Non-real-time communication mode was used. Browsing WWW and downloading video or other information are examples of typical asymmetric, non-real-time applications. Videotelephony would be an example of a symmetric, real-time application. IMT-2000 is also able to adapt the bearer service bit rate, quality, and radio link parameters depending on the link quality, traffic, network load, and radio conditions within the negotiated service limits.

The mobile office is an IMT-2000 application offering e-mail, WWW access, fax, and file transfer services. These services can already be provided by the second-generation networks. However, the higher bit rate of IMT-2000 makes them more attractive to the end user. A typical engineering document might exceed 1 Mbyte (8 Mbit) if it contains a large number of pictures. It would thus take approximately 4 seconds to transfer it with IMT-2000 but 80 seconds with a 100-Kbps GPRS connection. The higher user bit rate of IMT-2000 will increase the total traffic because users will now use applications that they would not have if only lower bit rates were available. A higher carrier bit rate will enable more efficient sharing of resources between the users and higher multiplexing gains.

Applications adapt their bit rates according to the information that needs to be transmitted. In the second-generation systems, speech service uses variable bit rate transmission. In the third-generation networks, many more services, such as video and packet data, will produce variable rate data. Consequently, the network will need the capability to take advantage of this variability to utilize the radio spectrum more efficiently.

Figure 3.5 depicts the classification of quality of service (QoS) categories. The classification is based on the delay sensitivity of the traffic.

3.7   INTERNET

A typical World Wide Web session could be like the one shown in Figure 3.6 [2]. After the session has started, a user browses Web pages by clicking one page at a time when it is downloaded to his terminal. One page (file) corresponds to a burst of packets, which can be of variable length. Note that these packets are upper-layer packets, which can be segmented for the transmission over the air interface. Between clicking the pages, the user might pause to read a page. The reading time corresponds to the longer delays between bursts of packets. Typically, the World Wide Web traffic is highly asymmetric (i.e., the user is downloading data from the network, and only some acknowledgments or commands are transmitted from the terminal to the network).

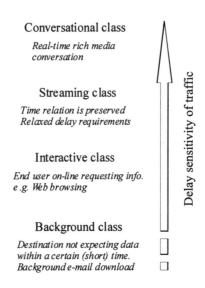

Figure 3.5   Classification of QoS categories.

While browsing the Web, a user could download a file using the file transfer protocol (FTP). This would establish a parallel session with the Web browser but with different traffic characteristics. A file transfer can be modeled as one long burst of packets. Transferring e-mail with large attachments generates a similar type of traffic.

Understanding the nature of the traffic is critical for a good system design. Between the transmission pauses, the system can either keep the radio connection, or release it and then establish it again when there is new data. This is a tradeoff between spending radio resources to keep the connection alive and spending radio resources to reestablish the connection. Furthermore, the access time will be longer if the connection is released and then reconnected. The question is how the timers should be set to minimize the consumption of radio resources and the access delay for a user.

The average transmission rate depends on the packet length and the time intervals between packets (see Figure 3.6). If the packets are large and the time interval between packets is short, capacity requirements will be much higher. The same applies for reading time between Web pages. The reading time depends on the contents of the page. For example, a page that needs some information to be filled in (e.g., address and credit card information for on-line shopping) requires more time than a page that just contains a simple greeting and a command to continue to the next page.

Figure 3.6   World Wide Web traffic characteristics.

The WWW traffic and file transfer are examples of packet services that do not need a specific minimum bit rate; rather they can utilize the spare capacity of the network. This is illustrated in Figure 3.7. The available bit rate, however, might have an effect on the traffic characteristics of a WWW user. In case the available bit rate is very small, the user might strip off unnecessary features. For example, he could disable the retrieval of pictures and download only text files, thus reducing the required bit rate.

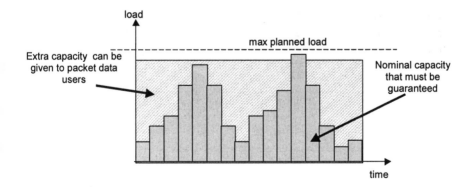

Figure 3.7   Guaranteed versus spare capacity.

3.8   WIRELESS VIDEO

Figure 3.8 illustrates a wireless video concept, which is one of the predicted IMT-2000 applications. Earlier in this chapter, we described a user watching a preview of a movie, which is a *retrieval-type* service. Other examples of such services are video messaging (photos, drawings) and news delivery through video [3]. *Conversational-type* services (e.g., videoconferencing and videotelephony) involve interaction between two or more end users. The retrieval-type and conversational-type video services have very different transmission requirements. In the retrieval-type service, transmission delay may be longer, on the order of 0.5 to 1 second, and encoding can be performed off-line, resulting in a higher compression ratio and thus a lower bit rate [3]. For the conversational services, transmission delay should not exceed 200 to 300 ms in order to avoid the annoying effect of users speaking simultaneously. The delay has a direct impact on the required signal-to-interference ratio. A longer delay allows for longer interleaving and retransmission of erroneous frames and, subsequently, smaller signal-to-interference ratio and higher overall capacity. The packet error ratio for an acceptable video quality is on the order of 10^{-3} to 10^{-4}, and the bit error ratio in the order of 10^{-6} to 10^{-7}. The conversational services will require more capacity than retrieval services because of lower delays, resulting in higher signal-to-interference ratio requirements to achieve these error rates.

Figure 3.8   IMT-2000 wireless video terminal (courtesy of Nokia).

ITU-T/SG16 (q11 and q15) and the International Standardization Organization (ISO) are working to set standards for low bit rate video coding. ITU-T/SG15 has produced recommendation H.263, which specifies a coded presentation for compressing the moving picture component of audio-visual signals at low bit rates [4]. Currently, SG16 is developing H.263+ with more efficient compression technology and H.263L (long term standard) aimed at achieving a dramatic increase in compression efficiency.

The ISO-MPEG4 (Motion Picture Experts Group) activity is aimed at a generic audio-visual coding system for multimedia communications. It will focus on providing functions for editing the content depicted in video scenes and for coding scenes containing natural video and computer graphics.

Bit rate requirements for videotelephony depend on the video codec used, picture quality required, and frame rate per second required as well as on the actual video sequence. It has been estimated that using an H.263 video codec, the minimum acceptable quality for low-motion video (e.g., a simple head-and-shoulders scene) requires a transmission bit rate of over 20 Kbps, and for generic video telephony over 40 Kbps [3]. If there is more bandwidth available, a higher quality can be achieved. The H.263 and MPEG use a hierarchical scalable video encoding mechanism where the quality of the compressed video can be traded for the compression ratio.

The instantaneous bit rate of video codecs varies greatly depending on the video sequence complexity. The peak rate requirement compared to the average bit rate can be relatively high. A buffer can be used to smooth the bit rate variability of the video encoder output. This variability of the video bit stream can be used to utilize radio resources more efficiently. The spare capacity can be utilized for services without quality of service guarantees (especially delay guarantees), as shown in Figure 3.7.

Figures 3.9 and 3.10 present examples of the bit rate variability of a low bit rate H.263 video codec. The first sequence is called Claire, which is a good quality video clip with a head and shoulders picture and little movement. From Figure 3.9 it can be seen that total bit rate varies between 10 and 17 Kbps dependent on the picture. The highest change in bit rate/second is about 5 Kbps/s (second 10).

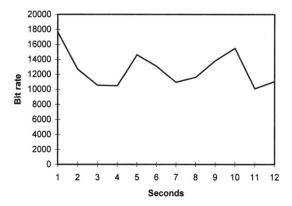

Figure 3.9   Bit rate/second requirement for Claire sequence.

The second example is called Carphone, which is a moving object with a moving background, and has a fair quality. From Figure 3.10 it can be seen that the total bit rate varies between 20 and 80 Kbps. The required bit rate can change almost 50 Kbps during 1 second (from second 8 to 9). This shows that there is large variation in the change of bit rate and air, which calls for efficient radio resource management to achieve good capacity. Variation in the bit rate allows statistical multiplexing of resources.

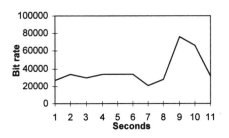

Figure 3.10   Bit rate/second requirement for Carphone sequence.

3.9   MULTIMEDIA SERVICES

Multimedia means transmission of several types of information (e.g., voice, video, and data) simultaneously. Bit rate and quality requirements of these components can be highly variable; for example, BER 10^{-3} and delay of 40 ms for speech and BER 10^{-7} with unconstrained delay for data. Multiplexing of different services is possible using existing multiplexing schemes such as H.223, which is part of the H.324 multimedia communication standard 324 [5], or H.250, which is part of the H.323 multimedia communication standard for packet-based multimedia communications. The H.323 multiplexing scheme converts the different data streams into one single stream for transmission over the network. It is more efficient, however, from the radio transmission point of view if the transmission parameters of each service are tailored to the requirements of the service. Therefore, multiplexing of different services should also be possible in the air interface. Third-generation networks will offer different transmission bearers for different types of services. The provision of these bearers in wideband CDMA is discussed in Section 5.8.

3.10   IMT-2000 BEARER SERVICES

Because the IMT-2000 applications are not yet known, we cannot specify the supported services based only on today's needs. We need a flexible and broad approach in order to have a future-proof radio interface. Therefore, based on the analysis of a large number of existing services and future needs, more generic radio requirements have been derived (data rate, BER, data rate variation, max delay). These application requirements can be mapped into generic radio bearers as shown in Table 3.1.

3.11   TRAFFIC MODELS

In order to analyze the capacity and design of access procedures, we need explicit models for the above services. In this section we classify services into real-time and non-real-time services. Examples of real-time services are speech and conversational video, and of non-real-time services WWW browsing and file transfer.

It should be noted that the models presented here are only one way of modeling these services. As new services emerge and applications develop, traffic characteristics will also change. One example is e-mail: attaching a file to an e-mail message has become very common during recent years, and consequently, the traffic characteristics of e-mail have changed. Therefore, it is important to carefully consider the used traffic model in order to be certain that it represents the main characteristics of the service.

Table 3.1
IMT-2000 Bearer Service Requirements

	Real time/constant delay		Non-real-time/variable delay	
Operating environment	Peak bit rate	BER / max transfer delay	Peak bit rate	BER / max transfer delay
Rural outdoor (terminal speed up to 250 km/h)	at least 144 Kbps (preferably 384 Kbps) granularity appr. 4 Kbps	delay 20 - 300 ms BER 10^{-3} - 10^{-7}	at least 144 Kbps (384 Kbps preferably)	BER = 10^{-5} to 10^{-8} Max transfer delay 150 ms or more
Urban/ suburban outdoor (terminal speed up to 150 km/h)	at least 384 Kbps (preferably 512 Kbps) granularity appr. 10 Kbps	delay 20 - 300 ms BER 10^{-3} - 10^{-7}	at least 384 Kbps (preferably 512 Kbps)	BER = 10^{-5} to 10^{-8} Max transfer delay 150 ms or more
Indoor/ Low range outdoor (terminal speed up to 10 km/h)	2 Mbps granularity appr. 200 Kbps	delay 20 - 300 ms BER 10^{-3} - 10^{-7}	2 Mbps	BER = 10^{-5} to 10^{-8} Max transfer delay 150 ms or more

Note: Currently there is no single specification of IMT-2000 or UMTS bearer services or requirements; this table has been composed from various sources and reflects the range of expected IMT-2000 bearer service parameters.

3.11.1   Real-Time Services

For real-time services, the traffic model is a traditional *birth-death process*. Speech and other real-time service users arrive to the system according to a *Poisson process* with an intensity λ_s [users/s].

The departure from the system is modeled by setting the *mean holding time* (*mht* [s]) of the users. That is, the session length for a user is exponentially distributed with an average of *mht*. The average holding time for all real-time users is 120 seconds. A minimum call size of 1 second should be assumed.

The activity of a service can be implemented as an on-off model where the lengths of active and silent bursts are exponentially distributed. The activity factor and the average length of both active and silent periods need to be defined. A typical activity factor for speech is 0.5, and the length of the active and silent periods is 0.7 seconds.

For variable rate video, a more realistic model should consider the different bit rates and their probability.

3.11.2   Non-Real-Time Services

As an example of non-real-time services, we present a traffic model for a WWW browsing session. The sessions are generated according to a Poisson distribution. A traffic model must take into consideration where in the system we want to model the behavior. The model presented describes how the data packets arrive to a buffer before the air interface. Figure 3.11 illustrates a downlink scenario where the arrow from the base station to the mobile station represents the traffic model. It should be noted that the flow due to one specific session is modeled, not the total flow from all sessions (i.e., there is one buffer per bearer service). Here, only the downlink model is presented. This model, however, can be easily generalized for the uplink, remembering that in the uplink, file size and idle period distributions are different.

Figure 3.11   The downlink packet flow to a user specific base station (BTS) buffer.

The definition of non-real-time traffic models follows [6]. The principles of this model are based on [7]. The following parameters define characteristics of the packet traffic as illustrated in Figure 3.12.

- Session arrival process;
- The number of packet bursts per session (N_{pc});
- The interarrival time between packet bursts (D_{pc}) (i.e., the reading time);
- The number of packets in a packet burst (N_d);
- The interarrival time between packets within a packet burst (D_d);
- The size of a packet (S_d).

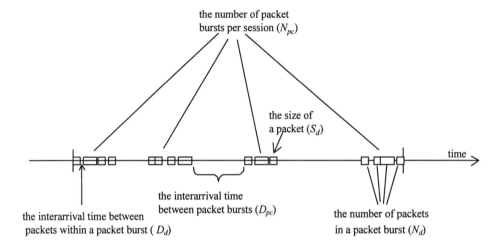

Figure 3.12 Packet service model.

The length of a session is modeled implicitly by the number of events during the session. Session arrival process defines how a session arrives to the system. The arrival of session setups to the network is modeled as a Poisson process. For each service there is a separate process.

N_{pc}, D_{pc}, N_d, and D_d are all geometrically distributed random variables with means μ_{Npc} [packets], μ_{Dpc} [seconds], μ_{Nd} [packets], and μ_{Dd} [seconds], respectively. The geometrical distribution is a discrete representation of the exponential distribution. It is used because the simulations are performed with a discrete time scale. The *geometrical distribution* for a random variable **x** with mean μ_x is defined by

$$P\{x = k\} = p(1 - p)^k \qquad k = 0, 1, \ldots \qquad (3.1)$$

where $p = \dfrac{1}{1 + \mu_x}$.

According to some studies, the packet size in WWW applications can be modeled as a *Pareto distribution* [8]. The probability density function (pdf) of the Pareto distribution is defined as [9]

$$f_x(x) = \frac{\alpha \cdot k^\alpha}{x^{\alpha+1}} \qquad \alpha, k \geq 0, \; x \geq k \tag{3.2}$$

and the cumulative distribution function (CDF) is defined as

$$F_x(x) = 1 - \left(\frac{k}{x}\right)^\alpha \qquad x \geq k \tag{3.3}$$

The average packet length is now obtained from the mean of the Pareto distribution [9]

$$\mu_x = \frac{k \cdot \alpha}{\alpha - 1} \qquad \alpha > 1 \tag{3.4}$$

Table 3.2 gives example mean values for the distributions of a typical WWW service. The average packet size can be calculated from (3.4) to be 896 bytes. The average requested file size is the average packet size multiplied by the average number of packets: 896 × 15 = 13.4 Kbytes. Different values for the average interarrival time between packets correspond to different average source speeds. For example, an average interarrival time of 0.02 seconds between packets corresponds to an average bit rate of 384 kbps. In order to have a finite variance for the distribution, the maximum packet size is limited to 67 Kbytes is [6]. Because different parameters of the traffic model will have a different impact on the system performance, it is also important to consider other parameter values as given here.

Table 3.2
Example Mean Values for the Distribution of a Typical WWW Service

Average number of packet bursts within a session (N_{pc})	Average reading time between packet bursts (D_{pc}) [s]	Average number of packets within a packet burst (N_d)	Average interarrival time between packets (D_d) [s]	Average bit rate [Kbps]	Parameters for packet size distribution
5	12	15	0.96	8	$k = 81.5$
			0.24	32	$\alpha = 1.1$
			0.12	64	
			0.05	144	
			0.02	384	
			0.004	2048	

REFERENCES

[1] van der Heijden, M., and M. Taylor (eds.), *Understanding WAP, Wireless Applications, Devices, and Services*, Norwood, Massachusetts: Artech House, 2000.

[2] Anderlind, E., and J. Zander, "A Traffic Model for Non-Real-Time Data Users in a Wireless Radio Network," *IEEE Communications Letters*, Vol. 1, No. 2, March 1997, pp. 37–39.

[3] Nieweglowski, J., and T. Leskinen, "Video in Mobile Networks," *European Conference on Multimedia Applications, Services and Techniques (ECMAST96)*, Louvain-la-Neuve, Belgium, May 28–30, 1996, pp. 120–133.

[4] Rijkse, K., "The H.263 Video Coding for Low-Bit-Rate Communication," *IEEE Communications Magazine*, Vol. 34, No. 12, December 1996, pp. 42–45.

[5] Lindbergh, D., "The H.324 Multimedia Communication Standard," *IEEE Communications Magazine*, Vol. 34, No. 12, December 1996, pp. 46–51.

[6] ETSI UMTS 30.03, "Selection Procedures for the Choice of Radio Transmission Technologies of the UMTS," *ETSI Technical Report*, 1997.

[7] Anagnostou, M. E., J.-A. Sanchez-P., and I. S. Venieris, "A Multiservice User Descriptive Traffic Source Model," *IEEE Transactions on Communications*, Vol. 44, No. 10, October 1996, pp. 1243–1246.

[8] Cunha, C. R., A. Bestavros, and M. E. Crovella, "Characteristics of WWW Client-Based Traces," *Technical Report TR-95-010*, Boston University Computer Science Department, June 1995.

[9] Mood, A. M., F. A. Graybilland, and D. C. Boes, *Introduction to the Theory of Statistics*, 3rd ed., New York: McGraw-Hill, Inc., 1974.

Chapter 4

RADIO OPERATING ENVIRONMENTS AND THEIR IMPACT ON SYSTEM DESIGN

4.1 INTRODUCTION

Radio propagation characterization is the bread and butter of communications engineers. Detailed knowledge of radio propagation is essential to develop a successful wireless system. A radio engineer has to acquire full knowledge of the channel to design a good radio communication system [1].

Therefore, knowledge of radio propagation characteristics is a prerequisite for the design of radio communication systems. Many measurements studies have been done to obtain information concerning propagation loss, spatial distribution of power when the environment is physically static, wideband and narrowband statistics concerning the random variables of received signals at a fixed location due to any surrounding movement, and delay spread. References [2–8] provide a good review of these measurements.

A radio channel is a generally hostile medium in nature. It is rather difficult to predict its behavior. Traditionally, radio channels are modeled in a statistical way using real propagation measurement data. In general, the signal fading in a radio environment can be decomposed into a large-scale path loss component together with a medium-scale slow varying component having a log-normal distribution, and a small-scale fast varying component with a Rician or Rayleigh distribution, depending on the presence or absence of the line-of-sight (LOS) situation between the transmitter and the receiver [7–17]. Accordingly, a three-stage propagation model should be appropriate to describe a wireless cellular environment: (1) area mean power depending on the path loss characteristics in the range from the transmitter to the area where the receiver is located; (2) local mean power within that area, which is slow varying, can be very satisfactorily represented by a log-normal distribution; and (3) superimposed fast fading instantaneous power, which follows a Rayleigh or Rician distribution. Figure 4.1 illustrates these three different propagation phenomena. An extreme variation in the transmission path between the transmitter and receiver can be found, ranging from direct LOS to severely obstructed paths due to buildings, mountains, or foliage. The phenomenon of decreasing received power with distance due to reflection, diffraction around structures, and refraction within them is known as path loss. Propagation models have been developed to determine the path loss and are known as large-scale propagation models because they characterize the received signal strength by averaging the power level over large transmitter-receiver separation distances, in the range of hundreds or thousands of meters.

Figure 4.1   Path loss, shadowing, and multipath fading.

Medium-scale propagation models determine the gradual changes in the local-mean power if the receiving antenna is moved over distances larger then a few tens or hundreds of meters. The medium-scale variation of the received signal power is called *shadowing*, and it is caused by obstruction by trees and foliage. The term local-mean power is used to denote the power level averaged over a few tens of wavelengths, typically 40λ [11, 13].

Small-scale propagation models characterize the fast variation of the signal strength over a short distance on the order of a few wavelengths or over short time durations on the order of seconds. Small-scale fast fading, also known as short-term fading or multipath fading, is due to multipath reflections of a transmitted wave by local scatterers such as houses, buildings, and man-made structures, or natural objects such as forests surrounding a mobile unit.

The accuracy of the models depends on their purpose. For detailed coverage and capacity analysis in a certain region, detailed models are needed. On the other hand, for rough capacity and range calculations needed in radio interface design, simple and easy-to-use models are sufficient. In the performance assessment, it is important to analyze the sensitivity of the result with respect to the propagation models used.

In addition to the modeling of the propagation environment, the mobile behavior of the wireless terminals needs to be understood in each radio environment. Mobility modeling is involved in the analysis of radio resource management, channel allocation, and handover performance. The accuracy of mobility models becomes essential for evaluation of design alternatives (e.g., different deployment strategies).

In this chapter, we first give an overview of the multipath channel and typical radio environments for third-generation wireless systems in Sections 4.2 to 4.4. Thereafter, detailed models for path loss, deployment, shadowing, and wideband radio channels are presented in Sections 4.5 and 4.6. In addition, the impact of two different sets of wideband channel models, ATDMA and CODIT channels, on the performance of narrowband and wideband CDMA system design is discussed. In the last section, mobility models for different radio environments are discussed.

4.2   MULTIPATH CHANNEL

The short-term fluctuations of the received signal caused by multipath propagation are called small-scale fading. The different propagation path lengths of the multipath signal give rise to different propagation time delays. Figure 4.2 shows an example of a *power-delay profile* of a multipath channel consisting of three distinct paths, which are also called *multipath taps*. Depending on the phase of each multipath signal when arriving at the receiver, they sum either constructively or destructively. Consequently, the power of each multipath tap is time varying, resulting in fading dips, as illustrated in Figure 4.3. The depth of the fading dips depends on the channel type. The distribution of the instantaneous power of the channel taps can be described by a distribution function, which depends on the radio environment. A so-called *Rayleigh fading* channel is the most severe mobile radio channel, and accordingly, the fading dips are deep. In a Rayleigh fading channel all multipath taps are independent and there is no dominant path. In a *Rician fading* channel the fading dips are shallower, which is due to a dominant path in addition to the scattered paths. A dominant path typically exists in microcell and picocell environments due to a LOS connection. The impact of different power delay profiles and fading characteristics on DS-CDMA performance will be discussed in Section 4.6.

In order to model the multipath fading channel, its behavior needs to be characterized. The concepts of *delay spread, coherence bandwidth, Doppler spread,* and *coherence time* are used to model the various effects of the multipath channel.

Figure 4.2   An impulse response of a multipath channel.

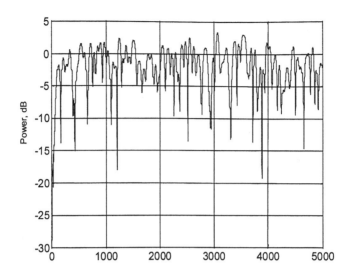

Figure 4.3   Fading channel.

The root-mean-square (rms) value of the delay spread is a statistical measure and tells the spread of the multipaths about the mean delay of the channel (for a mathematical definition of the rms delay spread, see Section 7.2.2). The maximum delay spread tells the delay difference between the first and the last multipath components in the power-delay profile. The coherence bandwidth is the maximum frequency difference for which the signals are still strongly correlated. The coherence bandwidth is inversely proportional to the delay spread (i.e., the smaller the delay spread, the larger the coherence bandwidth). If the transmission bandwidth of the signal is larger than the coherence bandwidth, the signal will undergo *frequency selective fading*. A *flat fading* channel results if the transmission bandwidth of the signal is smaller than the coherence bandwidth. As can be understood from the above descriptions, the delay spread and the coherence bandwidth are different views of the characteristics of the multipath fading channel. The coherence bandwidth is a measure of the diversity available to a RAKE or equalized receiver. A smaller coherence bandwidth means a higher order diversity. This is illustrated in Figure 4.4, where only part of the frequency carrier is fading. If the coherence bandwidth was as large as the transmission bandwidth, then the entire received spectrum would be observed to fade. The maximum delay spread can be used to calculate how many resolvable paths exist in the channel that could be used in the RAKE receiver (see Section 7.2.2).

The movement of the mobile station gives rise to a Doppler spread, which is the width of the observed spectrum when an unmodulated carrier is transmitted. If there is only one path from the mobile to base station, the base station will observe a zero Doppler spread combined with a simple shift of the carrier frequency (Doppler frequency shift). This is analogous to the change in the frequency of a whistle from a train perceived by a person standing on a railway line when the train is bearing down or receding from that person [6]. The Doppler frequency varies depending on the angle of the mobile station movement relative to the base station. The range of values when the Doppler power spectrum is nonzero is called the Doppler spread. The reciprocal of the Doppler shift is a measure of the coherence time of the channel. The coherence time is the duration over which the channel characteristics do not change significantly. The coherence time is related to the design of the channel estimation schemes at the receiver (see Section 5.8.1), error correction and interleaving schemes (see Section 5.5), power control schemes (see Section 5.12), and the frame length of the TDD schemes (see Section 9.4.1).

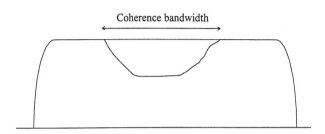

Figure 4.4   Coherence bandwidth and fading.

4.3   RADIO ENVIRONMENTS

There is a large number of environments where third-generation mobile radio systems are expected to operate. These include large and small cities, with variations in building construction, as well as tropical, rural, desert, and mountainous areas. In addition, the system construction (e.g., antenna design and height of antennas) impacts the radio environments. Because it is impossible to consider all possible radio environments in the design of a mobile radio system, more general models that will consider the essence of different radio environments are required. Therefore, the large number of possible radio environments has been condensed into three generic radio environments, according to [18–20]:

- Vehicular radio environment;
- Outdoor to indoor and pedestrian radio environment;
- Indoor office radio environment.

These radio environments correspond to the following cell types: macrocell, microcell, and picocell, respectively. In this section we describe the characteristics of these environments with respect the path loss attenuation, shadowing, and small-scale fading. Specific path loss attenuation, shadowing, and deployment models for each environment will then be presented in Section 4.5. In addition, Section 4.6 presents small-scale fading models. These models will be used in Chapters 9 and 10 for performance simulations. In addition to these models describing single radio environments, a mixed radio environment is described in Chapter 10 in conjunction with the discussion of hierarchical cell structures.

4.3.1 Vehicular Radio Environment

The vehicular environment is characterized by large macrocells and large transmit powers. It is also called the macrocell environment. The mobile speeds are also high because of fast-moving vehicles. There is typically no LOS component, implying that a received signal is composed of reflections. The average power of the received signal decreases with distance raised to some exponent, the *path loss exponent*. The path loss exponent varies depending on the environment but is typically between 3 and 5. The shadowing is caused by obstruction by trees and foliage, and the resulting medium-scale variation in the received signal power can be modeled with a log-normal distribution. The standard deviation varies considerably; for example, 10 dB can be used in urban and suburban areas. In rural and mountainous areas, a lower value can be used. Small-scale fading is Rayleigh. Typical delay spreads are on the order of 0.8 µs and the maximum value can be up to tens of microseconds.

4.3.2 Outdoor to Indoor and Pedestrian Radio Environment

Figure 4.5 illustrates the outdoor to indoor and pedestrian radio environment. This radio environment is characterized by small microcells and low transmit powers. The antennas are located below rooftops. Both line-of-sight and no line-of-sight (NLOS) connections exist. Indoor coverage can also be provided from a outdoor base station.

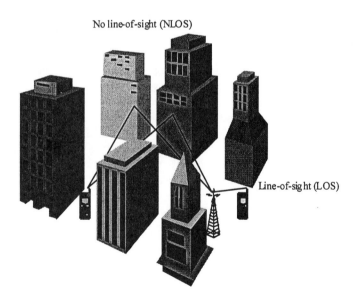

Figure 4.5   Outdoor to indoor environment.

The path loss attenuation in a microcell environment is shown in Figure 4.6. The path loss exponent has large variations, from 2 in areas with LOS, and up to 6 with NLOS due to trees and other obstructions along the path. Furthermore, a mobile station can experience a sudden drop of 15 to 25 dB when it moves around a corner. The standard deviation of the shadowing varies from 10 to 12 dB, and a typical building penetration loss average is 12 dB, with a standard deviation of 8 dB. The small-scale fading is either Rayleigh or Rician with typical delay spreads on the order of 0.2 µs.

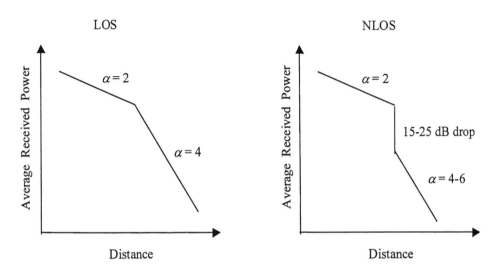

Figure 4.6   Microcell propagation in LOS and NLOS situations. (*After*: Wesley, E. K., *Wireless Multimedia Communications*, Reading, MA: Addison-Wesley, 1998.)

4.3.3   Indoor Office Radio Environment

In the indoor office radio environment (Figure 4.7), transmit powers are small and base stations and users are located indoors. Path loss attenuation exponent varies from 2 to 5 depending on the scattering and the attenuation by walls, floors, and metallic structures. The wall and floor penetration losses vary according to the material, from 3 dB for light textile to 13 to 20 dB for concrete block walls. A detailed overview of the penetration losses for different types of materials is presented in [7]. Shadowing is log-normal with a typical standard deviation of 12 dB. Fading ranges from Rician to Rayleigh. Rician fading is typically observed if the mobile station is in the same room as the base station and Rayleigh fading is observed when the mobile station is in another room or on a different floor than the base station. Typical average rms delay spreads are on the order of 50 to 250 ns. For a detailed overview, refer to [5]. The mobile station speeds are slow, ranging from a stationary user up to 5 km/h.

4.4   DISTRIBUTION FUNCTIONS

In this section, three important probability density functions (pdf) are given, namely, log-normal pdf used to model the shadowing, and Rayleigh and Rician pdf used to model the distribution of the instantaneous power in small-scale fading models.

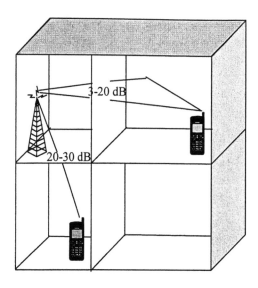

Figure 4.7   Indoor office radio environment.

The log-normal pdf is given by

$$f_L(p_o) = \frac{1}{p_o \sigma \sqrt{2\pi}} \left[-\frac{(\ln(p_o) - m)^2}{2\sigma^2} \right] \tag{4.1}$$

where p_o is the local mean power, and σ and $p_a = exp(m)$ are the logarithmic standard deviation of the shadowing and the mean, respectively.

The Rician pdf is given by

$$f_R(r) = \frac{r}{p_o'} \exp \left[-\frac{r^2 + s^2}{2p_o'} \right] I_0 \left(\frac{rs}{p_o'} \right) \qquad \text{for } s \geq 0, r \geq 0 \tag{4.2}$$

$$= 0 \qquad \text{for } r \leq 0$$

where r is the amplitude of the received signal, $I_0(\ )$ is the modified Bessel function of the first kind and zeroth order, s is the peak value of the specular radio signal, p_o' is the average power of the signal, $p_o' = p_o/(k+1)$, p_o is the total received local-mean power, and k is the Rician factor, which is defined as the ratio of the average specular power and the average fading power received over specular paths

$$k = \frac{s^2}{2p_o'} \tag{4.3}$$

When the direct signal does not exist i.e., when s becomes zero ($k = 0$), (4.2) reduces to the Rayleigh pdf

$$f_R(r) = \frac{r}{p_o} \exp\left[\frac{-r^2}{2p_o}\right] \qquad \text{for } 0 \le r \le \infty \tag{4.4}$$
$$= 0 \qquad \text{for } r < 0$$

4.5 PATH LOSS AND DEPLOYMENT MODELS

In this section, we describe path loss models for the radio environments presented previously. Typically, path loss models are derived using a combination of analytical and empirical methods. In the empirical approach, the measured data is modeled using curve fitting or analytical expressions. The validity of empirical models in other environments and frequencies can only be validated by comparing the model to data measured from the specific area and for the specific frequency. It should be noted that these models present only a snapshot of the real radio environment and are provided only to illustrate the modeling principles of the radio channel. They will be used in Chapters 9 and 10 for the system level simulations of a wideband CDMA system. As will be discussed in Chapter 12, for detailed network planning purposes, the propagation models need to be adjusted according to the real deployment environment.

4.5.1 Vehicular Radio Environment

The vehicular radio environment corresponds to a macrocell. The macrocell layout is typically modeled with a homogenous hexagonal grid, as depicted in Figure 4.8.

We use the path loss model presented in [19] where a Hata-like propagation model was proposed. The path losses are computed as a distance-dependent function $x^{-\alpha}$. Values ranging from 3.0 to 5.0 have been reported for the path loss exponent α depending on the environment.

In addition to this distance-dependent loss, there is the shadowing effect, which is modeled as a zero mean Gaussian random variable, ξ (dB). The standard deviation of the shadowing is assumed to be 10 dB, which is also used for the performance simulations in Chapter 7. The resulting path loss is computed as

$$L_{\text{macro}} = \xi + \alpha \cdot 10 \cdot \log(x) \quad \text{(dB)} \tag{4.5}$$

where x is distance in kilometers.

4.5.2 Outdoor to Indoor and Pedestrian Radio Environment

In this radio environment, the cell deployment model is of the Manhattan microcell type, as

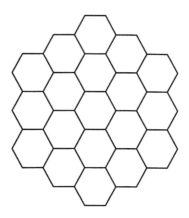

Figure 4.8 Macrocell layout.

depicted in Figure 4.9. The base station locations can be on street corners, as in Figure 4.9, or in the middle of street canyons. The number of base stations also has to be determined.

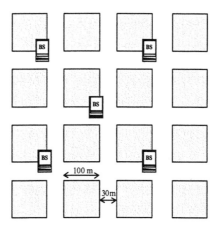

Figure 4.9 Manhattan microcell deployment model.

A three-slope model has been adopted for microcells [19]. A simplifying view has been adopted in the definition of this model, which is not suited for coverage analysis but rather for cellular capacity estimation.

A break point — located at a distance R_b from the transmitter, where the losses equal L_b (dB) — marks the separation between two LOS segments. The second of these LOS segments has a higher slope and predicts larger losses than the first one, as illustrated in Figure 4.6. On the other hand, turning a corner causes an additional loss, L_{corner}, in just a few meters but still causes an increase in the slope of this third segment. The model is mathematically given by the following expressions, where loss magnitudes are given in dB [18]:

$$L_{LOS_1} = L_b + 20 \cdot n_{LOS_1} \cdot \log\left(\frac{x}{R_b}\right) \qquad x \le R_b, \text{ LOS}$$

$$L_{LOS_2} = L_b + 40 \cdot n_{LOS_2} \cdot \log\left(\frac{x}{R_b}\right) \qquad x > R_b, \text{ LOS} \qquad (4.6)$$

$$L_{NLOS} = L_{LOS}(x_{corner}) + L_{corner} + 10 \cdot n_{NLOS} \cdot \log\left(\frac{x}{x_{corner}}\right) \qquad \text{NLOS}$$

where n_{LOS1}, n_{LOS2} and n_{NLOS} denote each segment slope, and the receiver position is defined by the distance from the transmitter to the receiver measured along the street path, x, and, if in a NLOS situation, the distance from the transmitter to the corner, x_{corner}.

The parameters are derived from the expressions in [19]

$$R_b = \frac{4 \cdot h_b \cdot h_m}{\lambda}$$

$$L_b = \left| 20 \cdot \log\left(\frac{\lambda^2}{8 \cdot \pi \cdot h_b \cdot h_m}\right) \right| \qquad (4.7)$$

$$L_{corner} = -0.1 \cdot w_s + 0.05 \cdot x_{corner} + 20$$

$$n_{NLOS} = -0.05 \cdot w_s + 0.02 \cdot x_{corner} + 4$$

For the microcell scenario $w_s = 30$m and assuming a wavelength of $\lambda = 0.15$m and the MS and BS antenna heights to be in the ratio $h_b \cdot h_m = 11.25$m^2, the following are obtained

$$R_b = 300\text{m}$$

$$L_b = 82\text{dB}$$

$$L_{corner} = 17 + 0.05 \cdot x_{corner}$$

$$n_{NLOS} = 2.5 + 0.02 \cdot x_{corner} \qquad (4.8)$$

$$n_{LOS_1} = 1$$

$$n_{LOS_2} = 2$$

and the resulting expressions for estimating the path losses will be the following:

$$L_{\text{LOS}_1} = 82 + 20 \cdot \log\left(\frac{x}{300}\right) \qquad\qquad x \leq 300, \text{ LOS}$$

$$L_{\text{LOS}_2} = 82 + 40 \cdot \log\left(\frac{x}{300}\right) \qquad\qquad x > 300, \text{ LOS} \qquad (4.9)$$

$$L_{\text{NLOS}} = L_{\text{LOS}}(x_{\text{corner}}) + 17 + 0.05x_{\text{corner}} + (25 + 0.2x_{\text{corner}})\log\left(\frac{x}{x_{\text{corner}}}\right) \qquad \text{NLOS}$$

As in the case of macrocells, the shadowing effect is modeled by $10^{\xi/10}$, where ξ is again a Gaussian random variable with the following first order statistics:

$$< \xi > = 0\,\text{dB}$$
$$\sigma_\xi = 4\,\text{dB} \qquad\qquad (4.10)$$

4.5.3 Indoor Office Radio Environment

An indoor office is modeled by a homogeneous isolated building with several floors. Typically, no stairs or elevators are modeled. Other relevant parameters are the location and number of base stations, but also the penetration loss of walls and floors. An example of an indoor office model is given in Figure 4.10.

Figure 4.10 Indoor office model example.

The *Motley-Keenan* model is used to model indoor path losses. This is an empirical model that takes into account the attenuation due to walls and floors on the way from the transmitter to the receiver. The path loss (dB) predicted by this model is given by [19]

$$L_{\text{pico}} = L_0 + 10 \cdot n \cdot \log(x) + \sum_{j=1}^{J} N_{w_j} \cdot L_{w_j} + \sum_{i=1}^{I} N_{w_i} \cdot L_{w_i} \qquad (4.11)$$

where L_0 denotes the loss at the reference point (at 1m), n is the power decay index, and x represents the transmitter to receiver path length. N_{w_j} and N_{f_i} denote the number of walls and floors of different kinds that are traversed by the transmitted signal, and L_{f_i} (dB) and L_{w_j} (dB) represent their corresponding loss factors. The values proposed for these parameters are the following:

$$
\begin{aligned}
L_0 &= 37\ \text{dB} \\
L_f &= 20\ \text{dB} \\
L_w &= 3\ \text{dB} \\
n &= 2
\end{aligned}
\tag{4.12}
$$

This yields to the following formulation for the loss model:

$$
L_{\text{pico}} = 37 + 20 \cdot \log(x) + 3 \cdot N_w + 20 \cdot N_f \quad \text{(dB)}
\tag{4.13}
$$

Unlike in other radio environments, no shadowing is modeled in the indoor office radio environment.

4.6 SMALL-SCALE MODELS

The small-scale fading is included in wideband channel models. There has been a vast amount of literature about channel modeling. During recent years, ATDMA and CODIT projects have performed wideband channel modeling specifically for third-generation systems and their expected radio environments. The outcome of the wideband propagation activities performed by ATDMA is reported in [21]. A complete and detailed study of the work carried out on propagation matters within the CODIT project can be found in [22, 23]. The ATDMA models together with models from the joint technical committee (JTC) in the United States have been used as a basis for the ITU channel models [24, 25]. As will be noticed from the following sections, channel models intended to model the same environment can differ significantly.

A comparison of ATDMA and CODIT channel models was carried out by SIG5 [19]. In the following sections we use this comparison to highlight differences in propagation environments and to analyze the impact of these channels on the performance of narrowband and wideband CDMA air interfaces assuming 1.25- and 5-MHz channel bandwidths, respectively.

The approach taken by CODIT for wideband modeling is statistical-physical modeling [23], while ATDMA uses stored channels, which are sets of complex impulse responses corresponding to different radio environments [21]. For SIG5, the channels were transformed into the well-known COST 207 format, which is used here as well.

4.6.1 Vehicular Radio Environment

In the vehicular environment, macrocell channel models are used. Table 4.1 presents the CODIT channel model in COST 207 format and Table 4.2 presents the ATDMA macrocell channel model.

To illustrate the differences in the channel models, the impulse responses of both channels are given in Figure 4.11.

It can be observed from Figure 4.11 that in the ATDMA wideband model, the multipath spread is much wider than in the CODIT channel model. Also, in the ATDMA model the average amplitude values differ more than in the CODIT model. In the ATDMA the second path is already attenuated by 10 dB, which leads to a difficult channel since the first tap is also Rayleigh fading [19]. So, the ATDMA channel is almost equally difficult for narrowband and wideband CDMA, even though wideband CDMA can resolve more taps and thus obtains somewhat more diversity gain.

Table 4.1
CODIT Macrocell Channel Model

Tap	Relative delay (ns)	Relative power (dB)	Doppler spectra
1	100	−3.2	CLASS
2	200	−5.0	CLASS
3	500	−4.5	CLASS
4	600	−3.6	CLASS
5	850	−3.9	CLASS
6	900	0.0	CLASS
7	1050	−3.0	CLASS
8	1350	−1.2	CLASS
9	1450	−5.0	CLASS
10	1500	−3.5	CLASS

Table 4.2
ATDMA Macrocell Channel Model

Tap	Relative delay (ns)	Relative power (dB)	Doppler spectra
1	0	0	CLASS
2	380	−10.0	CLASS
3	930	−22.7	CLASS
4	1940	−24.7	CLASS
5	2290	−20.7	CLASS
6	2910	−22.1	CLASS

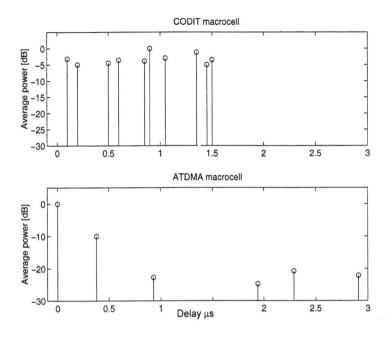

Figure 4.11 Impulse responses of the CODIT and ATDMA macrocell channel models.

4.6.2 Outdoor to Indoor and Pedestrian Radio Environment

The channel parameters for the microcell are given in Table 4.3 for CODIT and in Table 4.4 for ATDMA. The impulse responses for both channel models are illustrated in Figure 4.12.

Table 4.3
CODIT Microcell Channel Model

Tap	Delay (ns)	Power (dB)	Doppler spectrum	Ricean factor (dB)	Doppler shift (fi/fdop)
1	0	−2.3	RICE	−7.3	0.6066
2	0	0.0	RICE	−3.5	0.6066
3	0	−13.6	CLASS	−	−
4	50	−3.6	RICE	−3.5	0.6066
5	50	−8.1	CLASS	−	−
6	100	−10.0	CLASS	−	−
7	1700	−12.6	RICE	−2.2	0.6066

Table 4.4
ATDMA Microcell Channel Model

Tap	Delay (ns)	Power (dB)	Doppler spectra
1	0	−4.9	CLASS
2	230	0	CLASS
3	630	−11.4	CLASS
4	1110	−13.9	CLASS
5	1440	−16.1	CLASS
6	1840	−23.5	CLASS

From Figure 4.12 it can be observed that the first six taps in the CODIT microcell channel are very close to each other, and the receivers cannot resolve these paths. Even though some of these taps are Rice distributed, the total signal will be close to Rayleigh distributed, since there are two other paths with strong average power (−2.3 and −3.6 dB). This leads to almost the same situation as with a typical ATDMA macrocell, except that now there is one path at 1.7-μs relative delay, which provides some diversity. Both narrowband and wideband CDMA can only resolve the last tap, while the first six taps cannot be resolved.

The ATDMA microcell channel is also a rather harsh communications channel. The taps are now wider apart, but all taps are Rayleigh fading and the second strongest path is already −4.9 dB compared to the strongest one [19]. Wideband CDMA with 5-MHz bandwidth can resolve all six taps, while the one for narrowband CDMA can only resolve two groups of three taps each.

4.6.3 Indoor Office Radio Environment

The tap settings for CODIT picocell are given in Table 4.5, and for ATDMA picocell in Table 4.6. The impulse responses of both channels are given in Figure 4.13.

We notice from Figure 4.13 that the delay spread of CODIT picocell channel is very small. This, together with the fact that all the taps are Rayleigh fading, will lead to difficult flat fading channel [19]. The ATDMA indoor channel is much easier. The first strong Rice distributed tap should ensure better amplitude distribution than Rayleigh. Also, when using wideband CDMA, all the different paths should be resolvable.

Table 4.5
CODIT Picocell Channel Model

Tap	Delay (ns)	Power (dB)	Doppler spectra
1	0	−3.6	CLASS
2	50	0.0	CLASS
3	100	−3.2	CLASS

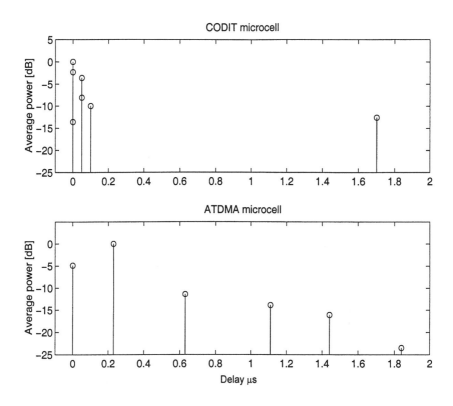

Figure 4.12 Impulse responses of the CODIT and ATDMA microcell channel models.

Table 4.6
ATDMA Picocell Channel Model

Tap	Delay (ns)	Power (dB)	Doppler spectra	Ricean factor (dB)	Doppler shift
1	0	0	RICE	−10.0	−
2	77	−3.3	RICE	−4.0	−
3	186	−9.3	RICE	−4.0	−
4	299	−14.3	RICE	−4.0	−
5	404	−20.3	RICE	−4.0	−
6	513	−26.8	CLASS	−	−

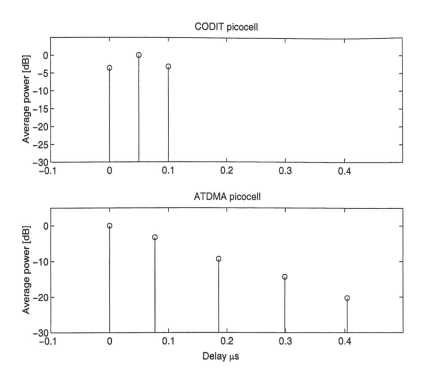

Figure 4.13 Impulse responses of the CODIT and ATDMA picocell channel models.

4.6.4   Spatial Channel Models

Deployment of adaptive/smart antenna techniques requires an understanding of the spatial properties of the wireless channel. Modern spatial channel models build on the classical understanding of fading and Doppler spread and incorporate additional concepts such as time delay spread, angle of arrival, and adaptive antenna geometries. For a more detailed treatment of spatial channel models, refer to [26] and references therein.

4.7   MOBILITY MODELS

Mobility modeling is involved with the analysis of aspects related to location management (location area planning, paging strategies), radio resource management (multiple access techniques, channel allocation strategies, handover rates), network signaling loads, and propagation (handover decision) [26, 28]. Different purposes require different types of mobility models. For example, location area planning requires the knowledge of the user location with an accuracy of a large-scale area, while analysis of radio resource schemes such as handover require medium-scale area accuracy (cell area) [27]. Mobility modeling can be based on simulation or analytical approaches. Typically, analytical models with simplified assumptions re used for analysis of critical network dimensioning parameters [27].

In this section we present analytical mobility models for a medium-scale area. The models are based on the SIG5 models [19], which were also used, with some changes, in the ETSI UMTS air interface evaluation process [20].

4.7.1 Vehicular Radio Environment

In the vehicular radio environment, users are completely free to move with a constant speed within the whole service area. The mobility model is a pseudo-random mobility model with semidirected trajectories [20]. Mobiles are uniformly distributed on the map, and their direction is randomly chosen at initialization. After every *decorrelation length,* there will be an update for the direction of the movement. The probability for the direction update is given in Table 4.7, and it is independent of the previous direction update event. If the direction update takes place, there is a uniform probability to change the direction of the movement between -45 and 45 degrees. The direction can also be changed within a given sector to simulate a semidirected trajectory.

Table 4.7
Parameters of the Vehicular Radio Environment Mobility Model

Parameter	Example value
Speed value	120 km/h
Probability to change direction at position update	0.2
Maximal angle for direction update	45°
Decorrelation length	20m

4.7.2 Outdoor to Indoor and Pedestrian Radio Environment

The deployment area is a regular grid of streets and buildings (see Section 4.5.2). The simulation area has a finite size with borders (the outer street tier). All buildings and streets have equal size. All users move only in the middle of the streets.

New users will be uniformly distributed over the possible coordinates in the service area. Each possible coordinate has an equal chance of being chosen. Mobiles move along streets (always in the middle of the streets) and may turn at crossings with a given probability (turning probability, TP), as indicated in Figure 4.14. The mobile's position is updated every m, and speed can be changed at each position update according to a given probability. The mobility model is described by the parameters shown in Table 4.8.

Table 4.8
Parameters of the Outdoor to Indoor and Pedestrian Radio Environment Mobility Model

Parameter	Example value
Mean speed	3 km/h
Standard deviation for speed	0.3 km/h
Probability to speed change at position update	0.2
Probability to turn at cross street	0.5

Mobiles are uniformly distributed in the streets, and their direction is randomly chosen at initialization. Users leaving the simulation area via the borders are lost. This loss influences the traffic density: it becomes inhomogeneous as it decreases from the middle to the borders. To keep the traffic density homogenous for the whole border, there must be an incoming probability equal to the leaving probability. In this model, this is achieved simply by generating a new subscriber when one leaves the area. This new subscriber is placed exactly at the border with the same probability of each point of the border. Because it is assumed that the leaving probability is the same for each part of the border, the resulting traffic density will be homogeneous.

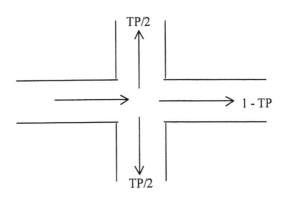

Figure 4.14 Turning point probability.

4.7.3 Indoor Office Radio Environment

In the indoor office radio environment, only one mobility class is assumed. The movement of the users is characterized as follows [20]:

- There is no mobility between the floors.
- Mobiles are either stationary or moving with constant speed from an office room to corridor or vice versa.
- If a mobile is in an office room, it has higher probability of being stationary.
- If a mobile is in the corridor, it has lower probability of being stationary.

Each mobile is either in the stationary or the moving state. The transition from the stationary state to the moving state is a random process. The time duration each mobile spend in the stationary state is drawn from the geometric (discrete exponential) distribution with different mean values depending on whether the mobile is in an office room or in the corridor

The transition from the moving state to the stationary state takes place when mobile reaches its destination.

When a mobile is in an office room and it is switched to the moving state, it moves to the corridor (see Figure 4.15), according to the following procedure:

1. Select the destination coordinates in the corridor with uniform distribution. (Each place in the corridor has equal probability of becoming the destination point.)
2. The mobile "walks" from its current location to the destination location so that first the vertical (y) coordinate is matched with the new coordinate, and next the horizontal (x) coordinate is matched with the destination coordinate. The speed is constant during the "walking."
3. When the mobile reaches the destination point, it is transferred into the stationary state.

Figure 4.15 The mobility model. Two example routes are drawn. On the left: from an office room to the corridor; and on the right: from the corridor to an office room.

By letting mobiles simply walk straight out from the office room, it is assumed that the door dividing each office room and corridor is as wide as the office room itself. When a mobile is in a corridor and it is switched to the moving state, it moves to any of the office rooms with equal probability. The following four-step procedure defines the movement along the corridor and from the corridor to an office room:

1. Select the destination office room by using discrete uniform distribution.
2. Select the destination coordinates with uniform distribution. (Each place in the corridor or in an office room has equal probability of becoming the destination point.)
3. The mobile "walks" from its current location so that first the horizontal (x) coordinate is matched with the new coordinate and next the vertical (y) coordinate is matched with the destination coordinate. The speed is constant during the "walking."
4. When the mobile reaches the destination point, it is transferred into the stationary state. At the stationary state, mobiles do not move at all.

Figure 4.16 represents the mobile station's movement with a state model. P(S,S) denotes the transition probability from the stationary state into the stationary state, P(S,M) is the transition probability from the stationary state into the moving state, P(M,S) denotes the transition probability from the moving state into the stationary state, and P(M,M) stands for the transition probability from the moving state into the moving state. To derive transition probabilities from the stationary state to the moving state, the following parameters must be set: ratio of mobiles at office rooms (r), mean office room stationary time (mr) and iteration time step (Δt). With these parameters, the transition probabilities per iteration time step (1 $\Delta t/mr$, 1-$\Delta t/mc$) and mean corridor stationary time (mc) can be derived so that flow to the office rooms equals the flow from the office rooms:

$$r \cdot \frac{\Delta t}{mr} = (1-r) \cdot \frac{\Delta t}{mc}$$ (4.14)

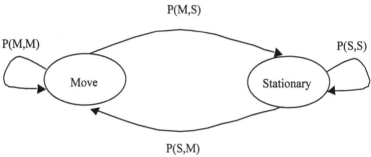

Figure 4.16 State transistion presentation of the mobile station movement.

With the parameters listed in Table 4.9, the following values are obtained:

Table 4.9
Parameters of the Indoor Office Mobility Model

Ratio of mobiles at office rooms	85%
Mean office room stationary time	30 seconds
Simulation time step	0.005 seconds
Number of office rooms	12
Mobile speed	1 m/s

- P(S,S) in an office room = 1−0.005/30 = 0.999833;
- P(S,M) in an office room = 0.005/30 = 0.0001667;

- P(S,S) in corridor room = 1−0.0009444 = 0.9990556;
- P(S,M) in corridor room = 0.005×85/(30×15) = 0.0009444;
- The average stationary time in the corridor becomes Δt/P(S,M) = 5.294 seconds.

P(M,S) and P(M,M) are not random. P(M,M) equals 1 if the destination is not reached,) otherwise; P(M,S) equals 0 if the destination is not reached, 1 otherwise.

REFERENCES

1] Prasad, R., "European Radio Propagation and Subsystems Research for the Evolution of Mobile Communications," *IEEE Communication Magazine*, Vol. 34, February 1996, p. 58.

2] Anderson, J. B., T. S. Rappaport, and S. Yoshida, "Propagation Measurements for Wireless Communications Channels," *IEEE Communication Magazine*, Vol. 33, January 1995, pp. 42–44.

3] Fleury, B. H., and P. E. Leuthold, "Radiowave Propagation in Mobile Communications: an Overview of European Research," *IEEE Communication Magazine*, Vol. 34, February 1996, pp. 70–81.

4] Molkdar, D., "Review on Radio Propagation into and Within Buildings," *IEE Proceedings*, Vol. 138, February 1991, pp. 61–73.

5] Hashemi, H., "The Indoor Propagation Channel," *Proc. IEEE*, Vol. 81, July 1993, pp. 943–968.

6] Steele, R., (ed.), *Mobile Radio Communications*, London: Pentech Press Publishers, 1992.

7] Rappaport, T. S., *Wireless Communication Principles & Practice*, Englewood Cliffs, New Jersey: Prentice Hall, 1996.

8] Prasad, R., *Universal Wireless Personal Communications*, Norwood, Massachusetts: Artech House, 1998.

9] Jakes, W. C., Jr., (ed.), *Microwave Mobile Communications*, New York: John Wiley & Sons, 1974.

10] Lee, W. C. Y., *Mobile Communications Engineering*, New York: McGraw-Hill, 1982.

11] Linnartz, J. P., *Narrowband Land-Mobile Radio Networks*, Norwood, Massachusetts: Artech House, 1993.

12] Lee, W. C. Y., *Mobile Cellular Telecommunications Systems*, New York: McGraw-Hill, 1989.

13] Lee, W. C. Y., *Mobile Communications Design Fundamentals*, New York: John Wiley & Sons, 1993.

14] Macario, R. C. V., (ed.), *Personal & Mobile Radio Systems*, London: Peter Peregrinus, 1991.

15] Stuber, G. L., *Principles of Mobile Communications*, Boston, Massachusetts: Kluwer Academic Publishers, 1996.

16] Mehrotra, A., *Cellular Radio Performance Engineering*, Norwood, Massachusetts: Artech House, 1994.

17] Hess, G. C., *Land-Mobile Radio System Engineering*, Norwood, Massachusetts: Artech House, 1993.

18] Recommendation ITU-R M.[FPLMTS.REVAL], "Guidelines for Evaluation of Radio

Transmission Technologies for IMT-2000/FPLMTS."

[19] Pizarroso, M., and J. Jiménez (eds.), "Common Basis for Evaluation of ATDMA and CODIT System Concepts," *MPLA/TDE/SIG5/DS/P/001/b1*, MPLA SIG 5, September 1995.

[20] ETSI UMTS 30.03, "Selection Procedures for the Choice of Radio Transmission Technologies of the UMTS," *ETSI Technical Report*, 1997.

[21] Strasser, G., (ed.), "Propagation Models Issue 1," *R2084/ESG/CC3/DS/P/012/b1*, April 1993.

[22] Jiménez, J., (ed.), "Final Propagation Model," *CODIT Deliverable, R2020/TDE/PS P/040/b1*, June 1994.

[23] Braun, W., and U. Dersch, "A Physical Mobile Radio Channel," *IEEE Trans. on Vehicular Technology*, Vol. 40, No. 2, May 1991, pp. 472–482.

[24] Xia, H. H., "Reference Models for Evaluation of Third-generation Radio Transmission Technologies," *Proceedings of ACTS Summit*, Aalborg, Denmark, October 1997, pp 235–240.

[25] Recommendation ITU-R M.1225, "Guidelines for Evaluation of Radio Transmission Technologies for IMT-2000," *Question ITU-R 39/8*, 1997.

[26] Special Issue on Smart Antennas, *IEEE Personal Communications*, Vol. 5, No. 1 February 1998.

[27] Markoulidakis, J. G., G. L. Lyberopoulos, D. F. Tsirkas, and E. D. Sykas, "Mobility Modeling in Third-Generation Mobile Telecommunications Systems," *IEEE Personal Communications*, Vol. 4, No. 4, August 1997, pp. 41–54.

[28] Jabbari, B., "Teletraffic Aspects of Evolving and Next-Generation Wireless Communications," *IEEE Personal Communications*, Vol. 3, No. 6, December 1996, pp 4–9.

Chapter 5

CDMA AIR INTERFACE DESIGN

5.1   INTRODUCTION

Chapter 2 presented the basic elements of DS-CDMA: RAKE receiver, power control, soft handover, and multiuser detection. This chapter discusses these elements in greater depth and presents the most important aspects to be considered when designing a wideband CDMA air interface. It starts by highlighting in Section 5.2 the system requirements and assumptions and their impact on the wideband CDMA air interface design. Secondly, logical and physical channel design are discussed in Sections 5.3 to 5.5. Thereafter, spreading codes and their properties are described in Section 5.6, along with the relevant criteria for selection and optimization. Section 5.7 on modulation covers not only data modulation but also spreading modulation and different spreading circuit types. Section 5.8 on control schemes gives the principles of forward error control (FEC) and automatic repeat request (ARQ). Next, the different data rates and quality of service requirements are described in Section 5.9, along with multirate scheme implementation. In addition, requirements and solutions for efficient transmission of packet data are given in Section 5.10. Transceiver design is illustrated from the receiver and transmitter perspective in Section 5.11. In addition to the RAKE receiver, an advanced receiver scheme, namely multiuser detection, is introduced in Section 5.12. Random access procedure is briefly discussed in Section 5.13. Section 5.14 describes handover and covers handover procedure, design requirements, and interfrequency handover. Design principles of power control are presented in Section 5.15. Finally, admission and load control are introduced in Section 5.16.

5.2   DEFINITION OF REQUIREMENTS AND GENERAL DESIGN PROCESS

The starting point for design of the air interface are the system requirements. Understanding these is critical for a good end result of the design process. Basic system requirements determine data rates, bit error rate (BER), and delay, as well as channel models, which define the radio environment. In Chapter 3, UMTS bearer service requirements were derived based on some service examples. In Chapter 4, path loss, shadowing, and wideband channel models for different radio environments were introduced. The radio environment is not in itself a system requirement, but it places constraints on the system design and therefore needs to be carefully modeled. A number of other issues, such as available frequencies, synchronization constraints, signaling requirements, and complexity aspects, have to be considered. The IMT-2000 system requirements have been specified in [1], and more detailed radio requirements are given in [2]. Corresponding ETSI documents for UMTS are [3, 4].

CDMA radio interface design involves several areas, as illustrated in Figure 5.1. Each area itself is very broad, and different parameters can be optimized alone. A good design, however, is always a tradeoff between several, often contradicting requirements. Good radio performance, for example, requires more advanced receivers resulting in higher costs and complexity, while another system requirement might be precisely the minimization of equipment cost. Spreading codes can be optimized alone or together with channel coding. The design of the multirate scheme for variable data rates requires careful joint-optimization of coding, modulation, and spreading codes. The handover synchronization scheme depends on the network synchronization (i.e., whether synchronous or asynchronous base stations are used). The preceding examples are just a few of the very large number of interdependencies among the different areas. Moreover, the final design depends on the way each optimization criteria is weighted. In the following sections, we highlight each area and their interrelationships in more detail.

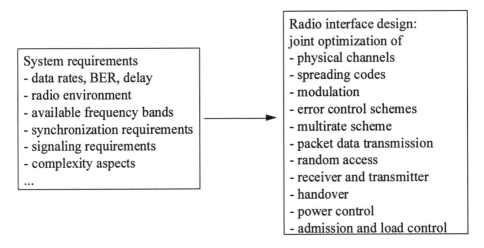

Figure 5.1 CDMA air interface design process.

5.3   LAYERED AIR INTERFACE STRUCTURE

As shown in Figure 5.2, the air interface functions are structured into protocol layers according to principles presented in [5]. The physical channel performs coding, modulation, and spreading for the physical channels. As shown in Figure 5.2, the link layer is further divided into two sublayers: medium access control (MAC) and link access control (LAC). The medium access layer coordinates the resources offered by the physical layer. The link access control performs the functions essential to set up, maintain, and release a logical link connection. The network layer contains call control, mobility management, and radio resource management functions. In this chapter we will restrict our discussion to the air interface dependent functions (i.e., to those function is in different protocol layers that depend on the underlying multiple access method, in this case, wideband CDMA). These layers include the physical layer, the MAC layer, and radio resource management in the network layer. The other functions are air interface independent and could be used for other types of multiple access schemes such as TDMA or FDMA.

The physical layer–related functions are discussed in Sections 5.5 to 5.12. The MAC layer is discussed as a part of packet data transmission in Section 5.10. Radio resource management schemes are discussed in Section 5.13 (random access), in Section 5.14 (handover), in Section 5.15 (power control), and in Section 5.16 (admission and load control). Some radio resource management functions such as handover, power control, and random access have an impact on the physical layer (measurement capabilities, impact on synchronization channels). Logical channel structure is an air interface independent function, but we discuss it in Section 5.4 in order to give background for the physical channel design.

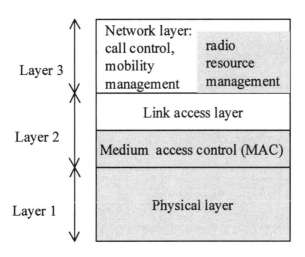

Figure 5.2   The layered structure of the air interface.

5.4   LOGICAL CHANNELS

Certain functions are needed in every cellular system to set up, release, and maintain connections. First, the mobile station acquires *synchronization* from the network. Thereafter, it obtains system information, such as cell identities, spreading codes, access channel, and neighboring cell lists, from the network. In case the call originates from the network, the mobile station is *paged*. After synchronization, the mobile station initiates connection setup by sending a *random access* message, and the network responds by sending an *access grant* message. Connection is established on a *traffic channel*. During the connection, control information carrying measurement and signaling data is transmitted.

The logical channels can be divided into control channels and traffic channels. A control channel is either common or dedicated. A *common control channel* is a *point-to-multipoint* control channel that carries connectionless messages, which are primarily intended to carry signaling information necessary for access management functions broadcast information (access request, access grant, paging messages, and user packet data) [5]. A *dedicated control channel* is a *point-to-point* bidirectional control channel, intended to carry signaling information such as handover measurements, service adaptation information, and power control information. Traffic channels carry a wide variety of user information streams. The procedures described lead to the following logical channel structure:

- Synchronization channel;
- Random access channel;
- Broadcast channel;
- Paging channel;
- Dedicated control channel;
- Traffic channel.

5.5   PHYSICAL CHANNELS

In general, the functions provided by logical channels need to be mapped into physical channels. The mapping depends on several aspects such as frame design, modulation method, and code design. In this section we discuss general principles for the design of different channels. These are discussed in detail later in this chapter and with example designs in Chapters 6, 7, and 8. The naming conventions for the physical channels of different wideband CDMA proposals vary considerably. Here we primarily use the same names as for the logical channels.

A specific function provided by a logical channel may be mapped into several physical channels. Or, vice versa, several fuctionalities may be mapped using, for example, time multiplexing into one physical channel. The different physical channels are separated by spreading codes.

A synchronization channel is used to provide the receiver with chip, bit, and frame synchronization. A pilot signal can be used as a reference signal for chip level synchronization and coherent detection, as discussed below. The mobile station uses the downlink synchronization channel for handover measurements and synchronization, as discussed in Section 5.14.3.

The broadcast channel transmits system-specific information. The system designer needs to determine the type and rate of information to be transmitted so that the broadcast channel data rate can be set. For the paging channel, the number of paging channels and the data rate need to be determined.

The random access channel structure depends on how fast the synchronization needs to be established and on the selected access strategy (see Section 5.13). In addition, the expected number of access attempts and synchronization time requirements determine the number of random access channels. Also, the amount of data transmitted during the access attempt is an important design aspect.

Typically, a dedicated control channel is bundled together with a traffic channel. Traffic and dedicated control channels can be time multiplexed, code multiplexed or I&Q multiplexed. The last alternative means that the user data is transmitted on the I channel and the control information on the Q channel. A pure code multiplexing without I&Q multiplexing leads to multicode transmission, which slightly increases the envelope variations of the transmitted signal, requiring a more linear power amplifier. Furthermore, with a code multiplexed dedicated control channel, the mobile station needs to receive two code channels instead of one, and thus, complexity is increased.

5.5.1 Frame Length Design

The frame length of the traffic channel depends on the service requirements and desired radio performance. In order to obtain good performance in a fading channel, the frame length of the traffic channel has to be long enough to support a reasonable interleaving depth. On the other hand, it cannot be too long and exceed the transmission delay requirement of service. Because the speech service is typically a dominant service, frame length is tailored according to it. As seen in Chapter 3, for third-generation wireless systems, the delay requirement for speech is specified to be 30 to 40 ms. Thus, a frame length of 10 ms is short enough, assuming interleaving is spanning over one frame. Some speech codecs introduce a 10-ms delay, and processing adds a few milliseconds. Thus, the total delay budget is between 20 and 25 ms. If a longer delay is allowed, a 20- ms frame could be used to improve the performance.

For data services, a longer interleaving period than for speech is often tolerable. Therefore, we would like to extend the interleaving period over one frame. This can be performed in multiples of the frame period and is called *interframe interleaving*. For packet services, a shorter frame length than 10 ms might be desirable to minimize the retransmission delay of erroneous packets.

The frame length control channel is typically assumed to be the same as for the traffic channel. The introduction of packet data, however, might motivate a smaller frame size for the control channel to decrease the channel setup time and to increase the traffic channel efficiency.

5.5.2 Measurement Signaling

Signaling requirements depend on the measurement needs (i.e., how often measurements for power control, handover, and load control are performed and how they need to be transmitted between mobile stations and the base station). When the rough transmission requirements for signaling have been defined, appropriate channel structures can be designed. The signaling traffic can either be transmitted on a dedicated control channel (outband signaling) or on a traffic channel by puncturing the user data (inband signaling). The selected method depends on the quality requirements for user data and mobile station complexity. Inband signaling degrades the transmission quality because user data bits are deleted. If the number of deleted bits is small, however, a power increase (with the same amount as for outband signaling) will compensate for the loss of quality.

5.5.3 Pilot Signals

Because of fading channels, it is hard to obtain a phase reference for the coherent detection of data modulated signal. Therefore, it is beneficial to have a separate pilot channel. Typically, a channel estimate for coherent detection is obtained from a common pilot channel. However, a common pilot channel transmitted with an omnidirectional antenna experiences a different radio channel than a traffic channel signal transmitted through a narrow beam. It has been noticed in second-generation systems that common control channels are often problematic in the downlink when adaptive antennas are used. The problem can be circumvented by user dedicated pilot symbols, which are used as a reference signal for channel estimation. In this case, a common pilot is only used for handover measurements and for channel estimation when the downlink beamforming is not used.

The user dedicated pilot symbols can either be time or code multiplexed. Figures 5.3 and 5.4 depict block diagrams of a transmitter and receiver for time multiplexed pilot symbols and a code multiplexed, parallel pilot channel, respectively. A drawback of the pilot channel transmitted on a separate code is that it requires extra correlators for despreading. The performance of both approaches is the same if the same amount of power is used for the pilot signals, assuming that optimal least mean squares channel recovery techniques are employed.

For time multiplexed pilot symbols, the ratio between the number of data symbols (L_d) and the number of pilot symbols (L_p) needs to be determined. For the parallel pilot channel, the power ratio of pilot and data channels needs to be determined. In the receiver, pilot symbols are averaged over a certain time period. The averaging length varies according to Doppler frequency. One possibility is to find a compromise by simulating the performance in different scenarios and by selecting a number of pilot symbols and an averaging period that results into reasonable performance over different operating conditions [6]. Another possibility is to have more flexibility and to have a variable number of pilot symbols and a variable averaging period. For a code multiplexed pilot channel, it is easier to adjust the pilot power.

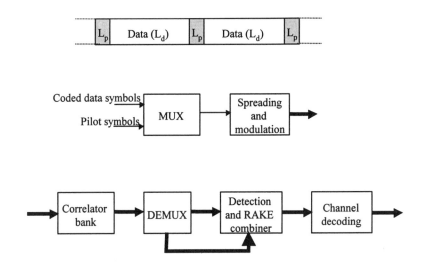

Figure 5.3   Time multiplexed pilot symbols.

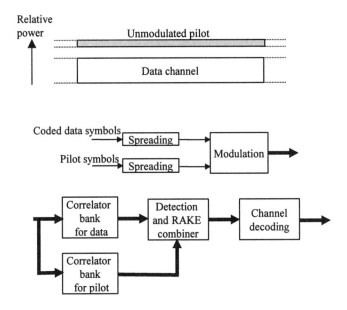

Figure 5.4   Code multiplexed, parallel pilot channel.

5.6   SPREADING CODES

Spreading codes[1] can be divided into *pseudo-noise* (PN) codes and *orthogonal* codes. PN codes are pseudo-random codes generated by a feedback shift register. The most commonly considered PN codes for DS-CDMA systems are generated using linear shift registers. The crosscorrelation between orthogonal codes is zero for a synchronous transmission. Orthogonal codes such as Walsh sequences are typically used for channel separation in DS-CDMA systems.

This section provides an engineer's perspective on the design and selection of spreading codes for DS-CDMA systems. For discussion of the fundamental properties of PN sequences and linear shift registers refer to [7, 8].

5.6.1   Basic Properties of Spreading Codes

In a DS-CDMA transmitter, the information signal is modulated by a spreading code, and in the receiver it is correlated with a replica of the same code. Thus, low crosscorrelation between the desired and interfering users is important to suppress the multiple access interference. Good autocorrelation properties are required for reliable initial synchronization, because large sidelobes of the autocorrelation function might lead to erroneous code synchronization decisions. Furthermore, good autocorrelation properties are important to reliably separate the multipath components. Because the autocorrelation function of a spreading code should resemble, as much as possible, the autocorrelation function of white Gaussian noise, the DS code sequences are also called pseudo-noise sequences. The autocorrelation and crosscorrelation functions are connected in such a way that it is not possible to achieve good autocorrelation and crosscorrelation values simultaneously. This can be intuitively explained by noting that having good autocorrelation properties is also an indication of good randomness of a sequence. Random codes exhibit worse crosscorrelation properties than deterministic codes [9]. Figure 5.5 illustrates auto- and crosscorrelation functions for a 31-chip-length M-sequence.

There exist two basic correlation functions: even and odd. For an *even* correlation function, the consequent information data bits are the same; and for an *odd* correlation function, consequent data bits are different. In an asynchronous CDMA system both functions exist with equal probability. Typically, however, only even crosscorrelation properties can be optimized.

Typically, PN sequences are generated with a linear feedback shift register generator, depicted in Figure 5.6. The outputs of the shift register cells are connected through a linear function formed by exclusive-or (XOR)-type logic gates into the input of the shift register. Nonlinear codes are used for military and cryptographic applications. Nonlinearity can be implemented, for example, with the aid of AND gates [9].

---

[1] Sometimes spreading codes are called spreading sequences. In this book, the terms spreading code and spreading sequence are used interchangeably.

Figure 5.5   Auto- and crosscorrelation functions.

Figure 5.6   Feedback shift register.

The symbol of a DS code can be classified as binary or nonbinary [9]. Nonbinary sequences are typically complex valued having equal signal amplitude. Binary or quaternary (4-phase) sequences are preferred in DS systems, because they fit in with BPSK, QPSK, and O-QPSK modulation methods typically used in DS-CDMA systems [9].

A spreading code can be either short or long.[2] Short codes span one symbol period, while long codes span several symbol periods. The reason to use short codes is either to attempt to control the correlation properties through a suitable selection of a short spreading code set or to minimize the complexity of multiuser detectors. Furthermore, the signal obtained by spreading with short codes is cyclostationary (periodic), which can be utilized in the design of the multiuser detection algorithm (see Section 5.12). Long codes facilitate a large number of spreading sequences, and they aim to randomize the interference.

---

[2] It should be noted that in the IS-95 system [10], the term *short code* is used (differently than in this book) for a relatively long sequence with length of 2^{15} used to spread several symbols, and the term *long code* is used for a code with length of 2^{42}.

A DS-CDMA system can be classified as synchronous or asynchronous.[3] In a synchronous system the transmission times of spreading codes are the same, whereas in an asynchronous system no time-control of transmission times between users exists. In a synchronous system (e.g., the downlink of a cellular system) completely orthogonal codes can be used. Multipath will result, however, in a partially nonorthogonal system. The uplink of a cellular system is asynchronous, although synchronization has been proposed for a wideband CDMA system in [11]. Short codes in an asynchronous system require a special synchronization scheme for handover because the time difference between base stations cannot be measured using spreading codes, unlike in a long code system [12].

5.6.2   Pseudo-Noise Sequences

Table 5.1 presents the number of M-sequences for different sequence lengths. As can be seen, the number of codes is rather limited. M-sequences, however, possess a certain property that has been used in the design of the IS-95 air interface. For each M-sequence, the time-shifted version of the code is a new code. Thus, the number of codes is the same as the length of the code. The propagation delay limits the minimum time separation between two codes, and therefore, long spreading sequences are needed. Assuming a chip rate of 3.6864 Mcps, one chip equals 0.27 µs. In IS-95 the minimum PN-offset is 64 chips corresponding to 15.6 km. To maintain the same PN-offset in kilometers between different pilot sequences as in IS-95, a PN-offset of 192 chips is required, and a code length of 2^{15} provides now 170 different pilots. Therefore, the use of the same M-sequence with different phase-offsets might lead to code planning [13] (see Chapter 12). An advantage of this code design is that crosscorrelation properties between the codes are now actually determined by the autocorrelation functions. The autocorrelation properties of M-sequences are very good, in contrast to their crosscorrelation properties. Because partial correlations are used, however, (i.e., correlation is performed over one symbol, not over the whole sequence), the good autocorrelation sequences of the full-length sequence are not fully utilized.

Table 5.1
M-Sequences

Sequence length	Number of codes
7	2
15	2
31	6
63	6
127	18
255	16
511	48

---

[3] It should be noted that synchronous/asynchronous CDMA in this context is different from the network synchronous/asynchronous scheme, which is related to synchronization between base stations.

In Table 5.2, a number of other well-known PN code families are presented. The number of Gold codes is approximately the same as the code length. Gold codes have two variations: even and odd degree codes. With an odd degree code (code lengths of 31, 127, 511, 2047, ...), a smaller value of maximum crosscorrelation relative to the code period is achieved [9]. Furthermore, there exists a code family called Gold-like sequences whose maximum crosscorrelation is slightly smaller than normal Gold codes. In practice, however, this reduction is so small that there is no performance difference.

There exist several Kasami families: small (S-Kasami), large (L-Kasami), and very large (VL-Kasami). In addition, Kasami sets can be further extended into the very very large family and so on. The larger the number of codes, the larger the maximum crosscorrelation value; thus, the overall performance of the system degrades. For BCH dual-codes, the number of codes and maximum crosscorrelation value depends on the error correction capability of the corresponding block code [9].

Recently, complex spreading sequences, called 4-phase sequences, have raised interest because of their possible improvement to the maximum crosscorrelation value [14]. Complex spreading sequences are suitable for quadrature modulation methods such as QPSK and O-QPSK. The most interesting code family is so-called Set A [14]. The code set size is the same as for the Gold sequences.

Table 5.2
Spreading Code Families

	Degree[a]	Number of codes	Maximum crosscorrelation
Gold	$n = 1 \bmod 2$	$2^n + 1$	$2^{\frac{n+1}{2}} + 1$
Gold	$n = 2 \bmod 4$	$2^n + 1$	$2^{\frac{n+2}{2}} + 1$
Gold like	$n = 0 \bmod 4$	2^n	$2^{\frac{n+1}{2}} - 1$
Gold like	$n = 0 \bmod 4$	$2^n + 1$	$2^{\frac{n+2}{2}} - 1$
S-Kasami	$n = 2 \bmod 4$	$2^{\frac{n}{2}}$	$2^{\frac{n}{2}} + 1$
L-Kasami	$n = 1 \bmod 2$	$2^{\frac{n}{2}}(2^n + 1)$	$2^{\frac{n+2}{2}} + 1$
	$n = 0 \bmod 4$	$2^{\frac{n}{2}}(2^n + 1) - 1$	$2^{\frac{n+2}{2}} + 1$
VL-Kasami	$n = 2 \bmod 4$	$2^{\frac{n}{2}}(2^n + 1)^2$	$2^{\frac{n+4}{2}} + 1$
Dual-BCH	$n = 0 \bmod 2$	$2^{\frac{n}{2}}$	$2^{\frac{n+2}{2}} + 1$
4-phase Set A	$n > 0$	$2^n + 1$	$2^{\frac{n}{2}} + 1$

[a] Code length is $2^n - 1$. *Source*: [9].

5.6.3 Orthogonal Codes

Orthogonal sequences are completely orthogonal for zero delay. For other delays, they have very bad crosscorrelation properties, and thus they are suitable only for synchronous applications. The downlink of a cellular system is synchronous without multipath. Due to multipath, however, an overlay code is required to suppress the multipath interference. Furthermore, an overlay code is required because of bad auto-correlation properties. The performance gain of the orthogonal Walsh codes depends on the channel profile, delay spread, and path loss. The better the isolation between cells and the smaller the multipath spread, the higher the performance gain [15].

The Walsh codes used in IS-95 are an example of orthogonal codes. Walsh codes can be despread easily by a Walsh-Hadamard transform. The length of a Walsh code is an even number of chips, and the number of codes is equal to the number of chips (e.g., there are 128 codes of length 128). A variable spreading factor scheme is one way of transmitting variable data rates (see Section 5.9). A Walsh code of length n can be divided into two codes of length $n/2$. All $n/2$ length codes generated from length n codes are orthogonal against each other. Furthermore, any two length $n/2$ and n codes are orthogonal except when one of the two codes is a mother code of the other. Walsh codes can be constructed according to (5.1).

$$H_0 = [0] \qquad H_n = \begin{bmatrix} H_{n-1} & H_{n-1} \\ H_{n-1} & \overline{H}_{n-1} \end{bmatrix} \qquad (5.1)$$

Another example of orthogonal codes are tree-structured orthogonal codes as proposed in [16]. Orthogonality between the different spreading factors can be achieved by tree-structured orthogonal codes whose construction is illustrated in Figure 5.7.

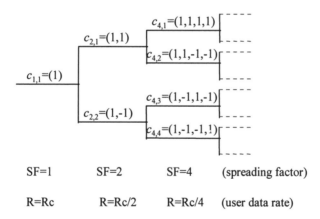

$$\begin{array}{llll} \text{SF=1} & \text{SF=2} & \text{SF=4} & \text{(spreading factor)} \\ \text{R=Rc} & \text{R=Rc/2} & \text{R=Rc/4} & \text{(user data rate)} \end{array}$$

Figure 5.7 Construction of orthogonal spreading codes for different spreading factors.

The tree-structured codes are generated recursively according to (5.2):

$$c_{2n} = \begin{pmatrix} c_{2n,1} \\ c_{2n,2} \\ \vdots \\ c_{2n,2n} \end{pmatrix} = \begin{pmatrix} \begin{pmatrix} c_{n,1} & c_{n,1} \\ c_{n,1} & -c_{n,1} \end{pmatrix} \\ \vdots \\ \begin{pmatrix} c_{n,n} & c_{n,n} \\ c_{n,n} & -c_{n,n} \end{pmatrix} \end{pmatrix} \tag{5.2}$$

where c_{2n} is an orthogonal code set of size $2n$. The orthogonality properties of these codes are similar to the variable length Walsh codes. Actually, this is just another way to generate Walsh functions. The order of the functions in the matrix is not the same as in the Hadamard matrix, but the functions themselves are the same. The generated codes within the same layer constitute a set of orthogonal functions and are thus orthogonal. Furthermore, any two codes of different layers are also orthogonal except when one of the two codes is a mother code of the other. For example, code $c_{4,4}$ is not orthogonal with codes $c_{1,1}$ and $c_{2,2}$.

5.6.4 Selection Criteria

The correlation properties and the number of codes are two important selection criteria when designing or selecting the most appropriate code family for a specific air interface design.

5.6.4.1 Correlation Properties

Because in an asynchronous system the codes can have any phase shift relative to each other, crosscorrelation properties of PN codes must be optimized. In addition, the optimization of the autocorrelation properties is important for acquisition performance. In what follows, we discuss different optimization criteria used for code design and their impact on system performance.

Several optimization criteria for spreading code design have been presented in the literature [9]. Typically, different variations of correlation optimization are used. Quite often, however, they lack a link to the final system performance, which is measured in terms of BER and average signal-to-noise ratio (SNR) [9].

The code design criterion used most often is minimization of the maximum peak even crosscorrelation value. For reasonably large code lengths (255), however, it has not been shown to give any difference between code families because both the odd and the even crosscorrelations impact the performance [9]. In addition, the minimization of the maximum correlation value does not minimize the overall multiple access interference, because in an asynchronous system all phase shifts impact the system's performance. Therefore, the distribution of crosscorrelation values is more important than the peak crosscorrelation value.

The maximum correlation value has more meaning when the SNR is large, the number of users is small, or when the codes are very short. This is the case in a variable spreading factor system for high data rates. Optimization of codes for this kind of situation, however, is difficult because the codes should be optimized against both codes of the same length and other spreading factors with different code lengths. If so-called subcode modulation is used [17], where the length of the spreading code is equal to the longest symbol corresponding to the lowest symbol rate, integration is not done over the full length of the sequence for higher symbol rates. Thus, the partial crosscorrelation properties determine the performance. In practice, however, the optimization of the partial crosscorrelations is difficult. Of course, in a situation when two high bit rate users with the same data rate coexist within the same bandwidth, a suitable selection of spreading codes would improve the performance.

Furthermore, the maximum correlation value has significance when users have a delay difference that produces this value. In this case, if the relative positions of the users do not change, then the observed performance degrades drastically. So, even if the system performs well on average, some users can experience low quality. This problem does not occur with long spreading codes in spite of occasional large crosscorrelation values, since correlation changes with every data bit. For spreading factors of 127 and 255, however, short spreading codes have a performance advantage of 0.5 to 1.0 dB over long spreading codes [9].

A more suitable optimization criterion than the maximum correlation value is the optimization of mean-square (MSQ) value of the crosscorrelation values. The MSQ value of even (periodic) and odd crosscorrelation values is almost equal for Gold, Kasami, and M-sequences [9].

From the performance point of view, it seems that any of the pseudo-noise codes presented in Section 5.6.2 could be used as the spreading codes for an asynchronous DS-CDMA system. Short spreading codes do have a small performance advantage over long codes [9].

5.6.4.2   Number of Codes

In the case of long codes, the number of codes is sufficient to accommodate enough users in a cell. On the other hand, one might sometimes run out of short spreading codes. In this section, different factors influencing the required number of spreading codes are discussed.

The required number of codes depends on the expected traffic load and spectrum efficiency. A large number of codes is required to support low bit rate speech users and also when a multicode transmission method is used (see Section 5.9.1). As an example, assume a spectrum efficiency of approximately 100 Kbps/MHz/cell for 5-MHz wideband CDMA [18]. With voice activity of 50% and voice codec rate of 8 Kbps, this would result into 125 users per cell ($= 5 \cdot 100/8 \cdot 0.5$).

In the downlink, orthogonal codes are usually used for channel separation. For example, there are 256 codes of length 256, and thus, even with a dedicated signaling channel using a different Walsh code than the traffic channel for every user, the number of codes would be enough in the above-described example. If the neighboring cells are empty, however, then one cell can accommodate a much higher number of users and more codes are required. Furthermore, mobiles in soft handover consume codes from at least two base stations. One way to increase the number of codes is to allocate the same set of orthogonal codes with different pilot codes for the same sector. This, however, would degrade the system performance because the two sets of codes would no longer be orthogonal. Also, quasi-orthogonal codes have been proposed to increase the number of orthogonal codes available [19].

In a variable data rate system, code allocation for different connections can be either static or dynamic [20]. The dynamic code allocation releases the spreading codes when the user is inactive. In static allocation, a connection is allocated a number of codes to support the maximum data rate. If the activity factor were very low, each connection would only use a very small part of the allocated codes. If there are several such users, then the base station might run out of codes. Therefore, the dynamic code allocation is the preferred method, especially in a packet data system with highly variable data rates.

In the uplink, each user has its own individual spreading code. If short codes and QPSK modulation are used, 250 codes are required for each cell assuming the same spectrum efficiency of 100 Kbps/MHz/cell as in the downlink example. Because the same codes cannot be reused in the neighboring cells, a much larger number of codes is required. Thus, it seems that for length 255 codes only the VL-Kasami set has enough codes (over one million codes). The L-Kasami set could be used as well, but then the code reuse factor is 16 (= 4112/250), which might not be enough.

5.6.4.3   Other Code Design Requirements

Orthogonal codes have an even number of chips in contrast to PN-sequences, which have an odd number of chips. If both types of codes are used in the system, it is beneficial for the spreading code length to be a multiple of a power of 2 in order to facilitate synchronization. Therefore, PN-sequences should be extended by one chip. This extra chip should be selected in such a way that the number of ones and zeros in the extended sequence is equal. In general, this extension of the code sequence does not change the correlation properties significantly.

For the access and synchronization channel, synchronization requirements might impact the spreading code selection. Acquisition time should be short, and thus, the number of codes cannot be too large or the codes too long. If long codes are used, however, then special synchronization schemes might be needed to reduce uncertainty in the code phase.

Receiver complexity impacts the selection of spreading codes. Multiuser detection, for example, might require the use of short spreading codes to reduce implementation complexity.

5.7   MODULATION

Figure 5.8 illustrates DS-CDMA transmission system functions related to modulation. The *data modulator* maps the m incoming coded data bits into one of $M (= 2^m)$ possible real or complex valued transmitted data symbols. The M-ary data symbols are fed to the *spreading circuit,* where the resulting signal is filtered and mixed with the carrier signal (i.e., spreading modulation is performed). Typical data modulation schemes are BPSK and QPSK. Spreading circuits can be binary, balanced quaternary, dual-channel quaternary, or complex DS-spreading. Typical spreading modulation schemes are

BPSK used with a binary spreading circuit, QPSK, and O-QPSK used with a quaternary or complex spreading circuit.

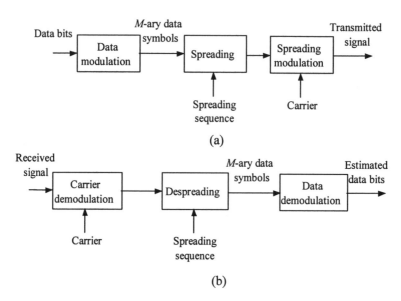

(a)

(b)

Figure 5.8   DS-CDMA transmission system: (a) transmitter and (b) receiver.

5.7.1   Data Modulation

Binary data modulation maps incoming data bits into transmitted data symbols with a mapping rule: $(1 \Rightarrow +1)$, $(0 \Rightarrow -1)$. Quaternary data modulation converts two consecutive data bits into one complex data symbol. Typically, Gray encoding is used, which has the following mapping rule: $(00 \Rightarrow +1+j)$, $(01 \Rightarrow -1+j)$, $(11 \Rightarrow -1+j)$, $(10 \Rightarrow +1-j)$. The resulting complex data symbol can be transmitted with quaternary or complex spreading. It should be noted that the quaternary data modulation does not give any advantage in a DS-CDMA system, as compared to binary data modulation. Actually, it is slightly more sensitive to phase errors due to crosscoupling of the I and Q channels.

Even higher order modulation methods such as 16-QAM have been proposed for CDMA to increase data throughput. However, the increase in data rate does not offset the higher carrier-to-interference ratio (C/I) requirement. With multiuser detection the difference is smaller because most of the interference is removed.

Demodulation can be noncoherent, differentially coherent, or coherent. With an ideal synchronization and without an external reference signal, the coherent reception gives a 3-dB advantage over noncoherent reception in an additive white Gaussian noise

(AWGN) channel. Coherent demodulation requires a phase reference, which can be obtained either from the data bearing signal or from an auxiliary source, such as external reference symbols or the pilot signal. Thus, the actual gain achieved against noncoherent detection depends on the amount of radio resources spent for reference symbols/pilot signal and performance loss due to errors in phase estimation.

5.7.1.1 M-ary Modulation

In addition to the above-described conventional M-ary data modulation, which converts incoming data symbols into complex valued symbols, the M-ary data modulator can convert m incoming data symbols into one of M ($M = 2^m$) possible data symbols represented by K channel symbols. These can then be transmitted using binary, balanced quaternary, quaternary, or complex spreading schemes. The M-ary modulator spreads the transmitted signal in the frequency domain. An example of such a scheme is the IS-95 uplink, where orthogonal 64-ary modulation is used (i.e., six consequent data symbols are used to select one Walsh-Hadamard symbol length of 64) [10]. In this way, the energy of six symbols is collected before a quadratic combiner at the receiver and the performance is improved when compared to binary modulation. The choice of the code rate depends on implementation complexity and channel coherence time, and performance improves as long as the carrier phase remains constant over the duration of a given M-ary symbol. Implementation of coherent detection is more difficult for this kind of scheme but has been proposed in [21–23] and is discussed further in Section 5.8.7.

5.7.2 Spreading Circuits

The binary DS-spreading circuit has a simple transmitter structure, and only one PN-sequence is needed for spreading. Its drawback is that it is more sensitive to multiple

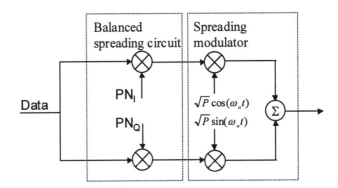

Figure 5.9 Balanced quaternary transmitter.

access interference than balanced quaternary spreading [24].

Balanced quaternary DS-spreading spreads the same data signal into I and Q channels using two spreading sequences. Figure 5.9 illustrates the balanced quaternary DS transmitter.

In the dual-channel QPSK spreading circuit depicted in Figure 5.10, the symbol streams on the I and Q channels are independent of each other. They can even have different powers. Similar to the balanced quaternary spreading circuit, two spreading sequences are needed.

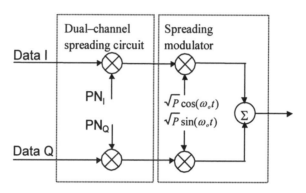

Figure 5.10 Dual-channel quaternary transmitter.

Figure 5.11 depicts a complex spreading scheme. Complex spreading uses either a complex spreading sequence or two real spreading sequences. Complex spreading reduces the peak to average ratio of the signal. It should be noted that with complex spreading, offset QPSK should not be used, because it will actually increase the peak-to-average power ratio. For despreading architectures, see Section 11.4.1.3.

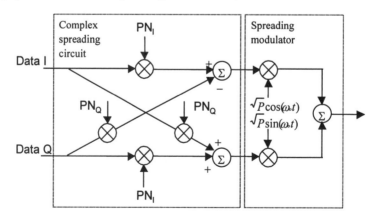

Figure 5.11 Complex spreading transmitter.

5.7.3 Spreading Modulation

Typically, linear modulation methods (BPSK, QPSK, and offset QPSK) have been proposed for wideband CDMA because they offer good modulation efficiency. Band-limited linear modulation methods with nonlinear power amplifiers result in spectrum regrowth (i.e., spectrum leakage to adjacent carriers). Thus, especially in the uplink, the modulated signal should tolerate nonlinear power amplifier effects as much as possible, since a nonlinear power amplifier is more power efficient than a linear one. In practice, after the modulation method has been fixed, pulse shaping determines the final spectrum properties of the modulation scheme. As discussed in Chapter 10, the roll-off factor depends on power amplifier linearity and adjacent channel attenuation requirements.

BPSK spreading modulation can be used with the binary spreading circuit. Non-filtered BPSK has a constant envelope and an infinite spectrum. In practice, however, the BPSK signal is filtered, resulting in a nonconstant envelope and higher linearity requirements.

QPSK and offset QPSK are quadrature types of modulation methods. The offset QPSK has a one-half chip delay in the Q channel, which prevents phase transitions via zero, as can be observed in Figure 5.12. Therefore, filtered O-QPSK has fewer envelope variations than QPSK, thus reducing the linearity requirements on the power amplifier. This is especially important in the uplink. If complex spreading is used, however, then QPSK should be used instead of offset QPSK.

The scattering diagrams of BPSK, QPSK, and offset QPSK modulation methods are shown in Figure 5.12.

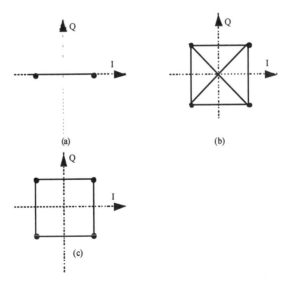

Figure 5.12 Scattering diagrams: (a) BPSK, (b) QPSK, and (c) O-QPSK.

5.8   ERROR CONTROL SCHEMES

Error control schemes can be classified in two basic categories: forward error control (FEC) and automatic repeat request (ARQ) schemes [25]. In addition, a combination of FEC and ARQ, a hybrid scheme, can be used [25]. In FEC, an error correcting code, sometimes referred to as the channel code, is used to combat transmission errors in a fading radio channel. In ARQ, an error detecting code together with a retransmission protocol is used. In the hybrid scheme, FEC reduces the need for retransmission by improving the error rate before ARQ.

5.8.1   Selection of Error Control Scheme

The choice of the error control scheme depends on the required quality of service (data rate, delay, and BER) and the radio channel. Raw channel error rates for transmission in a fading channel are typically on the order of 10^{-1} to 10^{-2}. For speech, the resulting error rate after demodulation must be on the order of 10^{-2} to 10^{-3} or better. This can be achieved, for example, with convolutional codes, which is the standard coding for all cellular systems. For data transmission, a bit error rate of 10^{-6} or better is required. This can be achieved with a concatenated coding scheme using a convolutional and a block code—typically a Reed-Solomon code [26]. Furthermore, nontransparent data services can utilize an additional ARQ scheme to improve the error performance. Maximum transmission delay of a service determines the maximum interleaving depth. Since the application requirements (BER, bit rate, and delay) vary greatly, the coding scheme should be adaptable.

5.8.2   Convolutional Codes

For convolutional codes, constraint length and code rate need to be selected. The constraint length should be as large as possible to obtain good performance [27]. The complexity of the decoder, however, increases with increasing constraint length. State-of-the-art VLSI (very large scale integration) implementations have been obtained for constraint length 9 convolutional codes. The code rate depends on the interleaving depth and the coherence time of the channel.

5.8.3   Concatenated Reed-Solomon/Convolutional Coding

This coding scheme employs an outer Reed-Solomon (RS) code and an inner convolutional code in conjunction with outer and inner interleavers [28]. The use of an inner bit interleaver/deinterleaver pair renders the fading channel memoryless and allows the convolutional code to mitigate multiuser interference more effectively [26]. The burst errors at the output of the Viterbi decoder are randomized by the outer symbol interleaver. The concatenated coding for mobile radio systems is designed in a way that the convolutional code produces an error rate on the order of 10^{-3}, and the RS code then improve this to a level of 10^{-6}. Of course, the actual improvement depends on the channel and on interleaving lengths. In addition, the error detection capability of the RS code can be used to provide an indication for automatic retransmission request for nontransparent data services.

5.8.4 Turbo Codes

Recently, turbo coding has received attention as a potential FEC coding scheme for cellular applications [29–32]. Turbo codes are parallel or serial concatenated recursive convolutional codes, whose decoding is carried out iteratively [29]. They have been demonstrated to closely approach the Shannon capacity limit on both AWGN and Rayleigh fading channels. The turbo code encoding structure is based on a combination of two or more weak error control codes. The information bits are interleaved between two encoders, generating two parity streams. These streams are then multiplexed and possibly punctured. The whole process results in a code that has powerful error correction properties.

5.8.5 Hybrid ARQ Schemes

The type-I hybrid ARQ scheme uses a code designed for error detection and correction [25]. Thus, it requires more parity check bits (overhead) than a code used for error detection only. Type-II hybrid ARQ codes the first transmission of a message with parity check bits used solely for error detection [25]. If an error is detected, the erroneous word is saved in the receiver buffer and a retransmission is required. The retransmission is a block of parity check bits formed from the original message and an error-correcting code. These parity check bits are then applied to the stored code word in the receiver. In case there is still an error, the next retransmission might contain either the original code word or an additional block of parity check bits.

5.8.6 Interleaving Schemes

A radio channel produces bursty errors. Because convolutional codes are most effective against random errors, interleaving is used to randomize the bursty errors. The interleaving scheme can be either block interleaving or convolutional interleaving [27]. Typically, block interleaving is used in cellular applications.

The performance improvement due to interleaving depends on the diversity order of the channel and the average fade duration of the channel. The interleaving length is determined by the delay requirements of the service. Speech service typically requires a shorter delay than data services. Thus, it should be possible to match the interleaving depth to different services.

5.8.7 Combined Channel Coding and Spreading

Very low rate coding schemes, which involve a combination of channel coding and spectrum spreading, have been presented by Hui and Viterbi [21–22]. In these schemes, signal spreading is performed using a very low rate convolutional code, and users are separated with PN-sequences, which have the same rate as the channel symbol rate (i.e., spectra is not spread by the PN-sequence at all but it is used to randomize the users' signals against each other). Hui actually proposed a combination of a very low rate convolutional code and short spreading sequences. The encoder consists of a binary convolutional encoder, a simple shift register of length k (k = constraint length of the code), and a M-ary orthogonal signal mapper [23]. The k code bits are used to select one of the M orthogonal code words (e.g., Walsh symbol set). The code rate is now $1/M$. It is also possible to use bi-orthogonal signal sets, as presented in [23].

5.9   MULTIRATE SCHEMES

Multirate design means that different services with different quality of service requirements are multiplexed together in a flexible and spectrum-efficient way. The provision of variable data rates with different quality of service requirements can be divided into three subproblems:

- How to map different bit rates into the allocated bandwidth;
- How to provide the desired quality of service;
- How to inform the receiver about the characteristics of the received signal.

The first problem concerns issues like multicode transmission and variable spreading. The second problem concerns coding schemes, and the third problem concerns control channel multiplexing and coding.

In addition to the above basic requirements, the design of a multirate solution needs to consider a number of other requirements and constraints. The uplink and downlink directions set different requirements for the multirate solution. Mobile station power amplifier requirements should not be too stringent, in order to facilitate use of power-efficient power amplifiers. In the downlink, the multirate solution should allow the mobile station receiver to save processing power.

5.9.1   Accommodation of Higher Data Rates

The increase of data rate in a CDMA system can be implemented by two basic schemes: *variable spreading factor* (VSF) and *multicode*. In a VSF scheme, the spreading ratio is reduced as the data rate increases. In a multicode scheme, additional parallel codes are allocated as the data rate increases. It is also possible to have combinations of these. Figure 5.13 illustrates the multicode transmission scheme. The high bit rate data stream is split into N parallel channels, each with data rate R_b. With multicode transmission, the mapping of simultaneously transmitted services into frames can be performed in two different ways. In the first way, services are transmitted simultaneously in different frames and in different codes. In the second way, services are mapped into the same frame, whose bits are then mapped into different codes. The advantage of this flexibility in service mapping will be discussed further in conjunction with power control (see Section 5.9.1.4).

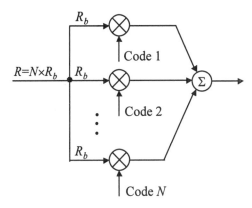

Figure 5.13   Multicode transmission scheme.

In the selection of a multirate transmission scheme, the following criteria need to be considered:

- Performance in the presence of multipath;
- Multiple access interference characteristics;
- Forward link orthogonality;
- Power control requirements;
- Complexity of the receiver and transmitter.

5.9.1.1   Performance in the Presence of Multipath

The performance of multicode and VSF schemes in the presence of multipath is considered. In the case of no multipath, there is no self-interference, assuming fully orthogonal spreading sequences [20]. When multipath exists, the self-interference is the same because in the case of multicode, there is N times less interference from other codes due to the higher spreading factor but there are N interfering codes instead of only one, as in the single code approach. Thus, in this respect, there seems to be only a small difference in the performance, assuming RAKE detection for BER greater than 10^{-6}. A drawback of the VSF scheme is that a very short code duration leads to intersymbol interference (ISI). For example, a data rate of 500 Kbps corresponds to a symbol duration of 2 µs. Thus, with a delay spread of 10 µs, there is ISI over five symbols.

5.9.1.2   Multiple Access Interference Characteristics

A fundamental assumption behind a matched filter detection approach is that the multiple access interference is Gaussian. The VSF approach may violate this assumption because there may be a high power user degrading the performance of the low rate users. In addition, for the highest data rates the spreading factor of the VSF scheme is very small; thus, a high maximum crosscorrelation relative to autocorrelation peak might degrade the performance. Therefore, in some instances it might be better to implement higher bit rates with a multicode scheme.

5.9.1.3 Forward Link Orthogonality

Because the forward link is synchronous, it is possible to maintain orthogonality between the codes. For multicode transmission we can select a set of orthogonal codes. With VSF, it is also possible to maintain orthogonality between different spreading factors if we impose the following constraint between different rates:

$$R = R_c/2^n \qquad n = 0, 1, 2, \ldots \qquad (5.3)$$

This can be achieved with the variable length Walsh sequences or with the three structured orthogonal codes [16] discussed in Section 5.6.3.

5.9.1.4 Power Control Requirements

In order for the receiver to be able to estimate the path loss for power control purposes, transmission power should not vary. Otherwise, the receiver has to first know the rate of a traffic channel. Another possibility is to use a fixed rate control channel for the power measurements. With multicode transmission, each of the parallel channels has a fixed power. For VSF, the power varies according to transmission rate, and thus, explicit transmission power needs to be signaled to the receiver. One further advantage of multicode transmission is that in case of several simultaneous services, the QoS can be adjusted using power control. Parallel service with similar power requirements can be time multiplexed, while code multiplexing is used for parallel services with different power requirements.

5.9.1.5 Complexity

Power amplifier linearity requirements should be as low as possible to allow the use of power-efficient power amplifiers. The multicode scheme results in larger envelope variations than the single code transmission. An additional drawback of the multicode scheme is that it requires as many RAKE receivers as there are codes. Each RAKE receiver, however, may be less complex, since a higher spreading factor might facilitate the use of fewer bits for quantization [20].

5.9.2 Granularity of Data Rates

Granularity means the minimum possible bit rate change. In order to maintain high flexibility, it is desirable to have as fine a granularity as possible. For multicode, granularity is R/M Kbps. If one code has the capability of smaller quantization of data rates, then better granularity can be achieved. For the VSF scheme, granularity varies according to the spreading factor and is finest for the low bit rates [20]:

$$\Delta R = \frac{R_c}{SF-1} - \frac{R_c}{SF} = \frac{R_c}{\frac{R_c}{R}-1} - R = \frac{R^2}{R_c - R} \tag{5.4}$$

where SF is the spreading factor, R_c is the chip rate, and R is the user bit rate.

For low bit rate services, this would most likely be sufficiently fine granularity. If we restrict the spreading factors according to (5.2), however, then the granularity is too coarse. Furthermore, if we assume for multicode transmission that one code carries 10 Kbps of traffic, the basic granularity is 10 Kbps. This is too large for some services such as low rate speech codecs, which typically require better granularity. The problem of granularity can also be considered from the viewpoint that, given the coded user bit rate, it is possible to match it with the given chip rate. For example, if one code carries 32 ksymbol/s and the source data rate is 9.6 Kbps coded with a 1/3 convolutional code, it results in a gross bit rate of 28.8 Kbps. Thus, we need to match the 28.8 Kbps to 32 Kbps.

One alternative to matching the bit and chip rate is the use of rate compatible punctured codes (RCPC) [33]. The basic idea of RCPC coding is to divide the bit stream of the mother code into blocks whose bits are either punctured or repeated according to a perforation/repetition matrix. The drawback of this approach is that a fixed number of rates have to be selected, because not all service rates can be directly matched to the available chip rate. Furthermore, for RCPC, good codes are only known up to the constraint length of 7.

Repetition coding, or puncturing, is another possibility for rate matching. Even unequal repetition coding (i.e., only some symbols in the frame are repeated to implement the desired symbol rate within the frame) can be used [17].

5.9.3 Transmission of Control Information

In order to vary the data rate or other service parameters, the receiver needs to know the structure of the received signal. This can be achieved either by transmitting explicit control information or by blind rate detection, for example, from the CRC information [34]. In case explicit control information is transmitted, the following issues need to be solved:

- Coding of the control information to achieve desired quality of service;
- Multiplexing of the control information;
- Position of the control information.

The control bits have to have a considerably lower error rate than the information bits, because, if a control word has an error, the whole frame is lost. Control information can either be coded together with the user data or independently from the user data. Furthermore, control information can be transmitted either code multiplexed or time multiplexed.

There are two possibilities for the position of control information: in the previous frame or in the same frame as the user data, as illustrated in Figure 5.14. Because the receiver is informed in advance of the transmission parameters, the processing of the data can be done "on-line." For services with a short delay requirement, however, the additional delay of one frame might be too long. A further drawback is that erroneous control information results in loss of the previous and the next frame, except if the next frame is transmitted with the same parameters as the previous frame.

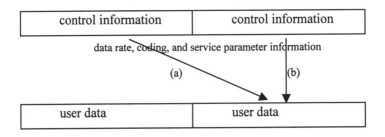

Figure 5.14 Control information transmitted (a) in the previous frame and (b) in the same frame as user data.

If the control information is transmitted within the same frame as the user data, then the receiver needs to decode the control information first. Thus, the data need to be buffered. Memory requirements for the buffer depend on the spreading code solution. In case the spreading ratios are multiples of each other, it is possible to despread with the lowest spreading ratio and to buffer the subsymbols after despreading, instead of the samples before the despreading, which need to be used if arbitrary spreading ratios are used. For high data rates, however, buffering might be still a problem.

5.10 PACKET DATA

Because non-real-time packet data services are not delay sensitive, they use the retransmission principle implemented with ARQ protocol to improve the error rate. The retransmission protocol can be either implemented in layer 2 as part of MAC and RLP or in the physical layer (layer 1). If packet data retransmission is implemented as part of layer 2, the transmission of packet data in the physical layer does not differ from the transmission of circuit switched data. So the multirate aspects discussed above also apply to the transmission of packet data. If the physical layer ARQ is used, then the physical layer is modified depending on the ARQ scheme used (see Section 5.8 about ARQ schemes). In both cases, the access procedure and handover for packet data services have certain special implications, which are discussed in the next subsections.

5.10.1 Packet Access Procedure

The packet access procedure in CDMA should minimize the interference to other users. Because there is no connection between the base station and the mobile station before the access procedure, initial access is not power controlled and thus the information transmitted during this period should be minimized. There are three scenarios for packet access:

1. Infrequent transmission of short packets containing little information;
2. Transmission of long packets;
3. Frequent transmission of short packets.

Because the establishment of a traffic channel itself requires signaling and thus consumes radio resources, it is better to transmit small packets within the random access message without power control. For long and frequent short packets, a dedicated traffic channel should be allocated.

If a dedicated channel has been reserved and there is nothing to send, the mobile station either cuts off the transmission or keeps the physical connection by transmitting power control and reference symbols only. In the former case, a virtual connection (higher layer protocols) is retained in order to rapidly reestablish the link in case of a new transmission. Selection between these two alternatives is a tradeoff between resources spent for synchronization and power control information, and resources spent for random access.

5.10.2 MAC Protocol

The task of the medium access protocol is to share the transmission medium with different users in a fair and efficient way. Sometimes, multiple access protocols such as FDMA, CDMA, and TDMA are also classified as medium access protocols. As already discussed in the beginning of this chapter, however, the medium access protocol is part of the link layer, while the multiple access scheme is part of the physical layer. The medium access protocol has to resolve contention between users accessing the same physical resource. Thus, it also manages the packet access procedure described in the previous section. Because the third-generation systems offer a multitude of services to customers at widely varying quality of service requirements, the MAC needs to offer capabilities to manage the access demands of different users and different service classes. This can be performed using reservation and priority schemes. Services with delay constraints can use a reservation scheme to reserve capacity to guarantee the quality of service. Priority schemes can be used to prioritize the requests from different services.

5.10.3 Packet Data Handover

Because CDMA operates with a reuse factor of one, it needs efficient and fast handover in order to avoid excessive interference with the other cells. This has been realized with soft handover in the case of circuit switched connections. Soft handover also improves performance through increased diversity. For packet connections, and especially for short packets, there may be no need to establish soft handover even if the user is at the cell edge. There is still the need, however, to route packets via the base station that provides the best connection. This is more important with frequent packet transmissions than with infrequent transmission. This can be implemented with frequent rescheduling of packet data transmission (i.e., if the transmission time exceeds a predefined timer, rescheduling is performed and the base station offering the best connection is selected).

If soft handover is used, then the implementation of ARQ depends on where it is terminated (i.e., where in the network the retransmission decision is made). If the ARQ is terminated after the soft handover combining device typically situated in the BSC, then ARQ is implemented as without soft handover. The only drawback of this solution is that it increases the delay for retransmission because of the fixed network transmission between the BTS and the BSC.

If ARQ is terminated in the BTS before the handover combining unit, a more complicated protocol is required. The mobile station transmits a packet in the uplink direction. Because all base stations involved in the handover receive the packet and each of them makes an independent decision whether there was an error or not in the packet, the problem is how to transmit and coordinate the retransmission requirements. The mobile station should retransmit only if both cells require retransmission. A return channel would be required for both links. Furthermore, it would not be possible to combine the information from the two links like the power control information, because acknowledgments are coded. Therefore, this scheme looks impractical.

5.11 TRANSCEIVER

In this section, the receiver and transmitter functions of the mobile and base stations and their relation to the overall system design are covered. Receiver algorithms for detection, code tracking, and acquisition are introduced and design examples are given. Based on this, their impact on different radio interface parameters is discussed. On the transmitter side especially, the power amplifier design and its impact on system parameters are explained.

5.11.1 Receiver

Figure 5.15 depicts the block diagram of a RAKE receiver. A RAKE finger despreads the received signal with a correlator. For coherent demodulation, the despread signal is multiplied by a complex amplitude to correct phase error and to weight each finger according to the selected combining strategy (maximal ratio or equal gain combining). The impulse response (IR) measurement block continuously measures the multipath profile. Whenever the delays of the impulse response change, the IR measurement block allocates new code phases for the code tracking block that tracks the small changes. After combining the signals from different RAKE fingers, deinterleaving and decoding of the channel code is performed. In addition, a searcher, not shown in the figure, is continuously scanning the neighboring cell pilots in order to provide pilot measurements for handover.

The number of RAKE fingers depends on the channel profile and the chip rate. The higher the chip rate, the more resolvable the paths. More RAKE fingers, however, are needed to catch all the energy from the channel to maintain good performance (see

Section 9.2.2). A very large number of RAKE fingers leads to combining losses. Practical implementation architecture for a RAKE receiver is discussed in Chapter 11.

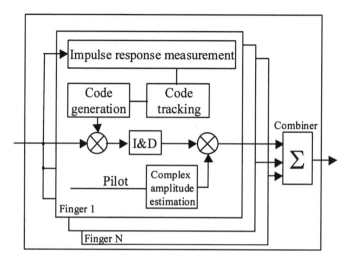

Figure 5.15   Block diagram of CDMA RAKE receiver.

5.11.1.1   Impulse Response Measurement

The impulse response measurement correlates the received signal with different phases of the pilot code to find the multipath components. The required measurement speed for IR measurement depends on the mobile speed and the radio environment. The faster the mobile station is moving, the faster the measurements need to be performed in order to catch the best multipath components for the RAKE fingers. Furthermore, in a long delay spread environment, the scanning window needs to be wider (see Chapter 12). In addition to measurement functions, this block performs RAKE finger allocation (i.e., assigning multipath components for RAKE fingers). For the code allocation, different strategies can be applied. Allocation can be done after the whole impulse response has been measured or immediately after a strong enough multipath component has been found.

5.11.1.2   Code Acquisition

Code acquisition is performed prior to the acquisition of system synchronization. The mobile station scans through the pilot signals. A priority order for the pilot signals may be set based upon the last pilot and neighboring pilots. If the connection is lost for some reason, scanning starts from the highest priority pilots. In the case of high interference, code acquisition could become a system bottleneck. A matched filter depicted in Figure 5.16 can be used for fast code acquisition. The predefined parallel data corresponds to the spreading code chips.

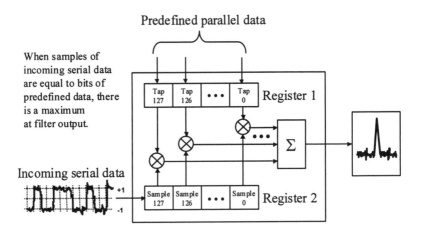

Predefined parallel data

When samples of
incoming serial data
are equal to bits of
predefined data, there
is a maximum
at filter output.

Incoming serial data

Figure 5.16 Matched filter.

5.11.1.3 Code Tracking

Typical implementation of a code tracking loop is an early-late delay locked loop. It consists of two correlators (early and late), which are allocated half a chip apart from the on-time correlator. Depending on the correlation result, the code phase is adjusted. Tracking loop performance depends on loop bandwidth. If the update is faster than a multipath component moves in delay, then synchronization errors are negligible, but loop noise increases. The requirements also depend on the detection strategy (i.e., whether conventional or multiuser detection is applied). The impact of code tracking errors on conventional detectors (matched filters) have been studied in [35, 36]. A comprehensive study of code tracking and acquisition can be found in [37].

5.11.1.4 Complex Amplitude Estimation

The complex amplitude estimation includes both amplitude and phase estimation. In the case of maximal ratio combining, the signal is weighted with the complex conjugate of the complex amplitude. If equal gain combining is performed, only the phase error is corrected, and each RAKE finger is considered with equal weight. The complex amplitude estimate needs to be averaged over a reasonably long period, while coherence time sets the upper limit for the averaging time (i.e., the channel should not change during the estimation).

5.11.1.5 Searcher

A searcher scans other cell pilots. During a call, the mobile station is scanning through the pilot signals, as well as measuring the downlink interference and possibly receiving uplink interference results. Because the number of pilots is very large, it might take a long time before a rising pilot in a neighboring set is noticed. Thus, searcher time can limit the system performance especially in a micro cellular environment, where a new base station can become active very fast due to the corner effect (see Section 12.5.1). One possibility of reducing the required hardware is to have flexible allocation between RAKE and searcher fingers. This would increase the scanning effectiveness in a low multipath environment. The number of fingers for searching depends on the desired speed for pilot scanning. Search window planning is discussed in Chapter 12 and in [38].

5.11.2 Transmitter

Figure 5.17 is a block diagram of a DS-CDMA transmitter. The data coming from the source is channel encoded, interleaved, and possibly different services are multiplexed. The resulting frames are passed through spreading and filtering, D-A converted, and passed trough IF and RF parts. The transmission parameters are controlled by the TX control block.

The most important component from the point of view of complexity and cost is the power amplifier in the RF section, especially for the mobile station. A nonlinear power amplifier can be operated near the compression point, and thus, it has good power efficiency resulting in low battery consumption. The linearity requirements are set by the sensitivity of the transmitted signal waveform for the nonlinearities and adjacent channel requirements. Multicode transmission is more sensitive to power amplifier nonlinearities than single code transmission because its peak-to-average ratio is higher because of the summing of several signals. If orthogonal Walsh sequences are used, the signal should be scrambled to avoid a high peak-to-average ratio. This is because if Walsh functions modulated with all ones or zeros are summed, then they form an impulse, which places high linearity requirements on the power amplifier. Nonlinear power amplifiers are discussed in more detail in Chapters 10 and 11.

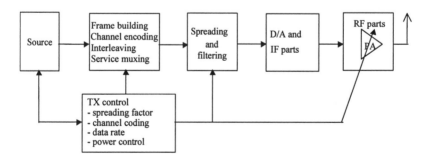

Figure 5.17 Transmitter block diagram.

5.11.2.1   Transmit Diversity

The forward link performance can be improved by transmit diversity. So far in CDMA systems, delay transmit diversity has been applied. It has, however, two main drawbacks: the loss of orthogonality that results in self-interference; and the requirement for additional RAKE fingers in the mobile station. Therefore, several more advanced transmit diversity concepts have been developed for third-generation CDMA systems.

Time division transmit diversity (TDTD) can be implemented by time-orthogonal signaling, by using a pseudo-random antenna hopping sequence. In a more optimal implementation, the transmit antennas are determined using closed loop antenna selection (CLAS). This results in the selective transmit diversity (STD) concept [39]. In order to be effective, the STD concept requires that the control loop delay is kept within the channel coherence time.

In code division transmit diversity (CDTD), the coded bits are distributed to more than one antenna, and all antennas transmit simultaneously to a given user. In orthogonal transmit diversity (OTD), different orthogonal codes for each antenna are used for spreading eliminating self-interference in flat fading.

The performance of TSD, TDTD, and OTD schemes has been compared in [39]. Power control with TDTD schemes has been studied in [40].

In multicarrier CDMA, transmitter diversity can be implemented by transmitting the carriers on multiple antennas. One or more carriers are mapped to one antenna. Because signals transmitted through different antennas experience independent fading, the performance is improved.

5.12   MULTIUSER DETECTION

Multiuser detection (MUD) and interference cancellation (IC) seek to improve performance by canceling the intracell interference and thus increasing the system capacity. The actual capacity increase depends on the efficiency of the algorithm, radio environment, and the system load. In addition to capacity improvement, MUD and IC alleviate the near-far problem typical to DS-CDMA systems.

Figure 5.18 depicts a system using multiuser detection/interference cancellation. Each user is transmitting data bits, which are spread by the spreading codes. The signals are transmitted over a Gaussian multiple access channel. In the receiver, the received signal is correlated with replicas of the user spreading codes. The correlator consists of a multiplier and an integrate and dump function. A matched filter can also be used. Multiuser detection processes the signal from the correlators jointly to remove the unwanted multiple access interference from the desired signal. The output of a multiuser detection block is the estimated data bits.

5.12.1 Capacity and Coverage Improvement

The performance of multiuser detection depends on the captured energy and impact of phase and code tracking errors [41]. The fraction of captured energy is a function of

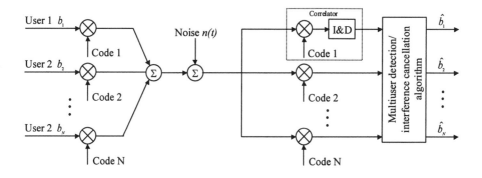

Figure 5.18 System model of multiuser detection.

chip duration multiplied by the number of RAKE branches divided by the delay spread. Furthermore, because typically the interference from a single cell (i.e., intracell interference) is canceled, the intercell interference limits the achievable capacity. If only intracell interference is canceled and intercell interference amounts to a fraction f of the intracell interference, then the boundary for capacity increase is $(1+f)/f$. With a propagation power law of 4, intercell interference is 55% of intracell interference; therefore, ideally, MUD can improve the capacity of the system by a factor of 2.8, compared to a system without MUD [42]. However, in practice, the MUD efficiency is not 100% but depends on the detection scheme, channel estimation, delay estimation, and power control error.

Typically, MUD is studied mainly for the uplink. One of the main motivations, in addition to complexity considerations, to justify the study of multiuser detection for the uplink has been the claim that the capacity of DS-CDMA systems is uplink limited. This claim, however, comes from the studies for IS-95. The noncoherent uplink of IS-95 with antenna diversity has normally worse performance than the coherent downlink with orthogonal codes. Even for IS-95, however, the downlink is sometimes the limiting direction [43]. All third-generation wideband CDMA systems use coherent detection also in the uplink and thus, together with antenna diversity, the uplink has better performance than the downlink [11, 44]. One way of improving downlink performance would be canceling the worst interferers from other cells.

As will be shown in Chapter 9, multiuser detection improves coverage and makes it more independent of the system load. Because MUD efficiency varies in different radio environments, this improvement is not fixed. This impact of MUD on coverage introduces a new variable to the network planning process, because MUD efficiency needs to be taken into account in the coverage design. On the other hand, the load factor is no longer crucial for the coverage area design because cell size does not shrink as fast with an increasing load as in the conventional detector. Because of orthogonal codes, however, the impact might be different in the uplink and downlink, especially if MUD is applied only in the uplink. This is because in low delay spread radio environments, forward link coverage does not depend heavily on load.

5.12.2 Development of Multiuser Detection

The idea of multiuser detection was first mentioned in 1979 by Schneider [45]. In 1983, Kohno et al. published a study on multiple access interference cancellation receivers [46]. In 1984, Verdú proposed and analyzed the optimal multiuser detector and the maximum likelihood sequence detector, which, unfortunately, is too complex for practical implementation since its complexity grows exponentially as a function of the number of users [47]. Consequently, Verdú's work inspired multiple researchers to find suboptimal multiuser detectors, which could achieve a close to optimal performance with reasonable implementation complexity.

The first studies concentrated on finding suboptimal detectors for the AWGN channel [48]. The following wave in the research of multiuser detection were studies on suitable detectors for fading multipath channels, first in slowly fading channels and then in relatively fast fading channels. Both flat fading and multipath channels have been considered. Because the multiuser detector parameters such as amplitude, phase, and crosscorrelations between users change over time, studies of adaptive multiuser detectors that self-tune the detector's parameters based on the received signal were started [49]. In the real world, the parameters required for signal detection are never perfect. Consequently, studies taking into account the impact of nonideal estimation of phase, amplitude, and delays have been initiated [35, 41]. Channel coding is an essential part of all proposed wideband CDMA air interfaces. So far, however, relatively few studies have considered multiuser detection together with channel coding [50–53]. MUD with channel coding can be either partitioned or integrated. Partitioned approaches treat the multiuser interference equalization problem and decoding problem separately. Integrated approaches perform the equalization and decoding operations together [50]. Partitioned approaches can be further divided into hard and soft decision approaches. In the soft decision approach a multiuser detection algorithm feeds the soft symbols after detection into a channel decoder. As discussed in Section 5.9, third-generation wideband CDMA systems will have multiple data rates. Recently, multiuser detectors for multirate CDMA systems have been proposed in [53–55].

5.12.3 Multiuser Detection Algorithms

Multiuser algorithm classification is depicted in Figure 5.19. Multiuser detection can be either optimal or suboptimal. Suboptimal multiuser detection algorithms can be classified into linear and interference cancellation type algorithms. Some detectors can also be classified into both categories. Furthermore, few other schemes not fitting this classification have been proposed. These include partial trellis-search algorithms [51], [52] and neural networks proposed for AWGN channels [56, 57].

Another proposed way to classify multiuser detection algorithms is by linear and nonlinear algorithms [58]. In this classification, however, the same algorithm can belong to both categories depending on the implementation.

Optimal multiuser detection consists of a matched filter followed by a maximum likelihood sequence detector implemented via a dynamic programming algorithm (e.g., Viterbi algorithm). The formulation of optimal multiuser detection for an AWGN channel, the maximum likelihood sequence estimator (MLSE), was published in 1986 by Verdú. Presentations of an optimal detector in a Rayleigh fading channel can be found in [59, 60]. Because the complexity of the optimal multiuser detector is exponential as a function of the number of users, a large number of studies to find suboptimal detectors have been carried out.

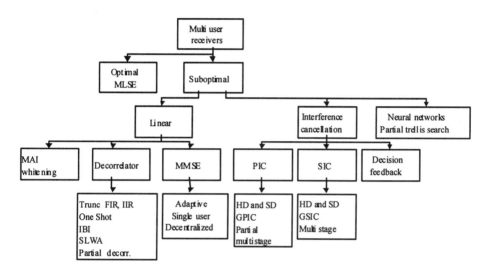

Figure 5.19 Multiuser detection classification.

5.12.3.1 Linear Detectors

The decorrelator (also called zero-forcing detector) proposed by Lupas and Verdú in [48] multiplies the matched filter outputs by the inverse of the crosscorrelation matrix. An advantage of this detector is that the received signal amplitudes do not have to be known. On the other hand, the received noise process is also filtered with the inverse matrix and hence increases the noise power. This is proportional to the users' mutual crosscorrelations. Because the decorrelator is a sequence detector, the detection process cannot be started until the whole transmitted sequence is received at the receiver. In practice this is not feasible and would result in a very long delay. Therefore, several finite delay decorrelator schemes have been proposed: isolation bit insertion (IBI), multishot sliding window algorithm (SLWA), hard decision method (HDM) [61, 62], improved one-shot detection scheme [63], and finite length FIR and IIR schemes [59]. The partial decorrelator forces the MAI due to past symbols to zero, while the MAI due to future symbols might be suppressed, for example, by interference cancellation. The partial decorrelator belongs to the family of nonlinear decision feedback (DF) detectors, which are typically characterized by a linear feedforward filter and a nonlinear feedback filter [64].

Another linear receiver is the linear minimum mean square error (LMMSE) [65], which, unlike the decorrelator, doesn't enhance the noise. Even though LMMSE has also been proposed for centralized receivers in AWGN and known fading channels, the main interest has been in decentralized adaptive implementations, which are especially attractive for implementation in mobile stations [66]. A modified LMMSE receiver for fading channels has been presented in [67]. MAI-whitening filters modeling the multiple access interference as colored noise are also examples of decentralized linear receivers [68].

5.12.3.2   Interference Cancellation

The idea of interference cancellation is to estimate the multiple access and multipath induced interference and then to subtract the interference estimate [69]. Successive interference cancellation (SIC) cancels the interference estimate on a user by user basis, while parallel interference cancellation (PIC) cancels the interference of all users simultaneously [70]. Group-wise algorithms detect symbols within a given group and cancel the interference on that group from other users. In wideband CDMA, the grouping can be performed based on data rates [53]. Multistage interference cancellation algorithms improve the interference estimates iteratively. If tentative data decisions are used, the scheme is called hard decision (HD) interference cancellation. If tentative data decisions are not used, the scheme is called soft decision (SD) interference cancellation. Complete subtraction of the MAI in PIC has been shown to result in biased decision statistics [71]. Therefore, partial cancellation has been proposed to reduce the effect of bias in SD-PIC [71] and HD-PIC [72]. A sliding window PIC receiver has been presented in [73].

5.12.4   System Model and MUD Algorithm Formulation

In the following paragraphs, we present a standard model for the DS-CDMA uplink with K users. BPSK modulated data of each user are spread by multiplying the data modulated signal by a pseudo-random sequence. The composite signal after passing through the channel and entering the base station can be modeled as

$$r(t) = \sum_{k=1}^{K} s_k(t - \tau_k) + n(t) \tag{5.5}$$

where τ_k is the delay of kth user's transmitted signal, and $n(t)$ is the complex AWGN with two-sided power spectral density σ^2. The kth user's signal is given as

$$s_k(t - \tau_k) = \sqrt{2P_k} \sum_i b_k(i) a_k(t - iT - \tau_k) e^{j\phi_k} \tag{5.6}$$

where P_k is the power of the the kth user, $b_k(i)$ is the BPSK modulated data for the kth user transmitted at time i, ϕ_k is the received signal phase of the kth user, and $a_k(t)$ is the spreading waveform given by

$$c_k(t) = \sum_{i=1}^{G} c_k^{(i)} p(t - iT_c) \tag{5.7}$$

where G is the number of chips, $c_k^{(i)} \in \pm 1$, and $p(t)$ is the pulse shape of a chip with duation T_c.

The conventional detector (matched filter or correlator receiver) can be modeled as follows. The output of the matched filter for user k is

$$y_k(i) = \int \mathrm{Re}\left\{ r(t)e^{-j\hat{\phi}_k} \right\} c_k(t - iT - \hat{\tau}_k) dt \tag{5.8}$$

where $\hat{\phi}_k$ and $\hat{\tau}_k$ are the estimated time delay and phase, respectively.

The estimated bits are then given as

$$\hat{b}_k(i) = \mathrm{sgn}(y_k(i)) \tag{5.9}$$

5.12.4.1 Decorrelator

For the decorrelator, the output of the matched filters can be expressed in matrix form using the block-based model [48]:

$$\mathbf{y} = \mathbf{RWb} + \mathbf{n} \tag{5.10}$$

where

$$\mathbf{y} = \left[\mathbf{y}^T(1), \mathbf{y}^T(2), ..., \mathbf{y}^T(M) \right]^T \tag{5.11}$$

$$\mathbf{y}(i) = \left[y_1(i), y_2(i), ..., y_{K1}(i) \right]^T \tag{5.12}$$

$$\mathbf{b} = \left[\mathbf{b}^T(1), \mathbf{b}^T(2), ..., \mathbf{b}^T(M) \right]^T \tag{5.13}$$

$$\mathbf{b}(i) = \left[b_1(i), b_2(i), ..., b_{K1}(i) \right]^T \tag{5.14}$$

where M is the number of symbols in the block. The correlation matrix is given by

$$\mathbf{R} = \begin{bmatrix} \mathbf{R}(0) & \mathbf{R}(-1) & \mathbf{0} & \cdots & \mathbf{0} & \mathbf{0} \\ \mathbf{R}(1) & \mathbf{R}(0) & \mathbf{R}(-1) & \mathbf{0} & \cdots & \mathbf{0} \\ \vdots & \ddots & \ddots & \ddots & \ddots & \vdots \\ \mathbf{0} & \cdots & \mathbf{0} & \mathbf{R}(1) & \mathbf{R}(0) & \mathbf{R}(-1) \\ \mathbf{0} & \mathbf{0} & \cdots & \mathbf{0} & \mathbf{R}(1) & \mathbf{R}(0) \end{bmatrix} \tag{5.15}$$

where the entries in the correlation matrix $R(i)$ are given by

$$R_{kiR}(i) = \cos(\phi_k - \phi_i) \int a_k(t - \tau_k) a_i(t + iT + \tau_i) dt \qquad (5.16)$$

The noise vector **n** is additive Gaussian and colored. The amplitude matrix **W** is a diagonal matrix containing the amplitudes of all users. Now the decorrelator output decision is given by

$$\hat{\mathbf{b}} = \text{sgn}(\hat{\mathbf{R}}^{-1}\mathbf{y}) \qquad (5.17)$$

5.12.4.2 MMSE

The MMSE receiver performs a linear transformation on the matched filter outputs that minimizes the mean square error (MSE). The detected bits are obtained from

$$\hat{\mathbf{b}}(i) = \text{sgn}\left(\left(\hat{\mathbf{R}} + \sigma^2 \mathbf{W}^{-2} \right)^{-1} \mathbf{y} \right) \qquad (5.18)$$

5.12.4.3 Parallel Interference Cancellation

The PIC receiver detects all users at the same time and then cancels interference simultaneously. It applies the multistage principle, and the received signal at stage n is given by

$$r_k^{(n)}(t) = r^{(n-1)}(t) - \sum_{l=1}^{K} \hat{s}_l(t - \hat{\tau}_l) \qquad (5.19)$$

5.12.4.4 Serial Interference Cancellation

The SIC receiver cancels one user at a time. It ranks the users according to their powers cancelling first the higher power users. The received signal for user k' after canceling k'-1 users is given by

$$r_{k'}(t) = r(t) - \sum_{l=1}^{k'} \hat{s}_l(t - \hat{\tau}_l) \qquad (5.20)$$

5.12.5 Design Aspects for Multiuser Detection

5.12.5.1 Uplink Versus Downlink

The uplink and downlink have different characteristics, which impact the design of the MUD scheme. The uplink is asynchronous (i.e., the users' transmission times are independent of each other). Propagation delays for cell ranges of 1 to 30 km are from 3 to 100 µs, introducing a significant amount of intersymbol interference between the symbols of different users. The downlink is synchronous, and, in contrast to base stations, a mobile station needs to detect only its own signal. Typically, orthogonal codes are used in the downlink, but multipath propagation partially destroys the orthogonality. The additional gain from multiuser detection depends on the multipath profile. Multipath spread in outdoor systems ranges from a few microseconds up to 20 µs.

Because a mobile station is only interested in demodulating its own signal, a different multiuser detection strategy than in the base station could be applied to reduce the complexity in the mobile station, while still improving the performance over conventional detection. Many of the proposed multiuser detectors can also be modified for single user detectors.[4] For multirate users, one possibility is to cancel only the self-interference from different parallel codes of the same user. In the downlink, the common control channels also cause interference. Since a mobile station has to detect the pilot signal, it could cancel the pilot signal, as proposed in [74].

5.12.5.2   Multiuser Parameter Estimation

Coherent multiuser detectors require the knowledge of the complex channel coefficients, the carrier phases, and the propagation delays. Even though conventional methods for parameter estimation can be used, better results would be achieved with methods that take interference from other users into account. Large parameter estimation errors might offset the gain from the multiuser detection.

The optimal MLSE receiver estimates the received complex amplitudes and multiplies the matched filter bank output with it [75]. In suboptimal implementations, data detection and complex channel coefficient estimation are decoupled from each other (i.e., multiuser detection is applied to data), while the amplitude and phase for each user are recovered independently from each other [59]. In this case, the optimal channel estimator would be LMMSE [59]. But because the channel covariance matrix required by the LMMSE estimator changes according to fading rate and signal-to-noise ratios, it is very difficult to implement in practice. It can be approximated, however, by an adaptive channel estimator filter such as FIR or recursive IIR predictors or smoothers [59]. Channel estimation can be either decision feedback (decision directed) or feedforward channel estimation (data aided). The latter uses only so-called pilot or reference symbols to estimate the channel. Because for wideband CDMA either pilot channel or symbols are used, data aided estimation schemes can be used.

Multiuser delay estimation methods include optimal ML estimation [76], subspace methods (MUSIC algorithm) [77], the hierarchical maximum likelihood method [78], and PIC-based delay estimators [79]. Subspace methods, however, seem not to be suitable for fast fading, multipath channels with an SNR less than 10 dB [77].

---

[4] Multiuser detectors that make a joint detection of the symbols of different users are also called centralized multiuser detectors. Single user detectors, also called decentralized multiuser detectors, demodulate a signal of one desired user only [59].

5.12.5.3   System Design Choices

If MUD is to be part of the next standard, some minimum performance requirements have to be specified. Because multiuser detection is very complex, one might have to make a tradeoff between complexity and performance. For example, how many users at maximum are involved in the joint detection process? As we saw in the previous section, MUD research is still in a phase that would not justify making it a mandatory feature for wideband CDMA standards.

Even though multiuser detection is a receiver technique, it might have an impact on the system design because of its large complexity, while a proper system design might ease the implementation of the multiuser detection. If short codes are used, then the implementation of multiuser detection is easier. For example, the receiver can exploit the cyclostationarity (i.e., the periodic properties) of the signal to suppress interference without knowing the interfering codes. This would be especially useful in the mobile station.

5.12.6 Choice of Multiuser Detector Algorithm

When considering which algorithm is the most appropriate choice for a practical multiuser detection algorithm, the following criteria should be considered:

- Complexity;
- Performance.

5.12.6.1   Complexity

The final implementation complexity of a multiuser detector receiver depends on the selected architecture and cannot be estimated without a detailed analysis of algorithms down to the ASIC and DSP implementation level. Rough estimates, however, of the implementation complexity can be obtained by estimating the number of arithmetic operations per second and the number of clock cycles required by synchronous DSP [59]. It should also be noted that an arithmetic operation could be implemented using either an integer or a floating point operation depending on the requirements of the selected algorithm. This might cause further significant differences to the implementation complexities of the algorithms. When assessing the required number of arithmetic operations, we follow the methodology in [59] and consider only the dominant term.

The ideal implementation of the linear detectors such as decorrelator and MMSE has cubic dependence on the number of the users times the number of multipath components [58]. Therefore, iterative algorithms such as the conjugate gradient (CG) method have been proposed to implement them [80]. The preconditioned conjugate gradient algorithm (PCG) seems to be one of the simplest algorithms for linear multiuser detection [59].

Tables 5.3 and 5.4 summarize the computational complexities of the algorithms. Table 5.4 has been calculated following the example in [59] where parameters of the FMA2 wideband CDMA scheme have been used. The number of users is assumed to be

$K = 150$. The number of multipaths is $L = 4$, the number of samples per chip is $N_s = 4$, and the code length is $N_c = 256$. The symbol rate after channel coding is 20 ksymbols/s (i.e., the symbol interval is $T = 50$ μs). The number of iterations for the PGC algorithm is $I = 124$, and the observation window size is $N = 13$. The number of cancellation stages in the PIC algorithm is $M = 2$ and in the SIC algorithm $M = 1$.

The iterative PCG algorithm is most complex and hardly feasible from an implementation point of view, especially if long spreading codes are used [81]. An alternative implementation for a decorrelator, an approximate decorrelator, has been proposed to relieve the processing requirement [82]. Even though it does not require the matrix inversion, the crosscorrelations between different user codes have to be calculated with every symbol requiring $O[(2(KL)^2 N_c N_s)/T] \approx 15$ Tflops. The symbol level PIC is the least complex multiuser detection algorithm. For long spreading codes, however, the correlation matrix needs to be updated for every symbol, thereby increasing complexity much further. It should be noted that the crosscorrelations could be calculated in parallel. The regenerative PIC is only five times more complex than the conventional matched filter (MF) detector and is well suited for long spreading codes. The regenerative SIC algorithm also has low complexity. The high clock rate, however, limits the number of users in the cancellation process. The high clock rate can be reduced by increasing the detection delay. The architecture presented in [83] would, however, introduce one symbol delay for each user in the detection process, making, for example, fast power control impossible for a large number of users. Thus, from a complexity point of view, the PIC receiver seems best suited for systems with long spreading codes.

Table 5.3
Implementation Complexity Requirements

	Operations/s	Clock cycles
Conventional	$O[(4N_cN_s KL)/T]$	$2/T$
PCG	$O[(12IN(KL)^2)/T]$	$16M/T$
PIC (symbol level)	$O[(4M(KL)^2)/T]$	$4M/T$
PIC (regenerative)	$O[(8MN_cN_s KL)/T]$	$5M/T$
SIC (regenerative)	$O[(8M N_cN_s KL)/T]$	$4MK/T$

Table 5.4
Implementation Complexity

	Operations/s	Clock cycles
Conventional (MF)	49 Gflops/s	40 kHz
PCG	139 Tflops/s	40 MHz
PIC (symbol level)	57 Gflops/s	160 kHz
PIC (regenerative)	196 Gflops/s	200 kHz
SIC (regenerative)	98 Gflops/s	12 MHz

5.12.6.2 Performance

In this section we discuss the performance of the decorrelator, MMSE, PIC, and SIC detectors in AWGN and Rayleigh fading channels. Furthermore, the impact of imperfect phase and delay estimation is discussed. Special attention is paid to near-far performance.

According to [58, 84] the decorrelator, MMSE, and SD-PIC receivers have almost equal performance in AWGN channel with bit energy-to-noise density ratio (E_b/N_0) less than 10 dB. The MMSE detector is slightly better than the decorrelator due to the noise enhancement property of the decorrelator [84]. The SD-SIC is considerably worse than the other schemes. The HD-PIC performs better than the SD-PIC receiver does. This is due to more reliable tentative decision of HD-PIC and thus less residual interference.

The impact of phase errors alone is not very large on any multiuser detection receiver and is much less severe than the delay errors [58]. Furthermore, it seems that the performance of all detectors deteriorates equally as the phase error increases [84].

In flat Rayleigh fading the performance of all detectors is almost equal [58]. The performance of the SIC scheme has improved as compared to the AWGN channel because the instantaneous powers of the users are different. To obtain the performance improvement requires, of course, power ranking of the users on a symbol by symbol basis.

In frequency selective Rayleigh fading the performance of the SD-PIC and SD-SIC receivers is slightly worse than the performance of the linear receivers. The reason is the additional multipath-induced multiple access interference, which degrades the estimation of the channel gains [58]. The decorrelator and MMSE receivers do not need separate estimation of the channel gains. Because the HD-PIC can use separate estimates of the channel gains, its performance is not degraded as much. Actually, for moderate channel gain estimation errors, the hard decision PIC still outperforms the decorrelator receiver [59].

The sensitivity of the decorrelator, MMSE, SD-PIC, and SD-SIC receivers for delay estimation errors has been compared in [58, 84]. In an AWGN channel all detectors degrade equally fast as the delay error increases. In a Rayleigh fading channel, however, the performance of the decorrelator and MMSE receivers degrades faster up to the standard deviation of 0.1 chip. Thus, the performance of the receivers becomes nearer and for larger delay errors the performance is almost equal. Sensitivity of the optimum receiver and HD-PIC for channel mismatch has been analyzed in [41].

In order to interpret the sensitivity result, we need to know how large an error the delay tracking scheme will produce. The most commonly used delay tracking device in the conventional CDMA receiver is the delay locked loop (DLL). The variance of delay tracking error can be assumed to be less than 0.01 chip [85]. Of course, in the near-far situation, the variance would be larger. A variance of 0.01 (standard deviation of 0.1 chips) still leads to significant degradation of multiuser detector performance in [58, 84]. For the conventional detector the degradation is smaller. This can be easily explained by the fact that the performance of the conventional detector deteriorates only because of reduced energy in the correlation process. In the multiuser detection algorithms, the error also impacts the multiple access interference estimates, and thus, the delay error of one user impacts all other users' signals as well.

There exist two issues concerning the near-far performance of multiuser detectors. For the VSF scheme, different data rates have different powers and thus there exists a constant near-far situation. For the multicode scheme, imperfect power control leads to variations in the received user powers.

Some conclusions for performance of multiuser detectors in the VSF system can be drawn from the results presented in [58, 84]. One high power user degrades the performance of the SD-PIC scheme significantly. The reason for this is that the cancellation of the weak user is inaccurate at the first stage due to the interference from the high power user [58]. Juntti observed similar behavior for the HD-PIC receiver in real multirate simulations [53]. The SIC scheme benefits from unequal powers. The SIC algorithm, however, is sensitive not to the power of the strongest user but to the power of the second strongest user [58]. This is because if there are two very strong users, the second strongest user degrades the estimation of the first user. This further reduces the attractiveness of the SIC algorithm. The decorrelator is insensitive to user powers as long as there is no delay estimation error. With imperfect delay estimation, however, it also loses its near-far resistance. In the presence of delay estimation errors, the near-far performance degrades and all schemes come closer to each other in performance.

The performance of different multiuser detection schemes for the VSF scheme has been studied in [53]. The most promising scheme was groupwise SIC (GSIC), where users are grouped according to their spreading factors and the PIC or decorrelator is applied within a group. Users with the lowest spreading factor were detected first, and their multiple access interference was subtracted from the MF outputs of the other users, which are then further detected by the PIC or decorrelator. The performance of the high bit rate users however, was not good because the undetected low bit rate users degrade the performance of high bit rate users.

5.12.6.3   MUD and Power Control

Third-generation wideband CDMA uses fast closed loop power control both in the uplink and the downlink. Since all multiuser detectors are in practice near-far limited, power control is required even if MUD is used. In the uplink, fast power control improves the performance in three ways: by equalizing the user powers, the detrimental near-far effect is mitigated; by compensating the channel fading, E_b/N_0 performance is improved; and by minimizing of the transmission power, the battery life of mobile stations is increased and intercell interference reduced. Thus, even with multiuser detection it is important to use power control.

In the downlink, power control also improves the performance against fading, but contrary to the uplink, it increases power differences between the received signals at the mobile station. Multiuser detection can facilitate a larger difference in the power levels and thus offers better compensation of deep fades [86].

All wideband CDMA schemes apply fast power control both in the uplink and the downlink. Fast power control tries to compensate for the effect of fast fading. Power control has the largest impact on the performance of the SIC scheme. The SIC scheme performs best when the user powers are different. This is obviously the case for instantaneous powers in Rayleigh fading. By ranking the users on a symbol basis, equal performance compared to the other schemes can be obtained. To relieve the implementation complexity of the SIC receiver with respect to the power ranking, it has been proposed to apply SIR-based power control [87]. The user that is detected first sees the interference from all other users, and the user that is detected last sees only noise and other cell interference. Consequently, given an equal quality goal, the power control will adjust the user powers so that the first user has the largest power and last user the smallest power. The performance of the first user would be equal to the conventional detector. Further gain can be obtained by taking into account the spatial distribution of the users within a cell. In order to minimize the intercell interference, the first user in the detection process should be closest to the base station [87]. For high mobility users this kind of assignment would be difficult.

It is well known that imperfect power control degrades the performance of the conventional detector. Imperfect power control has two effects on the receiver performance. Because fading is not compensated perfectly, the performance is degraded. Furthermore, nonequal powers due to imperfect power control lead to a near-far situation. As was discussed earlier, none of the multiuser detection receivers are near-far resistant when considering imperfect parameter estimation. Thus, imperfect power control will degrade also the performance of multiuser detectors. The power control error can be modeled as a log-normally distributed random variable. The combined power control error in the reverse link caused by the fast power control loop and open loop power control is on the order of 1.5 to 2.5 dB. Based on the studies so far it seems that all detectors will have similar performance degradation under imperfect power control. Furthermore, multiuser detectors require equally stringent power control as does the conventional matched filter detector.

5.12.6.4   Performance Versus Complexity

Because performance depends on how complex a multiuser detector we apply, we should weigh the complexity against performance. The one-stage SIC is not as good as others for equal powers. The two-stage SIC performs as well as the two- and three-stage PIC. The high clock rate and delay make it less attractive. The two-stage PIC (one cancellation stage) performs in most situations almost as well as the three-stage PIC, especially in the presence of parameter estimation errors [58, 84]. Given the reasonably low complexity of the PIC algorithm, however, it might be worthwhile to implement the three-stage PIC.

5.13   RANDOM ACCESS PROCEDURE

Random access is a process where a mobile station requests an access to the system, and the network answers the request and allocates a traffic channel for the mobile station. Random access is executed when the power of a mobile station is switched on or the synchronization is lost for some reason. Furthermore, random access is carried out for packet data transmission whenever there is something to transmit. As described in Section 5.3, before the random access can be performed, the following steps need to be performed:

- Code and frame synchronization;
- Retrieval of cell parameters such as random access code;

- Estimation of the downlink path loss and the initial power level for the random access.

Optimization criteria for the random access procedure are the speed of the process and low transmission power. Requirements for the speed of the random access procedure depend on the requirements for the initial synchronization time. The number of access channels depends on the anticipated access load. Information transmitted during the random access state will have an effect on this. Because excessive transmission power degrades CDMA system capacity, it is important to minimize the total power transmitted during the random access state. This is especially important because during the random access transmission power cannot be controlled by the fast closed loop power control. If initial transmission happens with a very low power, then access attempt can take a long time. On the other hand, high transmission power during the initial access results in fast synchronization but causes higher interference to other users during the synchronization period.

The minimum information that needs to be transmitted during a random access attempt is some kind of mobile identification. The actual traffic channel negotiation and service contract negotiation can then happen in a nonpublic channel. This can be performed in later state. A typical random access message consists of preamble, synchronization part, and data part, as illustrated in Figure 5.20. The data part contains at least the mobile station identification, while the preamble is an unmodulated wideband spread signal.

Preamble	Synchronization	Data

Figure 5.20 Random access message.

5.14 HANDOVER

In this section, we first define the different handover types that can be used in CDMA systems, including soft handover. In the second subsection, the handover procedure is defined. The implementation of the different phases of the soft handover procedure is studied in the third subsection. The third subsection also discusses macro diversity combining and other factors that impact the performance of soft handover. The fourth and fifth subsections describe softer and interfrequency handover, respectively.

5.14.1 Definitions

In soft handover a mobile station is connected to more than one base station simultaneously. *Idle handover* is a handover that is made when the mobile station is in idle state (i.e., does not have an active connection). *Hard handover* is a handover occurring when only one traffic channel is available at a time. *Softer handover* means a soft handover between sectors in the same cell. *Interfrequency handover* means handover between two different frequencies. *Intrasystem handover* is a handover between two systems (e.g., from a second-generation system to third-generation system).

Active set consists of the base stations involved in the soft handover with the given mobile station. If the active set is changed, an *active set update* occurs. *Candidate set* consists of the base stations that fulfill the criteria to be included in the active set but have not yet been included in the active set. *Neighbor set* contains the base stations that are likely candidates for soft handover. These are usually base stations whose geographical coverage areas are near the mobile station. *Remaining set* contains all base stations excluded from the other sets. *Discard set* is defined to be the set of base stations that belongs to the current active set but are going to be dropped from the active set because they no longer fulfill the criteria for the active set.

5.14.2 Handover Procedure

The handover procedure can be divided into three phases: measurement, decision, and execution phase, as illustrated in Figure 5.21. In the handover measurement phase, the typical downlink measurements performed by the mobile station are signal quality and signal strengths of its cell and neighboring cells. In the uplink, the base station measures signal quality. The measurement results are signaled to relevant network elements, to mobile stations, and to the BSC. In the handover decision phase, sometimes also called the evaluation phase, the measurement results are compared against predefined thresholds and it is decided whether handover should be initiated or not. In addition, admission control (see Section 5.16) is performed to verify that the new user can be accommodated into the new cell without degrading the quality of the existing users. In the execution phase, the mobile station enters the soft handover state, a new base station is added or released, or interfrequency handover is performed.

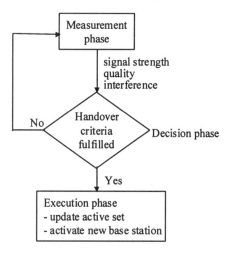

Figure 5.21 Handover phases.

5.14.3 Soft Handover

Soft handover is one of the most discussed features of CDMA. On the one hand it brings increased performance by increasing diversity, but on the other it is a necessity in a CDMA system for avoiding excessive interference from neighboring cells. The increased diversity comes at the expense of increased backhaul connections in the network.

5.14.3.1 Handover Decision

In IS-95, the handover decision is based on the pilot strength measurements of the downlink only. In wideband CDMA for third generation systems with asymmetric traffic, more decision parameters are needed. At least the following parameters can be identified:

- Distance attenuation;
- Uplink interference;
- Downlink interference.

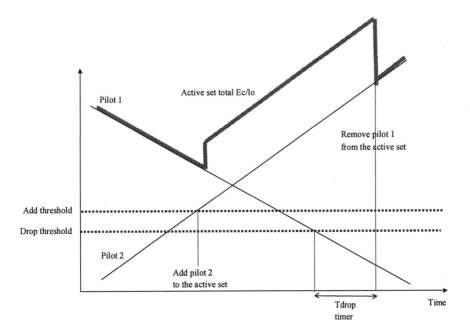

Figure 5.22 Handover parameters.

Figure 5.22 illustrates the handover parameters. In reality, there are delays associated with adding and removing the pilots from the active set. When the signal strength of base station 2 exceeds the *add threshold*, the mobile station enters the soft handover state. When the signal strength of base station 1 drops below the *drop threshold* and stays there until the drop timer has expired, base station 1 is removed from the active set.

The add and drop thresholds can also be adjusted dynamically to take into account situations where a weak pilot added to the active set would not bring any benefit. If there are already some dominant strong pilots, adding a new weak pilot does not improve performance. On the other hand, if all pilots are weak then even adding a new, weak pilot is beneficial. This type of *dynamic threshold* is specified in the IS-95B standard.

The selection of a cell with a high uplink load but with a low distance attenuation, instead of a cell with a low uplink load but a slightly higher distance attenuation, will result in a higher mobile station transmission power. Thus, it is important to take the uplink interference situation into account in the handover decision. Figure 5.23 depicts a situation where the high uplink load influences the handover decision. The mobile station is currently connected to BTS1 but should make a handover to BTS2 because the path loss attenuation is lowest because of the LOS connection.

Figure 5.23 High uplink load handover situation.

The high bit rate multimedia user, however, is producing a high interference level in the uplink, and thus, the required transmission power for the mobile station to connect to BTS2 is actually higher than to BTS3. Therefore, handover to BTS3 should be made.

In every moment, the mobile station tries to keep the best base stations available in the active set. The handover procedure, however, should avoid the so-called ping-pong effect, where new base stations have to be added and old ones released too often. This results in an excessive signaling reducing capacity and increased possibility of dropping the call.

The mobile station orders base stations according to the handover decision parameters, signal strength, and interference situation. The best base stations are included in the active set. If a base station is decided to be dropped from the active set, a timer is started. If the quality is still low after the timer expires, then the base station is dropped. The ping-pong effect is avoided by long enough timers. The rate of handover depends on users' mobility and cell size.

From a capacity perspective, it is beneficial if the downlink loads are balanced between the base stations. Thus, a base station that has a high load in the downlink should hand over traffic to other base stations. This can be done by either adjusting the pilot strength according to the downlink load or by the mobile station taking into account the downlink load in the handover decision.

There are a number of alternatives on how the uplink and downlink interference situation can be taken into account in the handover decision. The first and most straightforward alternative is to have constant pilot strength and to signal the uplink interference value to the mobile station. Furthermore, the downlink interference is measured in the mobile station. A drawback of this approach is the high signaling load in the downlink, which reduces the capacity. The second possibility is to adjust the pilot strength according to the uplink load so that the selection of the base station according to pilot strength also takes the interference conditions into account. This method avoids signaling the uplink interference value to the mobile station. The third possibility is that the pilot is adjusted according to downlink load. A fourth possibility is that the pilot is adjusted according to the uplink and downlink loads. A drawback of the pilot adjustment strategies is that they are slower than constant pilot strength plus signaling, and reducing the pilot power impacts the receiver performance of the mobile station. Some of the above strategies can be decided to be used by the network operator. Only the procedures that impact signaling needs and mobile station behavior need to be specified in the standard.

5.14.3.2 *Handover Measurements*

Handover measurements are performed to gather information for a handover decision. The averaging period for the measurement results depends on the mobile speed. Measurement results should be updated more frequently when the speed of the mobile station increases.

The different pilot sets need to be measured with different frequencies. The active set needs to be measured most frequently. Also the candidate set should be measured as often. The neighbor set is measured less often, and the remaining set is measured least often. On the other hand, it would be good to measure the neighbor set more frequently to avoid problems with rapidly rising pilots creating high interference.

5.14.3.3 Macro Diversity Combining

It has to be decided where and how the macro diversity combining is performed. There are several instances in which the macro diversity combining can be performed:

- In the RAKE receiver;
- In the channel decoder;
- After the channel decoder;
- After the source codec.

Combining in the RAKE receiver requires that the transmissions from different base stations are synchronized. Combining at this point provides the largest gain, and is most suitable for the mobile station, since the RAKE receiver sees the transmission from other base stations as additional multipath signals. In the network side, combining at the RAKE level or channel decoder is impractical. It would then need a large amount of signaling, since the soft bits should be transmitted to the soft handover termination point situated either in the BSC or in the switch. Thus, combining after the channel decoder seems to be the most practical alternative.

Maximal ratio, equal gain, and selection diversity are the principal combining techniques. In maximal ratio combining, the received signals are weighted according to their received SIR, while in equal gain combining, no weighting is performed. With selection diversity, the signal with the highest SNR is selected.

Figure 5.24 illustrates the different possibilities for the soft handover diversity combiner. There exist two basic possibilities: BSC or switch. According to the current UMTS network view, the radio access network comprising the base stations and base station controllers and the core network comprising the switch are separated. Thus, all radio-related functions should be placed in the access network, and consequently the diversity combiner should be situated in the BSC. This will also reduce the signaling load in the network, compared to a situation where the combiner is placed in the switch.

Soft handover between base stations belonging to different BSCs requires special arrangements (e.g., inter BSC links, as indicated in Figure 5.24). Furthermore, base stations can even belong to different switches.

5.14.3.4 Performance Considerations

Delays and errors in macro diversity measurements and execution lead to performance degradation. Some parameters have more significance in the overall performance.

Errors in SIR and signal strength measurements seem to have little impact on overall performance according to simulations performed in [88]. Standard deviation

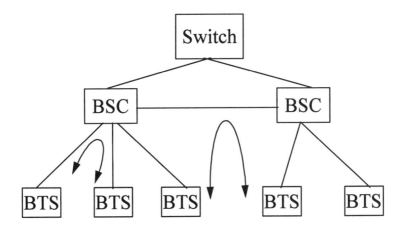

Figure 5.24 Placement of the diversity combiner in the network.

varied between 0 and 1.25 dB. Increased handover rate, however, and thus signaling, was not taken into account in the simulations.

Measurement delay is caused by the scanning procedure of the pilots. Measurement speed depends on the number of pilot signals and the number of searchers in the mobile station receiver. In IS-95, the worst case assumption is 200 ms. In [88], this was assumed to be 250 ms. Because the number of the pilots to be scanned is large, a sophisticated algorithm must be used to decide which base stations to scan, how often they should be scanned, and how averaging measurements would most likely improve the system performance [88].

After the decision to connect a new base station has been made, some time elapses before the setup of the new connection. This delay consists of signaling to the new base station, acquisition of the mobile station reverse link signal by the new base station, and starting the transmission in a new channel. Furthermore, the new base station has to signal the used traffic channel to the old base station, which in turn signals it to the mobile station. Thereafter, the mobile station receives the message and starts combining the two traffic channels. This process can take up to 200 to 300 ms.

Overall delay exceeding 300 ms starts to degrade the system capacity [88]. Furthermore, the downlink was noticed to be more sensitive to the delay due to the slow power algorithm. For third-generation systems using fast downlink power control, the impact of delays in the handover measurement and execution might be different.

The size of the active set depends on the radio conditions and number of RAKE fingers available. Two base stations often already give most of the achievable gain. The number of RAKE fingers decides whether the additional path diversity obtained through macro diversity can be utilized. In the forward link, the addition of a new base station to the active set can even degrade the performance, because transmission of additional signals creates more interference that could not be utilized because of a limited number of RAKE fingers. An active set update rate larger than 0.5 Hz (2 sec), preferably 1 Hz, is needed to not degrade the system quality [88]. Active set threshold (add threshold, see Figure 5.22) means that if the signal strength of a base station is within that window, for example 5 dB, it can be included in the active set. Too large a threshold will degrade the soft handover gain, because combining two base stations with large power differences does not produce very much diversity gain. On the other hand, too small an add threshold combined with a small drop threshold (Figure 5.22) increases the number of the handovers leading to excessive signaling due to the ping-pong effect.

5.14.3.5   Synchronization During Handover

Synchronization requirements and solutions for soft handover depend on the network synchronization and spreading code design. If the network is synchronized at chip level with an accuracy of a few microseconds as in IS-95, the handover procedure does not need to consider time differences between the base stations involved in the soft handover, with the exception of the propagation delay. Because it is desirable not to have external synchronization because of cost and complexity reasons, an asynchronous network is preferred. It will, however, place some special requirements on the handover synchronization, as explained below.

The mobile station has to be able to measure the time difference between the current base station and the new base station to be included in the active set. This time difference can exceed, in the worst case, the frame length. In an asynchronous network, code design further impacts the handover synchronization scheme. Long codes facilitate the measurement of this long time difference. With short codes, a special synchronization scheme is required that first measures the difference in frames with the help of, for example, the synchronization channel [12]. After the timing difference has been reduced to the frame level, it can be further reduced to the chip level. After the timing difference has been measured, the new base stations align the transmission of its forward link for the user in soft handover with the first base station.

5.14.4 Softer Handover

Softer handover occurs when the mobile station moves from one sector to another within one cell. The forward link resembles the soft handover situation (i.e., the mobile station receives the signals from the two sectors and combines them in the RAKE receiver). In the reverse link, in contrast to soft handover, the combining can also be performed in the RAKE receiver. In softer handover, there is no need for transactions between the base stations and, for example, the base station controller. Thus, softer handover between sectors can be established much faster than in soft handover, as fixed network signaling is not required. Therefore, it might be useful to implement street microcells as sectored cells to reduce the time for time-critical corner handover (see Chapter 12).

5.14.5 Interfrequency Handover

Seamless interfrequency handover is an important new feature in third-generation CDMA networks that have (1) hierarchical cell structures (i.e., microcells overlaid by a macrocell) and (2) hot spot cells with more carriers than surrounding cells, as illustrated in Figure 5.25. In the latter scenario, the interfrequency handover can be performed as an intracell handover within the hotspot cell. First, interfrequency handover is performed to frequency 1, and then a normal soft handover is carried out to the neighboring destination cell. For the first scenario, an efficient method to measure other carrier frequencies is required.

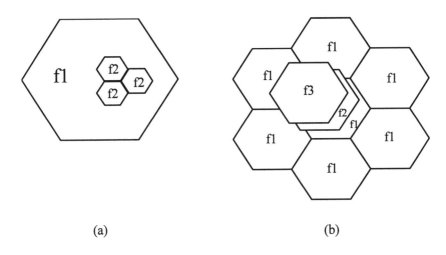

(a)   (b)

Figure 5.25   Interfrequency handover scenarios: (a) macro cell overlaying micro cells and (b) hot spot cell with several carriers.

For interfrequency handover, there are two implementation alternatives: compressed mode/slotted mode and dual receiver. The dual receiver has an expensive receiver but simple system operation. However, if there is already a diversity receiver (two receiver chains, and not selection diversity), then it is possible to use the other receiver for measurements on other carriers. Of course, diversity gain is lost during the measurement period. Compressed mode, on the other hand, has a simple receiver but a more complex system operation. The receiver algorithm has to operate with bursty signals during the slotted mode, resulting in poorer performance and more control effort.

For the compressed mode there exist a number of alternatives for implementation. In [89] four different methods are presented: variable spreading factor, code rate increase, multicode, and higher order modulation. Variable rate spreading and coding rate increment (puncturing) cause 1.5 to 2.5 dB loss in E_b/N_0 performance and higher order modulation causes an even higher loss (5 dB). This is due to breaks in power control and less coding. A further drawback of the variable spreading factor is that simple terminals have to be able to operate with different spreading ratios (e.g., a speech terminal that normally operates with a spreading ratio of 256, should now also operate with a spreading ratio of 128).

5.15 POWER CONTROL

In the design of a power control scheme, power control criteria, step size, dynamic range, and command rate need to be considered. Each of these aspects is discussed in the following subsections.

5.15.1 Power Control Criteria

As discussed in Chapter 2, all wideband CDMA proposals use open and fast closed loop power control methods. Depending on the power control criteria, several different algorithms can be derived. Most typical criteria are:

- Path loss–based power control;
- Quality-based power control.

Normally, the power control algorithm is a combination of these two basic criteria. Quality can be measured through, for example, SIR. Because different SIRs correspond to the same FER in different radio environments, we need to have a function that maps the desired FER into the required SIR target. This is performed by continuously measuring the FER and SIR, and then adjusting the SIR target to produce the desired FER.

5.15.2 Power Control Step Size

Power control step size defines how much a power control command changes the transmission power. Either a simple up/down adjustment or several power adjustment levels can be used. Typical step sizes are between 0.5 and 1 dB. It should be noted that power control adjustment is relative to the previous power setting, because an absolute power setting would require extremely accurate, and thus expensive, power control circuitry [90].

5.15.3 Dynamic Range Requirements

In the uplink, the open loop power control tries to equalize the powers of all mobile stations received by the base station to avoid the near-far effect. The dynamic range requirement is set by the maximum distance difference between two mobile stations and can be expressed as

$$\frac{P_{rx1}}{P_{rx2}} = \left(\frac{d_2}{d_1}\right)^{\alpha} \tag{5.21}$$

where

P_{rx1} = received signal power from the mobile station 1;
P_{rx2} = received signal power from the mobile station 2;
d_1 = distance between mobile station 1 and base station;
d_2 = distance between mobile station 2 and base station;
α = path loss attenuation factor.

In an extreme situation, if mobile station 1 is 100m from the base station and the mobile station 2 is 20 km away, and assuming a path loss attenuation factor of 3.5, the required dynamic range is 80 dB. Fast power control compensates slow and fast fading fluctuations. To compensate for these, the fast power control needs a dynamic range on the order of 10 to 30 dB around the setpoint set by the open loop control.

In the downlink, dynamics cannot be as high, because transmissions for different mobile stations come from a single source. Even if code channels, in theory, are orthogonal (when using orthogonal codes), they do not remain fully orthogonal in practice after the multipath channel and transmitter and receiver nonidealities. Thus, if a large dynamic range were used, power control would create a near-far situation for transmissions originating from the single transmitter. In the downlink, the dynamic range is on the order of 10 to 20 dB.

5.15.4 Power Control Command Rate

The second task for power control is to compensate for signal fluctuations caused by fast fading and to improve performance. It is very difficult to track fast fading at high mobile speeds due to measurement delay, signaling of the power control commands, and processing delay caused by extracting the power control command in the receiver. In addition, coding and interleaving are most effective in fast mobile speeds. According to a field test trial with IS-95, the "grey" area, where neither power control nor the coding and interleaving are effective, is from 20 to 40 km/h. For wideband CDMA, this area is most likely smaller. Therefore, low mobile speeds set the requirements for fast power control.

In the 900-MHz band, the maximum Doppler frequency is 25 Hz; while in the 2-GHz band, 55 Hz is the maximum for a mobile speed 50 km/h. Coherence times are 4 and 1.8 ms, respectively. Thus, the power control update time needs to be around 1 ms or longer. As we can see, the optimal power control rate depends on the mobile speed and carrier frequency, as the Doppler frequency for higher carrier frequencies is higher. Thus, it would be beneficial to have a variable power control rate according to the mobile speed.

Due to delay requirements, power control commands are transmitted uncoded. Therefore, the error probability of power control commands is very high. However, since power control commands are transmitted very often and the step size is small, the power control loop converges to the right value despite the high probability of errors.

The net capacity gain of power control depends on the power control command rate consuming radio resources, the increase of other cell interference (see Chapter 9), and the performance increase due to compensation of fast fading and the reduction of the near-far effect. Especially for low bit rate services, the power control command rate of, for example, 2 Kbps is a high percentage of the overall service bit rate. Thus, with a smaller command rate, the overall capacity could be better. The impact of power control on system performance will be discussed further in Chapter 9.

5.15.5 Power Control and Multiuser Detection

Quite often, multiuser detection is advocated to relax power control requirements, but in addition to tackling the near-far effect, power control improves performance in a fading channel. Thus, the basic requirements for the power control are the same with MUD as without MUD.

5.16 ADMISSION AND LOAD CONTROL

The purpose of admission control is to ensure that there are free radio resources for the intended call with required SIR and bit rate. The purpose of load control is to maintain in an objective manner the use of radio resources of the network within the given limits. Admission control is always performed when a mobile station initiates communications in a new cell, either through a new call or handover. Furthermore, admission control is performed when a new service is added during an active call. In general, the admission control procedure ensures that the interference created after adding a new call does not exceed a prespecified threshold.

The negotiation of a service contract is performed at the beginning of the call. Admission control can be involved because the user might accept, for example, a lower bit rate if it can be accommodated by the network.

The quantities used in admission and load control (e.g., SIR, total power) should be averaged to obtain stable values that are insensitive to power control errors. On the other hand, the averaging time cannot be too long, because then outdated information would be used in the admission and load control processes.

5.16.1 Load Factor

Load factor η is used to measure the network congestion. Load factor for the reverse link can be defined as follows:

$$1 - \eta = \frac{\text{SIR}_{loaded}}{\text{SIR}_{empty}} = \frac{\dfrac{S}{I_{tot}}}{\dfrac{S}{I_0}} = \frac{N}{I_{tot}} \Leftrightarrow \eta = 1 - \frac{N_0}{I_{tot}} \tag{5.22}$$

where

N_0 is the thermal noise spectral density;

I_{tot} is the total interference plus noise spectral density;

S is the received power at the base station from each user;

SIR is the signal-to-interference ratio.

From (5.5) we can see that the total received power at the base station can be used to measure the cell load. When the system is fully loaded then the load factor is one. Because a fully loaded system might get into an unstable state and drive the powers of all users to a maximum (see Chapter 9), a safety margin is required. Therefore, the load factor should be in the order of 0.4 to 0.8.

In the forward link, the load factor can be defined as the ratio of the maximum BTS transmission power to the predefined threshold value.

5.16.2 Admission Control Principles

Admission control needs to be done separately for uplink and downlink. This is especially important if the traffic is highly asymmetric. Typical criteria for admission control are call blocking and call dropping. Blocking occurs when a new user is denied access to the system. Call dropping means that a call of an existing user is terminated. Call dropping is considered to be more annoying than blocking. For detailed admission control algorithms, refer to [91–93].

5.16.3 Load Control Principles

The basic principle of load control is the same as admission control. While admission control is carried out as a single event, load control is a continuous process where the interference is monitored.

Load control measures the load factor, and, if the predefined load factor is exceeded, the network either reduces the bit rates of those users whose service contract allows it to be done, delays the transmission of those users without delay requirements, or drops low priority calls. If there is an underload, load control increases the bit rates of those users who can handle higher bit rates. The increase and decrease of the bit rates can be performed with priority order.

This kind of sophisticated control requires stability in all conditions. Thus, some time constants are required for the network not to react too fast and to prevent several users from simultaneously increasing their bit rates, resulting in too high a load.

REFERENCES

[1] Recommendation ITU-R M.687, "Future Public Land Mobile Telecommunication Systems (FPLMTS)," 1998.

[2] Recommendation ITU-R M.1034, "Requirements for the radio interface(s) for

Future Public Land Mobile Telecommunication Systems (FPLMTS)," 1998.

[3]   ETSI ETR 291, "System Requirements for UMTS," *ETSI Technical Report*, May 1996.

[4]   ETSI UMTS 21.01, "Requirements for the UMTS Terrestrial Radio Access system (UTRA)," 1997.

[5]   Recommendation ITU-R M.1035, "Framework for the radio interface(s) and radio sub-system functionality for Future Public Land Mobile Telecommunication Systems (FPLMTS)," 1998.

[6]   Higashi, A., T. Taguchi, and K. Ohno, "Performance of Coherent Detection and RAKE for DS-CDMA Uplink Channels," *Proceedings of PIMRC'95*, Toronto, Canada, September 1995, pp. 436–440.

[7]   Simon, M. K., J. K. Omura, R. A. Scholtz, and B. K. Levitt, *Spread Spectrum Communications Handbook*, revised edition, New York: McGraw-Hill, Inc., 1994.

[8]   Peterson, R. L., R. E. Ziemer, and D. E. Borth, *Introduction to Spread Spectrum Communications*, Englewood Cliffs, New Jersey: Prentice Hall, 1995.

[9]   Kärkkäinen, K., *Code Families and Their Performance Measures for CDMA and Military Spread-Spectrum Systems*, Ph.D. Dissertation, University of Oulu, Finland, 1996.

[10]   TIA/EIA/IS-95-B, "Mobile Station-Base Station Compatibility Standard for Dual-Mode Wideband Spread Spectrum Cellular System," 1998.

[11]   Koo, J. M., E. K. Hong, and J. I. Lee, "Wideband CDMA Technology for FPLMTS," *Proceedings of the 1st CDMA International Conference*, Seoul, Korea, November 1996.

[12]   Ojanperä, T., K. Rikkinen, H. Hakkinen, K. Pehkonen, A. Hottinen and J. Lilleberg, "Design of a 3rd Generation Multirate CDMA System with Multiuser Detection, MUD-CDMA," *Proceedings of ISSSTA'96*, Vol. 1, Mainz, Germany, September 1996, pp. 334–338.

[13]   Yang, J., D. Bao, and M. Ali, "PN Offset Planning in IS-95 Based CDMA System," *Proceedings of VTC'97*, Vol. 3, Phoenix, Arizona, May 1997, pp. 1435–1439.

[14]   Hammons, R. A., and V. J. Kumar, "On a Recent 4-Phase Sequence Design for CDMA," *IEICE Transactions on Communications*, Vol. E76-B, No. 8, August 1993, pp. 804–813.

[15]   DaSilva, V. M., and E. S. Sousa, "Effect of Multipath Propagation on the Forward Link of a CDMA Cellular System," *Wireless Personal Communications*, No. 1, Kluwer Academic Publishers, 1994, pp. 33–41.

[16]   Adachi, F., M. Sawahashi, and K. Okawa, "Tree-Structured Generation of Orthogonal Spreading Codes with Different Lengths for Forward Link of DS-CDMA Mobile Radio," *IEE Electronics Letters*, Vol. 33, No. 1, January 1997, pp. 27–28.

[17]   Hottinen, A., and K. Pehkonen, "A Flexible Multirate CDMA Concept with Multiuser Detection," *Proceedings of ISSSTA'96*, Vol. 1, Mainz, Germany, 1996, pp. 556–560.

[18]   Ojanperä, T., J. Skold, J. Castro, L. Girard, and A. Klein, "Comparison of Multiple Access Schemes for UMTS," *Proceedings of VTC'97*, Vol. 2, Phoenix, Arizona, May 1997, pp. 490–494.

[19] TTA, "Global CDMA I: Multiband Direct-Sequence CDMA RTT System Description," June 1998.

[20] Dahlman, E., and K. Jamal, "Wide-Band Services in a DS-CDMA Based FPLMTS System," *Proceedings of VTC'96*, Atlanta, Georgia, May 1996, pp. 1656–1660.

[21] Hui, J. Y. N., "Throughput Analysis for Code Division Multiple Accessing of the Spread Spectrum Channel," *IEEE Transactions on Vehicular Technology*, Vol. VT-33, No. 3, August 1984, pp. 98–102.

[22] Viterbi, A. J., "Very Low Rate Convolutional Code for Maximum Theoretical Performance of Spread-Spectrum Multiple-Access Channels," *IEEE Journal on Selected Areas in Communications*, Vol. 8, No. 4, May 1990, pp. 642–649.

[23] Rikkinen, K., "Comparison of Very Low Rate Coding Methods for CDMA Radio Communications System," *Proceedings of IEEE ISSSTA '94*, Oulu, Finland, 1994, pp. 268–272.

[24] Torrieri, D. J., "Performance of Direct-Sequence Systems with Long Pseudo-Noise Sequences," *IEEE Journal on Selected Areas in Communications*, Vol. 10, No. 4, May 1992, pp. 770–781.

[25] Lin, S., and J. Costello, Jr., *Error Control Coding: Fundamentals and Applications*, Englewood Cliffs, New Jersey: Prentice Hall, 1983.

[26] Cideciyan, R. D., E. Eleftheriou, and M. Rupf, "Concatenated Reed-Solomon/Convolutional Coding for Data Transmission in CDMA-Based Cellular Systems," *IEEE Transactions on Communications*, Vol. 45, No. 10, October 1997, pp. 1291–1303.

[27] Proakis, J. G., *Digital Communications*, 3rd ed., New York: McGraw-Hill, Inc., 1995.

[28] Wong, K. H. H., and L. Hanzo, "Channel Coding," in *Mobile Radio Communications*, R. Steele (ed.), New York: John Wiley & Sons Ltd, 1996, pp. 347–488.

[29] Berrou, C., A. Glavieux, and P. Thitimajshima, "Near Shannon Limit Error-Correcting Coding and Decoding: Turbo-Codes," *IEEE International Conference on Communications ICC'93*, Vol. 2, May 23–26, Geneva, Switzerland, pp. 1064–1070.

[30] Jung, P., J. Plechinger, M. Doetsch, and F. Berens, "Pragmatic Approach to Rate Compatible Punctured Turbo-Codes for Mobile Radio Applications," *Proceedings of 6th International Conference on Advances in Communications and Control: Telecommunications/Signal Processing*, Corfu, Greece, June 23–27, 1997.

[31] *Proceedings of International Symposium on Turbo-Codes and Related Topics*, Brest, France, September 3–5, 1997.

[32] Benedetto, S., D. Divsalar, and J. Hagenauer, "Concatenated Coding Techniques and Iterative Decoding: Sailing Toward Channel Capacity," Guest Editorial in *Journal on Selected Areas in Communications*, Vol. 16, No. 2, February 1998.

[33] Frenger, P., P. Orten, T. Ottosson, and A. Svensson, "Rate Matching in Multichannel Systems Using RCPC-Codes," *Proceedings of VTC'97*, Phoenix, Arizona, May 1997, pp. 354–357.

[34] Okumura, Y., and F. Adachi, "Variable Rate Data Transmission with Blind Rate Detection for Coherent DS-CDMA Mobile Radio," *IEE Electronics Letters*, Vol.

32, No. 20, September 1996, pp. 1865–1866.

[35] Parkvall, S., E. Ström, and B. Ottersen, "The Impact of Timing Errors on the Performance of Linear DS-CDMA Receivers," *IEEE Journal on Selected Areas in Communications*, Vol. 14, No. 8, October 1996, pp. 1660–1668.

[36] Sunay, O., and P. J. McLane, "Effects of Carrier Phase and Chip Timing Errors on the Capacity of Quadriphase Spread, BPSK Modulated DS CDMA System," *Proceedings of IEEE Globecom'95*, Singapore, November 1995, pp. 1114–1120.

[37] Glisic, S., and B. Vucetic, *Spread Spectrum CDMA Systems for Wireless Communications*, Norwood, Massachusetts: Artech House, 1996.

[38] Yang, S., *CDMA RF System Engineering*, Norwood, Massachusetts: Artech House, 1998.

[39] Hottinen, A., and R. Wichman, "Transmit Diversity by Antenna Selection in CDMA Downlink," *Proceedings of ISSSTA'98*, San City, South Africa, September 1998.

[40] Heikkinen, T., and A. Hottinen, "On Downlink Power Control and Capacity with Multi-Antenna Transmission," *Proceedings of VTC'98*, Ottawa, Canada, May 1998, pp. 475–479.

[41] Gray, S. D., M. Kocic, and D. Brady, "Multiuser Detection in Mismatched Multiple-Access Channels," *IEEE Transactions on Communications*, Vol. 43, No. 12, December 1995.

[42] Holtzman, J. M., "DS/CDMA Successive Interference Cancellation," in *Code Division Multiple Access Communications*, S. G. Glisic and P. A. Leppänen (eds.), Boston: Kluwer, 1995, pp. 161–180.

[43] Wallace, M., and R. Walton, "CDMA Radio Network Planning," *Proceedings of IEEE VTC'94*, Stockholm, Sweden, June 1994, pp. 62–67.

[44] Westman, T., and H. Holma, "CDMA System for UMTS High Bit Rate Services," *Proceedings of VTC'97*, Phoenix, Arizona, May 1997, pp. 830–834.

[45] Schneider, K. S., "Optimum Detection of Code Division Multiplexed Signals," *IEEE Transactions on Aerospace and Electronic Systems*, Vol. AES-15, No. 1, January 1979, pp. 181–185.

[46] Kohno, R., H. Imai, and M. Hatori, "Cancellation Technique of Co-Channel Interference in Asynchronous Spread-Spectrum Multiple Access Systems," *IEICE Transactions on Communications*, Vol. 65-A, May 1983, pp. 416–423.

[47] Verdú, S., *Optimum Multiuser Signal Detection*, Ph.D. Dissertation, University of Illinois, Urbana-Champaign, 1984.

[48] Lupas, R., and S. Verdú, "Near-Far Resistance of Multiuser Detectors in Asynchronous Code-Division Multiple Access Communications," *IEEE Transactions on Communications*, COM-38, April 1990, pp. 496–508.

[49] Verdú, S., "Adaptive Multiuser Detection," in *Code Division Multiple Access Communications*, S. G. Glisic and P. A. Leppänen (eds.), Boston: Kluwer, 1994, pp. 97–116.

[50] Giallorenzi, T. R., and S. G. Wilson, "Suboptimum Multiuser Receivers for Convolutionally Coded Asynchronous CDMA Systems," *IEEE Trans. Commun.*, Vol. 44, No. 9, September 1996, pp. 1183–1196.

[51] Giallorenzi, T. R., and S. G. Wilson, "Multiuser ML Sequence Estimator for Convolutionally Coded Asynchronous DS-CDMA Systems," *IEEE Transactions on Communications*, Vol. 44, No. 8, August 1996, pp. 997–1007.

[52] Fawer, U., and B. Aazhang, "Multiuser Receivers for Code-Division Multiple Access Systems with Trellis-Coded Modulation," *IEEE Journal on Selected Areas in Communications*, Vol. 14, No. 8, August 1996, pp. 1602–1609.

[53] Juntti, M., "Performance of Multiuser Detection in Multirate CDMA Systems," *Proceedings of ISSSTA*, July 1994, pp. 157–161.

[54] Saquib, M., and R. Yates, "A Two Stage Decorrelator for a Dual Rate Synchronous DS/CDMA System," *Proceedings of ICC'97*, Montreal, Canada, June 1997, pp. 334–338.

[55] Hottinen, A., H. Holma, and A. Toskala, "Multiuser Detection for Multirate CDMA Communications," *Proceedings of ICC'96*, Dallas, Texas, June 1996, pp. 1819–1823.

[56] Hottinen, A., "Self-Organizing Multiuser Detection," *Proceedings of ISSSTA'96*, Oulu, Finland, July 1996, pp. 152–156.

[57] Aazhang, B., B.-P. Paris, and G. Orsak, "Neural Networks for Multiuser Detection in Code-Division Multiple Access Systems," *IEEE Transactions on Communications*, Vol. 40, No. 7, July 1992, pp. 1212–1222.

[58] Buehrer, R. M., *The Application of Multiuser Detection to Cellular CDMA*, Ph.D. Dissertation, Virginia Polytechnic Institute and State University, Virginia, 1996.

[59] Juntti, M., *Multiuser Demodulation for CDMA Systems in Fading Channels*, Ph.D. Dissertation, University of Oulu, Finland, October 1997.

[60] Zvonar, Z., and D. Brady, "Multiuser Detection in Single-Path Fading Channels," *IEEE Transaction on Communications*, Vol. 42, No. 2/3/4, February/March/April 1994, pp. 1729–1739.

[61] Zheng, C., and S. K. Barton, "On the Performance of Near-Far Resistant CDMA Detectors in the Presence of Synchronization Errors," *IEEE Trans. on Commun.*, Vol. 43, No. 12, December 1995, pp. 3037–3045.

[62] Zheng, C., and M. Faulkner, "Power Control Requirements in Linear Decorrelating Detectors for CDMA," *Proceedings of VTC'97*, Phoenix, Arizona, May 1997, pp. 213–217.

[63] Peng, M., Y. J. Guo, and S. K. Barton "One-shot Linear Decorrelating Detector for Asynchronous CDMA," *Proceedings of Globecom'96*, 1996, pp. 1301–1305.

[64] Duel-Hallen, A., J. Holtzman, and Z. Zvonar, "Multiuser Detection for CDMA Systems," *IEEE Personal Communications,* April 1995, pp. 46–58.

[65] Madhow, U., and M. L. Honig, "MMSE Interference suppression for Direct-Sequence Spread-Spectrum CDMA," *IEEE Transactions on Communications*, Vol. 42, No. 12, December 1994, pp. 3178–3188.

[66] Zecevic, N., "Techniques and Adaptation Algorithms for Direct-Sequence Spread-Spectrum CDMA Single-User Detection," M.Sc. Thesis, Virginia Tech, July 1996.

[67] Latva-aho, M., and M. Juntti, "Modified Adaptive LMMSE Receiver for DS-CDMA Systems in Fading Channels," *Proceedings of IEEE International Symposium on Personal, Indoor, and Mobile Radio Communications (PIMRC)*, Helsinki, Finland, September 1997, pp. 554–558.

[68] Monk, A. M., M. Davis, L. B. Milstein, and C. W. Helstrom, "A Noise-Whitening Approach to Multiple Access Noise Rejection-Part I: Theory and Background," *IEEE Journal on Selected Areas in Communications*, Vol. 12, No. 5, June 1994, pp. 817–827.

[69] Moshavi, S., "Multiuser Detection for DS-CDMA Communications," *IEEE Communications Magazine*, October 1996, pp. 124–136.

[70] Prasad, R., *CDMA for Wireless Personal Communications*, Norwood, Massachusetts: Artech House, 1996.

[71] Correal, N. S., R. M. Buehrer, and B. D. Woerner, "Improved CDMA Performance through Bias Reduction for Parallel Interference Cancellation," *Proceedings of PIMRC'97*, Helsinki, Finland, September 1997, pp. 565–569.

[72] Divsalar, D., and M. Simon, "A New Approach to Parallel Interference Cancellation," *Proceedings of Globecom'96*, London, England, November 1996, pp. 1452–1457.

[73] Haifeng, W., J. Lilleberg, and K. Rikkinen, "A New Sub-Optimal Multiuser Detection Approach for CDMA Systems in Rayleigh Fading Channel," *Proceedings of Conference on Information Sciences and Systems (CISS)*, The Johns Hopkins University, Baltimore, Maryland, March 1997, pp. 276–280.

[74] I., C.-L., and C. A., Webb III, H. C. Huang, S. ten Brink, S. Nada, and R. D Gitlin, "IS-95 Enhancements for Multimedia Services," *Bell Labs Technical Journal*, Autumn 1996, pp. 60–87.

[75] Kailath, K., "Correlation Detection of Signals Perturbed by a Random Channel," *IRE Trans. Information Theory*, Vol. 6, No. 3, June 1960, pp. 361–366.

[76] Lilleberg, J., E. Nieminen, and M. Latva-aho, "Blind Iterative Multiuser Delay Estimator for CDMA," *Proceedings of PIMRC'96*, Taipei, Taiwan, October 1996, pp. 565–568.

[77] Luukkanen, P., and J. Joutsensalo, "Comparison of MUSIC and Matched Filter Delay Estimators in DS-CDMA," *Proceedings of PIMRC'97*, Helsinki, Finland, September 1997, pp. 830–834.

[78] Joutsensalo, J., J. Lilleberg, A. Hottinen, and J. Karhunen, "A Hierarchic Maximum Likelihood Method for Delay Estimation in CDMA," *Proceedings of VTC'96*, Atlanta, Georgia, April 1996, pp. 188–192.

[79] Latva-aho, M., and J. Lilleberg, "Parallel Interference Cancellation in Multiuser CDMA Channel Estimation," *Wireless Personal Communications*, Boston: Kluwer, Vol. 9, No. 2–3, August 1998, pp. 171–195.

[80] Juntti, M. J., B. Aazhang, and J. O. Lilleberg, "Iterative Implementation of Linear Multiuser Detection for Dynamic Asynchronous CDMA Systems," *IEEE Transactions on Communications*, Vol. 46, No. 4, April 1994, pp. 503–508.

[81] Ojanperä, T., "Overview of Multiuser Detection/Interference Cancellation for DS-CDMA," *Proceedings of ICPWC'97*, Mumbay, India, December 1997, pp. 115–119.

[82] Mandayan, N., and S. Verdú, "Analysis of Approximate Decorrelating Detector," *33rd Allerton Conference on Communication, Control and Computing*, 1997.

[83] Pedersen, K. I., T. E. Kolding, I. Seskar, and J. Holtzman, "Practical Implementation of Successive Interference Cancellation in DS/CDMA Systems," *Technical Report*, WINLAB, Rutgers University, 1996.

[84] Orten, P., and T. Ottoson, "Robustness of DS-CDMA Multiuser Detectors," *Proceedings of CTMC Globecom'97*, Phoenix, Arizona, November 1997, pp. 144–148.

[85] Keurulainen, J., "Delay-Locked Loop Performance in CDMA Receivers," M.Sc. Thesis, Helsinki University of Technology, 1994.

[86] Wichman, R., and A. Hottinen, "Multiuser Detection for Downlink CDMA Communications in Multipath Fading Channels," *Proceedings of VTC'97*, Vol. 2, Phoenix, Arizona, May 1997, pp. 572–576.

[87] Hatrack, P., and J. M. Holzman, "Reduction of Other-Cell Interference with Integrated Interference Cancellation," *Proceedings of VTC'97*, Phoenix, Arizona, May 1997, pp. 1842–1846.

[88] Andersson, T., "Tuning the Macro Diversity Performance in a DS-CDMA System," *Proceedings of VTC'94*, Stockholm, Sweden, June 8–10, 1994, pp. 41–45.

[89] Gustafsson, M., K. Jamal, and E. Dahlman, "Compressed Mode Techniques for Inter-Frequency Measurements in a Wide-band DS-CDMA System," *Proceedings of PIMRC'97*, Helsinki, Finland, September 1997, pp. 231–235.

[90] Fukasawa, A., T. Sato, Y. Takizawa, T. Kato, M. Kawabe, and R. E. Fisher, "Wideband CDMA System for Personal Radio Communications," *IEEE Communications Magazine*, October 1996, pp. 116–123.

[91] Knutsson, J., P. Butovitsch, M. Persson, and R. D. Yates, "Evaluation of Admission Control Algorithms for CDMA System in a Manhattan Environment," *Proceedings of the 2nd CDMA International Conference*, Seoul, Korea, October 1997, pp. 414–418.

[92] Huang, C. Y., and R. D. Yates, "Call Admission in Power Controlled CDMA Systems," *Proceedings of IEEE VTC'96*, Atlanta, Georgia, May 1996, pp. 1665–1669.

[93] Liu, Z., and M. E. Zarki, "SIR Based Call Admission Control for DS-CDMA Cellular System," *IEEE Journal on Selected Areas in Communications*, Vol. 12, 1994, pp. 638–644.

Chapter 6

WCDMA

6.1 INTRODUCTION

This chapter presents the WCDMA air interface, referred also as UMTS terrestrial radio access (UTRA), developed by the third-generation partnership project (3GPP). 3GPP has the goal to harmonize and standardize in detail the similar proposals from ETSI, ARIB, TTC, TTA, and T1.

Table 6.1 lists the parameters of WCDMA. WCDMA has two modes characterized by the duplex method: FDD (frequency division duplex) and TDD (time division duplex), for operating with paired and unpaired bands, respectively [1]. The TDD mode is described in Chapter 7.

The chip rate of the system is 3.84 Mcps. The frame length is 10 ms and each frame is divided into 15 slots (2560 chip/slot at the chip rate 3.84 Mcps). Spreading factors range from 256 to 4 in the uplink and from 512 to 4 in the downlink. Thus, the respective modulation symbol rates vary from 960 k symbols/s to 15 k symbols/s (7.5 k symbols/s) for FDD uplink. For separating channels from the same source, orthogonal variable spreading factor (OVSF) channelization codes are used. In the downlink, Gold codes with a 10-ms period (38400 chips at 3.84 Mcps) are used to separate different cells, with the actual code itself length $2^{18}-1$ chips. In the uplink, Gold codes with a 10-ms period, or alternatively short codes with a 256-chip period, are used to separate the different users.

For the channel coding three options are supported: convolutional coding, turbo coding, or no channel coding. Channel coding selection is indicated by upper layers. Bit interleaving is used to randomize transmission errors. The modulation scheme is QPSK.

The carrier spacing has a raster of 200 kHz and can vary from 4.2 to 5.4 MHz. The different carrier spacings can be used to obtain suitable adjacent channel protections depending on the interference scenario. Figure 6.1 shows an example for the operator bandwidth of 15 MHz with three cell layers. Larger carrier spacing can be applied between operators than within one operator's band in order to avoid inter-operator interference. Interfrequency measurements and handovers are supported by

171

WCDMA to utilize several cell layers and carriers.

This chapter is organized as follows. WCDMA specification structure is given in Section 6.2. Protocol structure and logical and transport channels are described in Section 6.3. Physical channels, spreading, multirate schemes (variable data rates), packet data, and handover are discussed in Sections 6.4–6.8. Section 6.9 describes the future evolution of the WCDMA covering release 2000 standards and beyond.

6.2 WCDMA SPECIFICATIONS

The air interface description in the following is based on the 3GPP wideband CDMA specifications as listed in Table 6.2. The physical layer is specified in TS25 series of 3GPP specifications.

Table 6.1
Parameters of WCDMA

Channel bandwidth	5 MHz
Duplex mode	FDD and TDD
Downlink RF channel structure	Direct spread
Chip rate	3.84 Mbps
Frame length	10 ms
Spreading modulation	Balanced QPSK (downlink) Dual-channel QPSK(uplink) Complex spreading circuit
Data modulation	QPSK (downlink) BPSK (uplink)
Channel coding	Convolutional and turbo codes
Coherent detection	User dedicated time multiplexed pilot (downlink and uplink), common pilot in the downlink
Channel multiplexing in downlink	Data and control channels time multiplexed
Channel multiplexing in uplink	Control and pilot channel time multiplexed I&Q multiplexing for data and control channel
Multirate	Variable spreading and multicode
Spreading factors	4–256 (uplink), 4–512 (uplink)
Power control	Open and fast closed loop (1.6 kHz)
Spreading (downlink)	OVSF sequences for channel separation Gold sequences 2^{18}-1 for cell and user separation (truncated cycle 10 ms)
Spreading (uplink)	OVSF sequences, Gold sequence 2^{41} for user separation (different time shifts in I and Q channel, truncated cycle 10 ms)
Handover	Soft handover Interfrequency handover

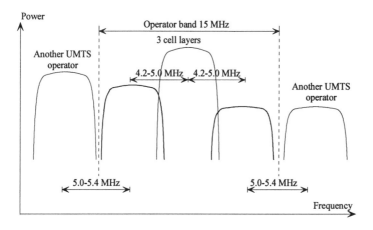

Figure 6.1 Frequency utilization with WCDMA.

6.3 PROTOCOL ARCHITECTURE

Figure 6.2 shows the air interface protocol architecture. The protocol architecture is similar to the current ITU-R protocol architecture, ITU-R M.1035. The air interface is layered into three protocol layers:

- The physical layer (layer 1, L1);
- The data link layer (layer 2, L2);
- Network layer (layer 3, L3).

The physical layer interfaces the medium access control (MAC) sublayer of layer 2 and the radio resource control (RRC) layer of layer 3. The physical layer offers different transport channels to MAC. A transport channel is characterized by how the information is transferred over the radio interface. Transport channels are channel coded and then mapped to the physical channels specified in the physical layer. MAC offers different logical channels to the radio link control (RLC) sublayer of layer 2. A logical channel is characterized by the type of information transferred.

Layer 2 is split into following sublayers: MAC, RLC, packet data convergence protocol (PDCP) and broadcast/multicast control (BMC). Layer 3 and RLC are divided into control and user planes. PDCP and BMC exist in the user plane only. In the control plane, layer 3 is partitioned into sublayers where the lowest sublayer, denoted as RRC, interfaces with layer 2. The RLC sublayer provides ARQ functionality closely coupled with the radio transmission technique used.

Table 6.2
3GPP RAN Specifications

Specification number	Name	Scope
TS 25.201	Physical layer – general description	Describes the contents of the layer 1 documents (TS 25.200 series); where to find information; a general description of layer 1
TS 25.211	Physical channels and mapping of transport channels onto physical channels (FDD)	Establishes the characteristics of the layer-1 transport channels and physical channels in the FDD mode, and specifies: • transport channels • physical channels and their structure • relative timing between different physical channels in the same link, and relative timing between uplink and downlink; • mapping of transport channels onto the physical channels.
TS 25.212	Multiplexing and channel coding (FDD)	Describes multiplexing, channel coding, and interleaving in the FDD mode and specifies: • coding and multiplexing of transport channels; • channel coding alternatives; • coding for layer 1 control information; • different interleavers; • rate matching; • physical channel segmentation and mapping.
TS 25.213	Spreading and modulation (FDD)	Establishes the characteristics of the spreading and modulation in the FDD mode, and specifies: • spreading • generation of channelization and scrambling codes; • generation of random access preamble codes; • generation of synchronization codes; • modulation.
TS 25.214	Physical layer procedures (FDD)	Establishes the characteristics of the physical layer procedures in the FDD mode, and specifies: • cell search procedures; • power control procedures; • random access procedure.
TS 25.215	Physical layer - measurements (FDD)	Establishes the characteristics of the physical layer measurements in the FDD mode, and specifies: • the measurements performed by layer 1; • reporting of measurements to higher layers and network; • handover measurements and idle-mode measurements.

Figure 6.2   Air interface protocol architecture.

6.3.1   Logical Channels

The MAC layer provides data transfer services on logical channels. A set of logical channel types is defined for different kinds of data transfer services as offered by MAC. Each logical channel type is defined by the type of information that is transferred. Logical channel types are depicted in Figure 6.3. Logical channels are classified into two groups:

- Control channels for the transfer of control plane information (Table 6.3)
- Traffic channels for the transfer of user plane information (Table 6.4).

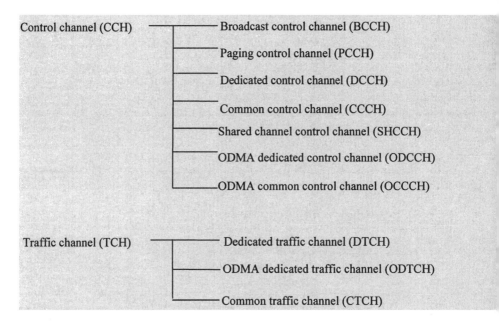

Figure 6.3 Logical channel structure.

Table 6.3
Logical Control Channels

Broadcast control channel (BCCH)	Downlink channel for broadcasting system control information.
Paging control channel (PCCH)	Downlink channel that transfers paging information and is used when: • Network does not know the location cell of the mobile station; • The mobile station is in the cell connected state (utilizing sleep mode procedures).
Common control channel (CCCH)	Bidirectional channel that transfers control information between network and mobile stations. This channel is used: • By the mobile stations having no RRC connection with the network; • By the mobile stations using common transport channels when accessing a new cell after cell reselection.
Dedicated control channel (DCCH)	Point-to-point bidirectional channel that transmits dedicated control information between a mobile station and the network. This channel is established through RRC connection setup procedure.
ODMA common control channel (OCCCH)	Bidirectional channel for transmitting control information between mobile stations.
ODMA dedicated control channel (ODCCH)	Point-to-point bidirectional channel that transmits dedicated control information between mobile stations. This channel is established through RRC connection setup procedure.

Table 6.4
Traffic Channels

Dedicated traffic channel (DTCH)	Point-to-point channel, dedicated to one mobile station, for the transfer of user information. A DTCH can exist in both uplink and downlink.
ODMA dedicated traffic channel (ODTCH)	Point-to-point channel, dedicated to one mobile station, for the transfer of user information between mobile stations. An ODTCH exists in relay link. A point-to-multipoint unidirectional channel for transfer of dedicated user information for all or a group of specified mobile stations.

6.3.2   Transport Channels

A transport channel is defined by how and with what characteristics data is transferred over the air interface. There exist two types of transport channels:

- Dedicated channels;
- Common channels, listed in Table 6.5.

There is one dedicated transport channel, the dedicated channel (DCH), which is a downlink or uplink transport channel. The DCH is transmitted over the entire cell or over only a part of the cell using beam-forming antennas. The DCH is characterized by the possibility of fast rate change (every 10 ms), fast power control, and inherent addressing of mobile stations.

6.3.2.1   Mapping Between Logical Channels and Transport Channels

Figure 6.4 shows the mapping between logical and transport channels. The following connections exist:

- BCCH is connected to BCH and may also be connected to FACH.
- PCCH is connected to PCH.
- CCCH is connected to RACH and FACH.
- SHCCH is connected to RACH and USCH/FACH and DSCH.
- DTCH can be connected to either RACH and FACH, to RACH and DSCH, to DCH and DSCH, to a DCH, a CPCH (FDD only).
- CTCH is connected to FACH.
- DCCH can be connected to either RACH and FACH, to RACH and DSCH, to DCH and DSCH, to a DCH, a CPCH to FAUSCH, CPCH.

Table 6.5
Common Transport Channels

Broadcast channel (BCH)	Downlink transport channel that is used to broadcast system- and cell-specific information. The BCH is always transmitted over the entire cell with a low fixed bit rate.
Forward access channel (FACH)	Downlink transport channel. The FACH is transmitted over the entire cell or over only a part of the cell using beam-forming antennas. The FACH uses slow power control.
Paging channel (PCH)	Downlink transport channel. The PCH is always transmitted over the entire cell. The transmission of the PCH is associated with the transmission of a physical layer signal, the paging indicator, to support efficient sleep mode procedures.
Random access channel (RACH)	Uplink transport channel. The RACH is always received from the entire cell. The RACH is characterized by a limited size data field, a collision risk and by the use of open loop power control.
Common packet channel (CPCH)	Uplink transport channel. The CPCH is a contention-based random access channel used for transmission of bursty data traffic. CPCH is associated with a dedicated channel on the downlink, which provides power control for the uplink CPCH.
Downlink shared channel (DSCH)	Downlink transport channel shared by several mobile stations The DSCH is associated with a DCH.

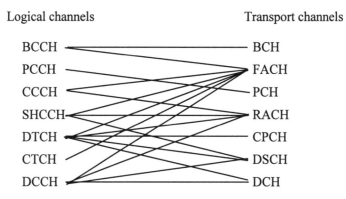

Figure 6.4   Mapping between logical and transport channels.

6.4   PHYSICAL CHANNELS

The transport channels are channel coded and matched to the data rate offered by physical channels. Thereafter, the transport channels are mapped on the physical channels. Physical channels consist of radio frames and time slots. The length of a radio frame is 10 ms and one frame consists of 15 time slots. A time slot is a unit, which consists of fields containing bits. The number of bits per time slot depends on the physical channel. Depending on the symbol rate of the physical channel, the configuration of radio frames or time slots varies. The basic physical resource is the code/frequency plane. In addition, on the uplink, different information streams may be transmitted on the I and Q branch. Consequently, a physical channel corresponds to a specific carrier frequency, code, and, on the uplink, relative phase (0 or p/2). In Section 6.4.1, the different physical channels and their structure are presented.

6.4.1   Uplink Physical Channels

There are two uplink dedicated physical and two common physical channels:

- The uplink dedicated physical data channel (uplink DPDCH) and the uplink dedicated physical control channel (uplink DPCCH);
- The physical random access channel (PRACH) and physical common packet channel (PCPCH).

The uplink DPDCH is used to carry dedicated data generated at layer 2 and above (i.e., the dedicated transport channel (DCH)). There may be zero, one, or several uplink DPDCHs on each layer 1 connection. The uplink DPCCH is used to carry control information generated at layer 1. Control information consists of known pilot bits to support channel estimation for coherent detection, transmit power-control (TPC) commands, feedback information (FBI), and an optional transport-format combination indicator (TFCI). The transport-format combination indicator informs the receiver about the instantaneous parameters of the different transport channels multiplexed on the uplink DPDCH, and corresponds to the data transmitted in the same frame. For each layer 1 connection there is only one uplink DPCCH.

Figure 6.5 shows the principle frame structure of the uplink dedicated physical channels. Each frame of length 10 ms is split into 15 slots, each of length T_{slot} = 2560 chips, corresponding to one power-control period.

The parameter k in Figure 6.5 determines the number of bits per uplink DPDCH/DPCCH slot. It is related to the spreading factor (SF) of the physical channel as SF = $256/2^k$. The DPDCH spreading factor may thus range from 256 down to 4. An uplink DPDCH and uplink DPCCH on the same layer 1 connection generally are of different rates and thus have different spreading factors.

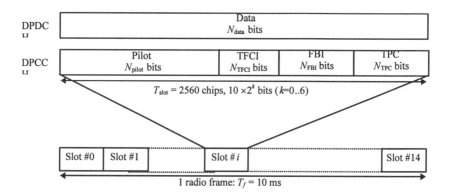

Figure 6.5 Frame structure for uplink DPDCH/DPCCH. (*Source*: [3], reproduced with permission from ETSI.)

Multiple parallel variable rate services (= dedicated logical traffic and control channels) can be time multiplexed within each DPDCH frame. The overall DPDCH bit rate is variable on a frame-by-frame basis. In most cases, only one DPDCH is allocated per connection, and services are jointly interleaved sharing the same DPDCH. Multiple DPDCHs can also be allocated, however. When multicode transmission is used, several parallel DPDCHs are transmitted using different channelization codes. There is only one DPCCH per connection, however.

The PRACH is used to carry the RACH. The random-access transmission is based on a slotted ALOHA approach with fast acquisition indication. The mobile station can start the transmission at a number of well-defined time-offsets, denoted access slots. There are 15 access slots per two frames and they are spaced 5120 chips apart. Figure 6.6 shows the access slot numbers and their spacing to each other. Information on what access slots are available in the current cell is given by higher layers. The structure of the random-access transmission is shown in Figure 6.7. The random-access transmission consists of one or several preambles of length 4096 chips and a message of length 10 or 20 ms. The mobile station indicates the length of the message part to the network by using specific signatures.

The preamble part of the random-access burst consists of 256 repetitions of a signature. There are a total of 16 different signatures, based on the Hadamard code set of length 16.

Figure 6.8 shows the structure of the random-access message part radio frame. The 10 ms message part radio frame is split into 15 slots, each of length $T_{slot} = 2560$ chips. Each slot consists of two parts, a data part that carries layer 2 information and a control part that carries layer 1 control information. The data and control parts are transmitted in parallel. A 20-ms-long message part consists of two consecutive message part radio frames.

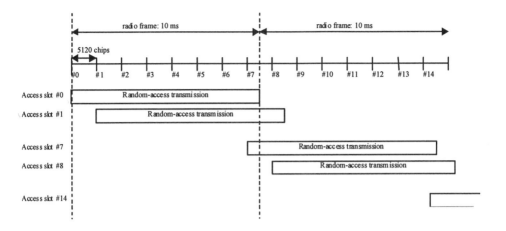

Figure 6.6 RACH access slot numbers and their spacing. (*Source*: [3], reproduced with permission from ETSI.)

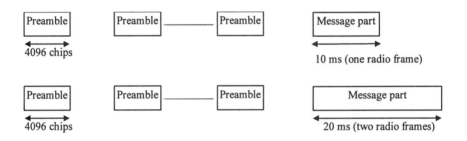

Figure 6.7 Structure of the random-access transmission. (*Source*: [3], reproduced with permission from ETSI.)

The data part consists of 10×2^k bits, where $k = 0, 1, 2, 3$. This corresponds to a spreading factor of 256, 128, 64, and 32, respectively, for the message data part.

The control part consists of eight known pilot bits to support channel estimation for coherent detection and two TFCI bits. This corresponds to a spreading factor of 256 for the message control part.

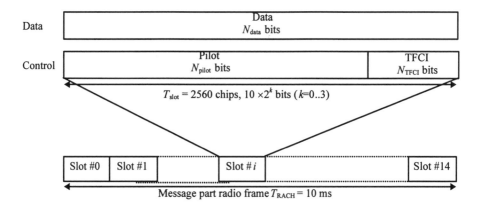

Figure 6.8 Structure of the random-access message part radio frame. (*Source*: [3], reproduced with permission from ETSI.)

The PCPCH is used to carry the CPCH transport channel. The CPCH transmission is based on DSMA-CD approach with fast acquisition indication. The mobile station can start transmission at a number of well-defined time-offsets, relative to the frame boundary of the received BCH of the current cell. The structure of the CPCH random access transmission is shown in Figure 6.9. The CPCH random-access transmission consists of one or several access preambles of length 4096 chips, one collision detection preamble (CD-P) of length 4096 chips, a DPCCH power control preamble (PC-P) (which is either 0 slots or 8 slots in length), and a message of variable length Nx10 ms.

Figure 6.9 Structure of the CPCH random-access transmission. (*Source*: [3], reproduced with permission from ETSI.)

6.4.2 Downlink Physical Channels

There is one downlink dedicated physical channel, one shared and five common control channels:

- Downlink dedicated physical channel (DPCH);
- Physical downlink shared channel (DSCH);
- Primary and secondary common pilot channels (CPICH);
- Primary and secondary common control physical channels (CCPCH);
- Synchronization channel (SCH).

Figure 6.10 shows the frame structure of the DPCH. On the DPCH, the dedicated transport channel is transmitted time multiplexed with control information generated at layer 1 (known pilot bits, power-control commands, and an optional transport-format combination indicator). DPCH can contain several simultaneous services when TFCI is transmitted or a fixed rate service when TFCI is not transmitted. The network determines if a TFCI should be transmitted.

When the total bit rate to be transmitted exceeds the maximum bit rate for a downlink physical channel, multicode transmission is employed (i.e., several parallel downlink DPCHs are transmitted using the same spreading factor). In this case, the layer 1 control information is put on only the first downlink DPCH.

Figure 6.10 Frame structure for downlink DPCH. (*Source*: [3], reproduced with permission from ETSI.)

Common pilot channel (CPICH) is a fixed-rate (30 Kbps, SF=256) downlink physical channel that carries a predefined bit/symbol sequence. There are two types of common pilot channels, the primary and secondary CPICH, as shown in Table 6.6.

The primary CCPCH is a fixed-rate (30 Kbps, SF=256) downlink physical channels used to carry the BCH. Common control physical channels are not inner-loop power controlled. Figure 6.11 shows the frame structure of the primary CCPCH. The primary CCPCH is not transmitted during the first 256 chips of each slot. Instead, primary and secondary SCHs are transmitted during this period.

Table 6.6
Primary and Secondary CPICH

Primary CPICH	<ul><li>Uses the same channelization code always;</li><li>Scrambled by the primary scrambling code;</li><li>One per cell;</li><li>Broadcast over the entire cell;</li><li>The primary CPICH is the phase reference for the SCH, primary CCPCH, AICH, PICH. It is also the default phase reference for all other downlink physical channels.</li></ul>
Secondary CPICH	<ul><li>Zero, one, or several per cell;</li><li>May be transmitted over only a part of the cell;</li><li>A secondary CPICH may be the reference for the secondary CCPCH and the downlink DPCH. If this is the case, the mobile station is informed about this by higher-layer signaling.</li></ul>

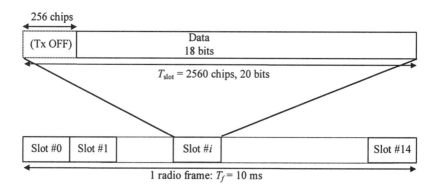

Figure 6.11   Frame structure for primary CCPCH. (*Source*: [3], reproduced with permission from ETSI.)

The frame structure of the secondary CCPCH is shown in Figure 6.12. The secondary CCPCH is used to carry the FACH and PCH. The main difference between the primary and secondary CCPCH is that the primary CCPCH has a fixed predefined rate while the secondary CCPCH can support variable rate. Furthermore, a primary CCPCH is continuously transmitted over the entire cell while a secondary CCPCH is only transmitted when there is data available and may be transmitted in a narrow lobe in the same way as a dedicated physical channel (only valid for a secondary CCPCH carrying the FACH).

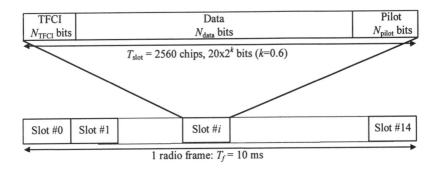

Figure 6.12   Frame structure for secondary CCPCH. (*Source*: [3], reproduced with permission from ETSI.)

Figure 6.13 depicts the structure of the synchronization channel (SCH) used for cell search. The SCH consists of two subchannels, the primary and secondary SCH.

The primary SCH consists of a modulated code of length 256 chips, the primary synchronization code (PSC) denoted c_p in Figure 6.13, transmitted once every slot. The PSC is the same for every cell in the system.

Figure 6.13   Structure of synchronization channel. (*Source*: [3], reproduced with permission from ETSI.)

The secondary SCH consists of repeatedly transmitting a length 15 sequence of modulated codes of length 256 chips, the secondary synchronization codes (SSC), transmitted in parallel with the primary SCH. The SSC is denoted $c_s^{i,k}$, where $i = 1, 2, ..., 64$ is the number of the scrambling code group, and $k = 0, 1, ..., 14$ is the slot number. Each SSC is chosen from a set of 16 different codes of length 256. This sequence on the secondary SCH indicates to which of the code groups the cell's downlink scrambling code belongs.

The physical downlink shared channel is used to carry the downlink shared channel. It is shared by users based on code multiplexing. The structure of the PDSCH is shown in Figure 6.14. As the DSCH is always associated with a DCH, the PDSCH is always associated with a downlink DPCH. For PDSCH the spreading factors may vary from 256 to 4. If the spreading factor and other physical layer parameters can vary on a frame-by-frame basis, the TFCI shall be used to inform the mobile stations of the instantaneous parameters of PDSCH.

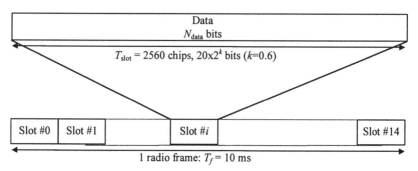

Figure 6.14 Frame structure for the PDSCH. (*Source*: [3], reproduced with permission from ETSI.)

The acquisition indicator channel (AICH) is a physical channel used to carry acquisition indicators, which correspond to signature s on the PRACH or PCPCH.

The page indicator channel (PICH) is a fixed-rate (SF=256) physical channel used to carry the page indicators. The PICH is always associated with a secondary CCPCH to which a PCH transport channel is mapped.

6.5 MULTIRATE USER DATA TRANSMISSION

WCDMA has a flexible multirate transmission scheme that enables transmission of different types of services using different data rates and quality of service parameters. For example, channel coding type, interleaving depth, and data rate can be varied to achieve the desired quality of service.

Figure 6.15 and Figure 6.16 show the multirate transmission and multiplexing schemes for the uplink and downlink, respectively. Data from transport channels is encoded and thereafter mapped to the physical channels and transmitted over the radio transmission link. The channel coding scheme is a combination of error detection, error correcting, rate matching, interleaving, and transport channels mapping onto physical channels.

Data arrives to the coding/multiplexing unit in the form of transport block sets once every transmission time interval, which is transport-channel specific and can be 10, 20, 40, or 80 ms. Multirate transmission consists of following steps:

- Addition of cyclic redudancy check (CRC) to each transport block;
- Concatenation of transport block and segmentation of code block;

- Channel coding;
- Rate matching;
- Insertion of discontinuous transmission (DTX) indication bits;
- Interleaving;
- Segmentation of radio frames;
- Multiplexing of transport channels;
- Segmentation of physical channel;
- Mapping to physical channels.

Error detection is provided on transport blocks through CRC. The CRC is 24, 16, 12, 8, or 0 bits, and higher layers signal what CRC length should be used for each transport channel.

After CRC addition, transport block concatenation and code block segmentation are performed. All transport blocks in are serially concatenated. If the number of bits in the transmission time interval is larger the maximum size of the used code block, then code block segmentation is performed after the concatenation of the transport blocks. The maximum size of the code blocks depends on whether convolutional coding, turbo coding, or no coding is used. The maximum code block sizes are:

- Convolutional coding: 504;
- Turbo coding: 5114;
- No channel coding: unlimited.

Radio frame size equalization is padding the input bit sequence in order to ensure that the output can be segmented in consecutive radio frames of the same size. Radio frame size equalization is only performed in the uplink. In the downlink, rate matching output block length is already suitable for radio frame segmentation.

When the transmission time interval is longer than 10 ms, the input bit sequence is segmented and mapped onto consecutive radio frames. This enables interleaving over several radio frames improving spectrum efficiency.

Because WCDMA provides flexible data rates, the number of bits on a transport channel can vary between different transmission time intervals. The rate matching adapts this resulting symbol rate to the limited set of possible symbol rates of a physical channel. Rate matching means that bits on a transport channel are repeated or punctured according to the defined rate matching attribute, which is semistatic and can only be changed through higher layer signaling.

In the downlink the transmission is interrupted if the number of bits is lower than maximum (i.e., DTX is used to fill up the radio frame with bits). The insertion point of DTX indication bits depends on whether fixed or flexible positions of the transport channels in the radio frame are used. It is up to the network to decide for each transport channel whether fixed or flexible positions are used during the connection. DTX indication bits only indicate when the transmission should be turned off, they are not transmitted.

One or more physical channels can be used to transmit the result. When more than one physical channel is used, physical channel segmentation divides the bits among the different channels. After the second interleaving, physical channel mapping is performed.

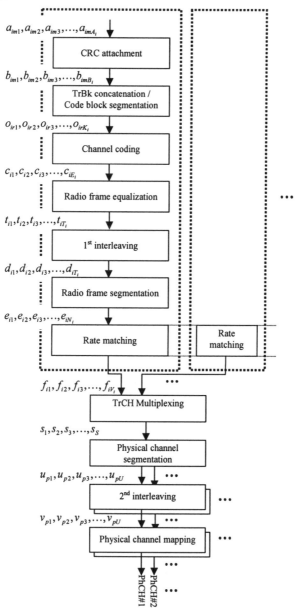

Figure 6.15   Multirate scheme for uplink. (*Source*: [4], reproduced with permission from ETSI.)

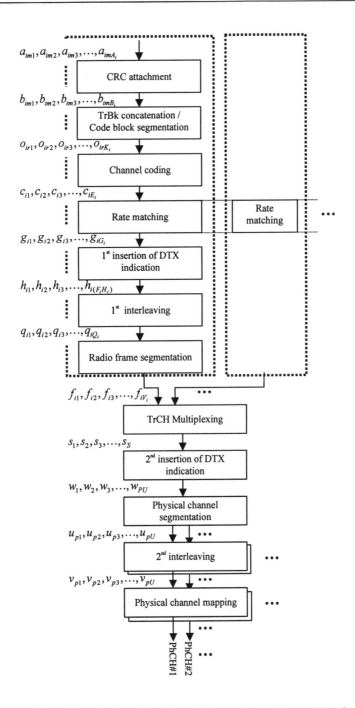

Figure 6.16   Multirate scheme for downlink. (*Source*: [4], reproduced with permission from ETSI.)

6.5.1 Transport Format Detection

Transport format detection can be performed both with and without transport format combination indicator (TFCI). If a TFCI is transmitted, the receiver detects the transport format combination from the TFCI. When no TFCI is transmitted, so-called blind transport format detection may be used (i.e., the receiver side detects the transport format combination using some information, for example, received power ratio of DPDCH to DPCCH or CRC check results).

6.5.2 Channel Coding

Table 6.7 lists the channel coding parameters for different transport channel types. The following channel coding schemes can be applied:

- Convolutional coding with constraints length 9 and coding rate 1/3 or 1/2;
- Turbo coding;
- No channel coding.

The first and second interleaving are both block interleavers with intercolumn permutations.

Table 6.7
Error Correction Coding Parameters

Transport channel type	Coding scheme	Coding rate
BCH	Convolutional code	1/2
PCH		
RACH		
CPCH, DCH, DSCH, FACH		1/3, 1/2
	Turbo code	1/3
	No coding	

The turbo coding scheme is a parallel concatenated convolutional code (PCCC) with eight-state constituent encoders.

The initial value of the shift registers of the PCCC encoder shall be all zeros (see Figure 6.17). The output of the PCCC encoder is punctured to produce coded bits corresponding to the desired code rate. For rate 1/3, none of the systematic or parity bits are punctured.

Figure 6.18 depicts the overall eight-state PCCC turbo coding scheme including turbo code internal interleaver. The turbo code internal interleaver consists of mother interleaver generation and pruning. For arbitrary given block length K, one mother interleaver is selected from the 134 mother interleavers set. After the mother interleaver generation, l-bits are pruned in order to adjust the mother interleaver to the block length K. Tail bits T1 and T2 are added for constituent encoders RSC1 and RSC2, respectively.

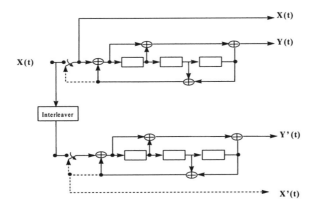

Figure 6.17 Structure of the eight-state PCCC encoder (dotted lines effective for trellis termination only). (*Source*: [4], reproduced with permission from ETSI.)

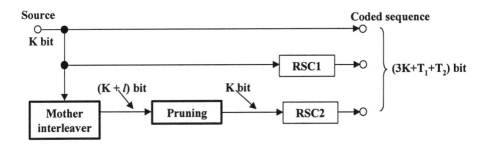

Figure 6.18 Overall eight-state PCCC turbo coding. (*Source*: [4], reproduced with permission from ETSI.)

6.6 SPREADING AND MODULATION

WCDMA applies a two-layered code structure consisting of a orthogonal spreading codes and pseudo-random scrambling codes. Spreading is performed using channelization codes, which transforms every data symbol into a number of chips, thus increasing the bandwidth of the signal. Orthogonality between the different spreading factors can be achieved by the tree-structured orthogonal codes whose construction was described in Chapter 5. Scrambling is used for cell separation in the downlink and user separation in the uplink.

6.6.1   Uplink Spreading

In the uplink, either short or long spreading (scrambling) codes are used. The short codes are used to ease the implementation of advanced multiuser receiver techniques; otherwise, long spreading codes can be used. Short codes are S(2) codes of length 256 and long codes are Gold sequences of length 2^{41}, but the latter are truncated to form a cycle of a 10-ms frame.

IQ/code multiplexing used in the uplink leads to parallel transmission of two channels, and therefore, attention must be paid to modulated signal constellation and related peak-to-average power ratio (crest factor). By using the complex spreading circuit shown in Figure 6.19, the transmitter power amplifier efficiency remains the same as for QPSK transmission in general.

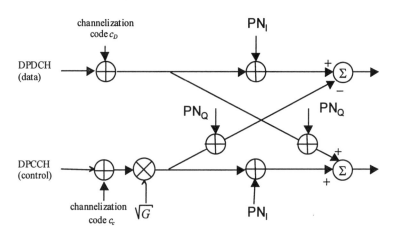

Figure 6.19   IQ/code multiplexing with complex spreading circuit.

Moreover, the efficiency remains constant irrespective of the power difference G between DPDCH and DPCCH. Thus, signal envelope variations are very similar to the QPSK transmission for all values of G.

The IQ/code multiplexing solution with complex scrambling results in power amplifier output backoff requirements that remain constant as a function of power difference. Furthermore, the achieved output backoff is the same as for one QPSK signal.

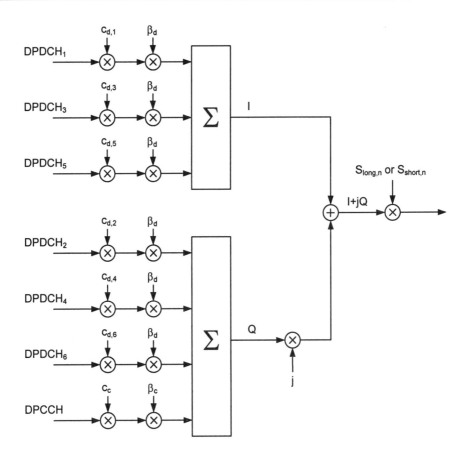

Figure 6.20 Spreading for uplink DPCCH and DPDCHs. (*Source*: [5], reproduced with permission from ETSI.)

Figure 6.20 shows the uplink spreading of DPCCH and DPDCHs, which are spread by different channelization codes. One DPCCH and up to six parallel DPDCHs can be transmitted simultaneously. After channelization, the real-valued spread signals are weighted by gain factors, which is different for DPCCH and DPDCHs but is the same for all DPDCHs. After transforming the signal from real to complex, it is then scrambled by the complex-valued scrambling code, which can be long or short.

6.6.1.1   PRACH Spreading

The PRACH consisted of preamble and message parts. The preamble is a complex-valued code and message part consists of data and control parts. Figure 6.21 shows spreading and scrambling of the PRACH message part. After channelization, the spread signals are weighted by gain factors. Scrambling is performed by a 10-ms-long complex-valued scrambling code.

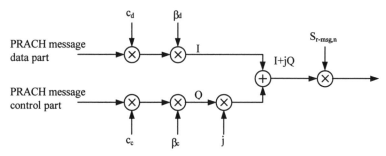

Figure 6.21 Spreading of PRACH message part. (*Source*: [5], reproduced with permission from ETSI.)

6.6.1.2 PCPCH

The PCPCH consists of preamble and message parts. The preamble part consists of a complex-valued code, and the message part consists of data and control parts. Figure 6.22 shows the spreading of the PCPCH message part. The control and data part are spread by different channelization codes After channelization, the spread signals are weighted by gain factors. Scrambling is performed by a 10-ms long complex-valued scrambling code.

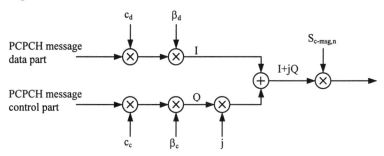

Figure 6.22 Spreading of PCPCH message part. (*Source*: [5], reproduced with permission from ETSI.)

6.6.2 Downlink Spreading

In the downlink, the same orthogonal channelization codes are used as in the uplink. For scrambling, Gold codes of length 2^{18} are used, but they are truncated to form a cycle of a 10-ms frame (i.e., 384,000 chips). To form a complex-valued code, the same truncated code is used with different time shifts in I and Q channels.

It is possible to generate 2^{18}-1 scrambling codes, but only 8191 of them are used. Each cell is allocated one primary scrambling code. In order to reduce the cell search time, the primary scrambling codes are divided into 512 sets. Thus, the mobile station needs to search at maximum 512 10-ms-long codes.

In addition to primary scrambling codes, there are 15 secondary scrambling code sets. Secondary scrambling codes are used when one set of orthogonal channelization codes is not enough. This can be the case when adaptive antennas are used in the downlink. It should be noted that use of the secondary scrambling code destroys the orthogonality between code channels. This is counterweighted, however, by the spatial isolation offered by adaptive antennas.

The primary CCPCH and primary CPICH are always transmitted using the primary scrambling code. The other downlink physical channels can be transmitted with either the primary scrambling code or a secondary scrambling code from the set associated with the primary scrambling code of the cell.

Figure 6.23 illustrates the spreading operation for all downlink physical channels except SCH (i.e., for P-CCPCH, S-CCPCH, CPICH, AICH, PICH, and downlink DPCH). The nonspread physical channel consists of a sequence of real-valued symbols. For all channels except AICH, the symbols can take the three values +1, -1, and 0, where 0 indicates DTX. For AICH, the symbol values depend on the exact combination of acquisition indicators to be transmitted.

Each pair of two consecutive symbols is first serial-to-parallel converted and mapped to an I and Q branch. The mapping is such that even and odd numbered symbols are mapped to the I and Q branch, respectively. For all channels except AICH, symbol number zero is defined as the first symbol in each frame. For AICH, symbol number zero is defined as the first symbol in each access slot. The I and Q branches are then spread to the chip rate by the same real-valued channelization code $C_{ch,SF,m}$. The sequences of real-valued chips on the I and Q branch are then treated as a single complex-valued sequence of chips. This sequence of chips is scrambled by a complex-valued scrambling code $S_{dl,n}$. In case of P-CCPCH, the scrambling code is aligned with the P-CCPCH frame boundary (i.e., the first complex chip of the spread P-CCPCH frame is multiplied with chip number zero of the scrambling code). In case of other downlink channels, the scrambling code is aligned with the scrambling code applied to the P-CCPCH. In this case, the scrambling code is thus not necessarily aligned with the frame boundary of the physical channel to be scrambled.

Figure 6.24 shows how different downlink channels are combined. Each complex-valued spread channel, corresponding to point S in Figure 6.23, is separately weighted by a weight factor G_i. The complex-valued P-SCH and S-SCH are also weighted separately. All downlink physical channels are then combined using complex addition.

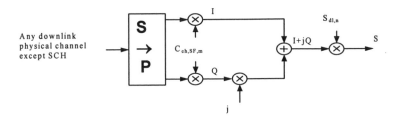

Figure 6.23 Spreading for all downlink physical channels except SCH. (*Source*: [5], reproduced with permission from ETSI.)

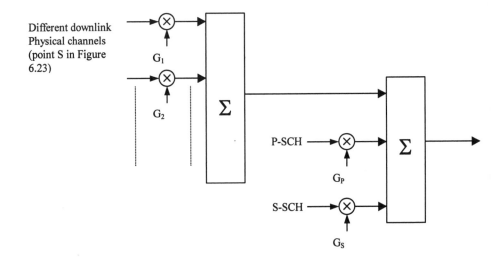

Figure 6.24 Spreading and modulation for SCH and P-CCPCH. (*Source*: [5], reproduced with permission from ETSI.)

6.6.2.1   Synchronization Codes

The primary synchronization code (PSC), CPSC is constructed as a so-called generalized hierarchical Golay sequence. The PSC is furthermore chosen to have good aperiodic autocorrelation properties.

6.6.3   Modulation

The complex-valued chip sequence generated by the spreading process is QPSK modulated. Figure 6.25 illustrates the modulation principle used in the uplink and downlink. The pulse shaping is root-raised cosine with roll-off factor 0.22 and is the same for the mobile and base stations.

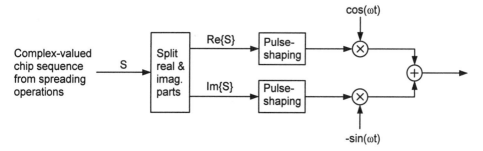

Figure 6.25   Modulation principle. (*Source*: [5], reproduced with permission from ETSI.)

6.7   TRANSMIT DIVERSITY

Two types of diversity mechanism can be used in the downlink of WCDMA: open loop and closed loop. The open-loop downlink transmit diversity employs a space/time block coding–based transmit diversity (STTD). Even though the STTD encoding is optional in the base station, STTD support is mandatory at the mobile station. Another type of open loop transmit diversity, time switched transmit diversity (TSTD), can be applied to the SCH channel. This diversity mechanism is also optional in the base station but mandatory in the mobile station.

Not all diversity methods can be used will all physical channels. Table 6.8 summarizes the possible application of open and closed-loop transmit diversity modes on different downlink physical channels. Simultaneous use of STTD and closed loop modes on DPCH and PDSCH is not allowed.

Table 6.8
Application of Tx Diversity Modes on Downlink Physical Channels

Channel	Open loop mode		Closed loop mode
	TSTD	STTD	
P-CCPCH	-	X	-
SCH	X	-	-
S-CCPCH	-	X	-
DPCH	-	X	X
PICH	-	X	-
PDSCH (associated with DPCH)	-	X	X
AICH	-	X	-

Note: "X" - can be applied, "-" cannot be applied.

6.8   AIR INTERFACE PROCEDURES

Different air interface procedures are required to make a radio system work (i.e., to establish communication and maintain it with a minimum consumption of radio resources). In this section the following air interface procedures are covered:

- Cell search operation;
- Handover;
- Power control;
- Uplink synchronous transmission scheme (USTS).

6.8.1   Cell Search

During the cell search, the mobile station searches for a cell and determines the downlink scrambling code and common channel frame synchronization of that cell. Because the radio frame timing of all common physical channels is related to the timing of P-CCPCH, it is enough to find the timing of P-CCPCH only.

The cell search is typically carried out in three steps: slot synchronization; frame synchronization and code-group identification; and scrambling-code identification. An example procedure from the 3GPP specification TS25.214 is described as follows:

- *Step 1: Slot synchronization.* During the first step of the cell search procedure, the mobile station uses the SCH's primary synchronization code to acquire slot synchronization to a cell. This can be done with a single matched filter matched to the primary synchronization code that is common to all cells. The slot timing of the cell can be obtained by detecting peaks in the matched filter output.
- *Step 2: Frame synchronization and code-group identification.* During the second

step of the cell search procedure, the mobile station uses the SCH's secondary synchronization code to find frame synchronization and identify the code group of the cell found in the first step. This is done by correlating the received signal with all possible secondary synchronization code sequences and identifying the maximum correlation value. Because the cyclic shifts of the sequences are unique, the code group and the frame synchronization are determined.

- *Step 3: Scrambling-code identification.* During the third and last step of the cell search procedure, the mobile station determines the exact primary scrambling code used by the found cell. The primary scrambling code is typically identified through symbol-by-symbol correlation over the CPICH with all codes within the code group identified in the second step. After the primary scrambling code has been identified, the primary CCPCH can be detected. And the system- and cell specific BCH information can be read.

In case the mobile station has received information about which scrambling codes to search for, steps 2 and 3 above can be simplified.

6.8.2   Handover

WCDMA has several different types of handovers:

- Soft, softer, and hard handover;
- Interfrequency handover;
- Handover between FDD and TDD modes;
- Handover between WCDMA and GSM.

The handover algorithm to make the handover decision needs different types of measurement information. Table 6.9 lists the measurements that can be carried for handover purposes. The actual handover algorithm implementation is left to equipment manufacturers.

Base stations in WCDMA need not be synchronized, and therefore, no external source of synchronization, like GPS, is needed for the base stations. Asynchronous base stations must be considered when designing soft handover algorithms and when implementing location services.

Before entering soft handover, the mobile station measures observed timing differences of the downlink SCHs from the two base stations. The structure of SCH is presented in Section 6.4.4. The mobile station reports the timing differences back to the serving base station. The timing of a new downlink soft handover connection is adjusted with a resolution of one symbol (i.e., the dedicated downlink signals from the two base stations are synchronized with an accuracy of one symbol). That enables the mobile RAKE receiver to collect the macro diversity energy from the two base stations. Timing adjustments of dedicated downlink channels can be carried out with a resolution of one symbol without losing orthogonality of downlink codes.

Table 6.9

Handover Measurements

Received signal code power (RSCP)	Received power on one code measured on the pilot bits of the primary CPICH. The reference point for the RSCP is the antenna connector at the mobile station.
TDD received signal code power	Received power on one code measured on the PCCPCH from a TDD cell. The reference point for the RSCP is the antenna connector at the mobile station.
Received signal code power after radio link combination	Received power on one code measured on the pilot bits of the DPCCH after radio link combination. The reference point for the RSCP is the antenna connector at the mobile station.
SIR	Signal-to-interference ratio, defined as: $(RSCP/ISCP) \times (SF/2)$, where ISCP = interference signal code power, the interference on the received signal measured on the pilot bits. Only the nonorthogonal part of the interference is included in the measurement. SF = the spreading factor used. The SIR shall be measured on DPCCH after RL combination. The reference point for the SIR is the antenna connector of the mobile station.
RSSI	Received signal strength indicator, the wideband received power within the relevant channel bandwidth. Measurement shall be performed on downlink carrier. The reference point for the RSSI is the antenna connector at the mobile station.
GSM carrier RSSI	Received signal strength indicator, the wideband received power within the relevant channel bandwidth. Measurement shall be performed on a GSM BCCH carrier. The reference point for the RSSI is the antenna connector at the mobile station.
CPICH Ec/No	The received energy per chip divided by the power density in the band. The Ec/No is identical to RSCP/RSSI. Measurement shall be performed on the primary CPICH. The reference point for Ec/No is the antenna connector at the mobile station.
Transport channel BLER	Estimation of the transport channel block error rate (BLER). The BLER estimation shall be based on evaluating the CRC on each transport block after radio link combination. Used in SIR target setting for fast power control in the outer loop power control.
Physical channel BER	The physical channel bit error rate (BER) is an estimation of the average bit error rate before channel decoding of the DPDCH data after radio link combination.
Mobile station transmitted power	The total mobile station transmitted power on one carrier. The reference point for the UE transmitted power shall be the mobile station antenna connector.

6.8.2.1   Interfrequency Handovers

Interfrequency handovers are needed for utilization of hierarchical cell structures; macro, micro, and indoor cells (see Chapter 10). Several carriers and interfrequency handovers may also be used for taking care of high capacity needs in hot spots. Interfrequency handovers will be needed also for handovers to second-generation systems like GSM or IS-95. In order to complete interfrequency handovers, an efficient method is needed for making measurements on other frequencies while still having the connection running on the current frequency. Two methods are considered for interfrequency measurements in WCDMA:

- Dual receiver;
- Compressed mode.

The dual receiver approach is considered suitable, especially if the mobile terminal employs antenna diversity. During the interfrequency measurements, one receiver branch is switched to another frequency for measurements, while the other keeps receiving from the current frequency. The loss of diversity gain during measurements needs to be compensated for with higher downlink transmission power. The advantage of the dual receiver approach is that there is no break in the current frequency connection. Fast closed loop power control is running all the time.

The compressed mode approach (also referred as slotted mode) depicted in Figure 6.26 is required for the mobile station without dual receiver. The information normally transmitted during a 10-ms frame is compressed either by code puncturing or by changing the FEC rate. For the performance of different interfrequency handover options, see Section 5.14.5.

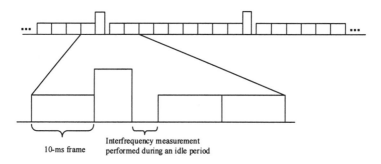

10-ms frame

Interfrequency measurement
performed during an idle period

Figure 6.26 Compressed mode structure.

6.8.2.2   Handover Between GSM and WCDMA

The handover between the WCDMA system and the GSM system, offering worldwide coverage today, has been one of the main design criteria taken into account in the WCDMA frame timing definition. The GSM-compatible multiframe structure, with the superframe being multiple of 120 ms, allows similar timing for intersystem measurements as in the GSM system itself. Apparently, the needed measurement interval does not need to be as frequent as for GSM terminal operating in a GSM system, as intersystem handover is less critical from an intrasystem interference point of view. Rather, the compatibility in timing is important, so that when operating in WCDMA mode, a multimode terminal is able to catch the desired information from the synchronization bursts in the synchronization frame on GSM carrier with the aid of a frequency correction burst. This way the relative timing between a GSM and WCDMA carriers is maintained similar to the timing between two asynchronous GSM carriers. The timing relation between WCDMA channels and GSM channels is indicated in

Figure 6.27, where the GSM traffic channel and WCDMA channels use a similar 120-ms multiframe structure. The GSM frequency correction channel (FCCH) and GSM synchronization channel (SCH) use one slot out of the eight GSM slots in the indicated frames, with the FCCH frame with one time slot for FCCH always preceding the SCH frame with one time slot for SCH, as indicated in Figure 6.27. Further details on GSM common channel structures can be found in [2].

A WCDMA terminal can do the measurements either by requesting the measurement intervals in a slotted mode form where there are breaks in the downlink transmission, or by performing the measurements independently with a suitable measurement pattern. With independent measurements the dual receiver approach is used instead of the slotted mode because the GSM receiver branch can operate independently of the WCDMA receiver branch.

For smooth interoperation between the systems, information needs to be exchanged between the systems, in order to allow the WCDMA base station to notify the terminal of the existing GSM frequencies in the area. In addition, more integrated operation is needed for the actual handover where the current service is maintained, naturally taking into account the lower data rate capabilities in GSM when compared to the UMTS maximum data rates reaching all the way to 2 Mbps.

Figure 6.27   Measurement timing relation between WCDMA and GSM frame structures.

The GSM system is likewise expected to be able to indicate the WCDMA spreading codes in the area to make the cell identification simpler. After that the existing measurement practices in GSM can be used for measuring the WCDMA when operating in GSM mode.

As the WCDMA does not rely on any superframe structure as with GSM to find out synchronization, the terminal operating in GSM mode is able to obtain the WCDMA frame synchronization once the WCDMA base station scrambling code timing is acquired. The base station scrambling code has 10-ms period and its frame timing is synchronized to WCDMA common channels.

6.8.3   Power Control

WCDMA has fast closed loop and open loop power control procedures. The fast power control operates with a rate of 1.5 KHz (i.e., one command per slot). The nominal power control step size is 1 dB, but multiples of the nominal step sizes can also be used. Power control commands can only be sent every second slot.

In soft handover state, power control commands from different base stations are combined, taking the reliability of each command into account. In the compressed mode, the fast power control uses a larger step size for a short period after a compressed frame in order to converge faster.

Open loop power control is used before initiating transmission on the RACH or CPCH. Power control accuracy is required to be ±9 dB.

6.8.4   Uplink Synchronous Transmission Scheme

Uplink synchronous transmission scheme is an alternative technology applicable for low mobility terminals. USTS can reduce uplink intracell interference by means of making a cell receive orthogonalized signals from mobile stations. To orthogonalize receiving signals from mobile stations:

- The same scrambling code is allocated to all dedicated physical channels in a cell;
- Different channelization codes are allocated to all dedicated physical channels across all mobile stations in a cell, and the spreading factor and code number of channelization code are delivered from network to each mobile station;
- The channelization codes for DPDCH and DPCCH in a mobile station are chosen from either the upper half part or the lower half part of the OVSF code tree in a mobile station to reduce the peak-to-average power ratio;
- Additional scrambling codes can be allocated if all channelization codes are occupied;
- The signal transmission time of each mobile station is adjusted.

The transmission time control is carried out by two steps. The first step is initial synchronization and the second is tracking:

1. *Initial synchronization:* Adjust transmission time through the initial timing control message over FACH. When the cell receives the signal from the mobile station over RACH, the cell measures the time difference between the received timing and the reference time in the unit of 1/8 chip duration. The message for initial synchronization, which contains the difference in

time, is delivered to mobile station via FACH.

2. *Tracking process (closed loop timing control):* Adjust the transmission time through the time alignment bit (TAB) over DPCCH. UE adjusts its transmission time according to the message. The cell periodically compares the reference time with the received signal timing from mobile station. When the received timing is earlier than the reference time, TAB = 0. When the received timing is later than the reference time, TAB = 1. TAB replaces the TPC bit every timing control period of 20 ms the last TPC bit of every two frames is replaced by TAB. At the mobile station, hard decision on the TAB shall be performed, and when it is judged as 0, the transmission time shall be delayed by 1/8 chip, whereas if it is judged as 1, the transmission time shall be advanced by 1/8 chip.

6.8.5 Packet Data

WCDMA has three different types of packet data transmission possibilities [2]:

- Common channels;
- Dedicated channel;
- Shared channels.

The advantage of using common channels for packet data transmission is the very short link setup time. Common channels RACH (uplink) and FACH (downlink) are used for short infrequent packets. Soft handover cannot be used for these channels. In addition, only open loop power control is in operation. Therefore, RACH and FACH packet transmission should be limited to short packets that only use a limited capacity. Figure 6.28 illustrates packet transmission on RACH.

The common packet channel (CPCH) can be used to transmit short and medium-sized data packets. The setup procedure is similar to RACH, but the channel can be shared in time division fashion between different users and fast power control can be used. Soft handover cannot be used for CPCH.

Common channel without fast power control

Figure 6.28 Packet transmission on the RACH channel.

Larger or more frequent packets are transmitted on a dedicated channel. Now, soft handover and fast power control are used and thus radio performance is better than transmission in the common channels. The data rate can be varied during the transmission. Consequently, the number of orthogonal codes in the downlink has to be allocated according to the highest bit rate. If several high bit rate dedicated packet access connections exist, there might not be enough orthogonal channels available.

To overcome the limitation of running out of orthogonal codes, downlink shared channel (DSCH) can be used. In this packet transmission mode, a single physical channel is shared using time division. With DSCH, fast power control but not soft handover can be used. Fast power control signaling is carried on a low bit rate dedicated channel.

6.9 WCDMA EVOLUTION – RELEASE 2000 AND BEYOND

WCDMA will continue to evolve based on research and new innovations. Release 2000 (now renamed release 4) specifications will provide efficient IP support enabling provision of services through an all-IP core network. Technical issues to be solved include header compression and QoS. The later 3GPP releases will provide even higher data rates up to approximately 10 Mps. A work item to specify high-speed packet access (HSPA) has been started. The following enhancements are proposed:

- Adaptive modulation and coding (AMCS);
- Hybrid automatic-repeat-request (HARQ);
- Smaller frame size;
- Position of scheduling mechanism;
- Fast cell site selection;
- Uplink DCH associated with a access control channel (ACCH).

REFERENCES

[1] Chaudry, P., M. Mohr, and S. Onoe, "The 3GPP Proposal for IMT-2000," *IEEE Communications Magazine*, Vol. 39, No. 12, December 1999, pp. 72–81.

[2] Holma, H., and A. Toskala (eds.), *WCDMA for UMTS*, Chichester, England: John Wiley & Sons, Ltd., 2000.

[3] TS 25.211, "Physical Channels and Mapping of Transport Channels onto Physical Channels (FDD)," 3GPP Technical Specification, Version 3.2.0, March 2000.[1]

[4] TS 25.212, "Multiplexing and Channel Coding (FDD)," 3GPP Technical Specification, Version 3.2.0, March 2000.

[5] TS 25.213, "Spreading and Modulation (FDD)," 3GPP Technical Specification, Version 3.2.0, March 2000.

---

[1] The present technical specifications are the property of 3GPP. No reproduction shall be made without ETSI authorization. The original version of these 3GPP technical specifications can be obtained from the publication office of ETSI.

Chapter 7

WCDMA TIME DIVISION DUPLEX

7.1   INTRODUCTION

The WCDMA air interface consists of two components: FDD for paired frequency bands and TDD for unpaired frequency bands. The main parameters of the two components have been harmonized in the 3GPP standardization process. While the nominal bandwidth for WCDMA is 5 MHz, for TDD there exists also a narrowband mode that uses 1.25 MHz bandwidth. This is mainly a result of harmonization with a TDD proposal from China. The 3GPP WCDMA TDD mode is originally based on the time division CDMA (TD-CDMA) concept [1, 2].

This chapter is organized as follows. Section 7.2 explains the TDD concept and its main characteristics. Section 7.3 lists the 3GPP specifications for TDD, and Section 7.4 presents the TDD system description.

7.2   TDD CONCEPT

In this section the basic TDD concept and its main characteristics such as asymmetric services, reciprocal channel in the uplink and downlink, and interference issues are studied.

In time division duplex (TDD), the uplink and downlink transmissions are time multiplexed into the same carrier, in contrast to frequency division duplex (FDD), where uplink and downlink transmissions occur in frequency bands separated by the duplex frequency. Figure 7.1 illustrates the principles of TDD and FDD. Because for IMT-2000 there exist two types of frequency bands, paired and unpaired, both FDD and TDD modes are required to fully utilize the IMT-2000 spectrum.

7.2.1   Asymmetric Services

Services such as the Internet, multimedia applications, and file transfers often set different capacity requirements for the uplink and downlink. The utilization of a TDD frequency band is not fixed between the uplink and downlink (unlike with FDD), and this flexibility in resource allocation can be used if the air interface design is flexible enough.

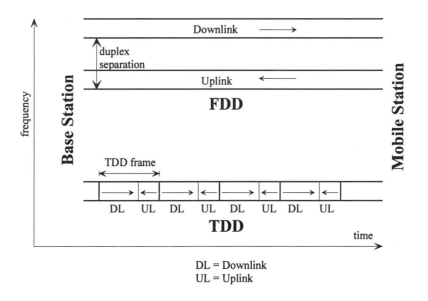

Figure 7.1 TDD and FDD principles.

7.2.2 Reciprocal Channel in Uplink and Downlink

In FDD operation, the uplink and downlink transmissions are separated by a duplex separation. Because fast fading due to multipath propagation depends on the frequency, it is uncorrelated between the uplink and downlink. The FDD transmitter cannot predict the fast fading that will affect its transmission.

In TDD operation, the same frequency is used for both the uplink and the downlink. Based on the received signal, the TDD transmitter is able to know the fast fading of the multipath channel. This assumes that the TDD frame length is shorter than the coherence time of the channel. This assumption holds if TDD mobiles are slowly moving terminals. The reciprocal channel can then be utilized for:

- Open loop power control;
- Spatio-temporal transmission diversity (adaptive antennas for transmission [3], pre-RAKE [4–6]).

With open loop power control, the need for power control signaling is reduced compared to closed loop power control. Closed loop power control signaling also introduces some delay and is subject to errors, which is not the case with open loop power control. In order to have fast enough open loop power control, the TDD frame must be short enough. According to [7], Doppler frequencies up to 80 Hz (43 km/h at 2-GHz carrier frequency) can be supported with a very small degradation if the uplink part of the TDD frame length is 1.5 ms. If the TDD system is intended only for slowly moving terminals, then longer TDD frames could also be used. With open loop power control, the interference situation at the receiver is not known by the transmitter, only the signal level is known. Changes in the interference level must be signaled.

Transmission diversity can be utilized with diversity antennas (space domain diversity) or with pre-RAKE (time domain diversity). In selection diversity combining, the receiver measures the received signal from diversity antennas and selects the best antenna for reception. Antenna diversity techniques are easily applied at the base stations but those receiver techniques are not suited for small handheld terminals. In order to achieve antenna diversity in the downlink, transmission diversity is utilized at the base station. Based on uplink reception, the best antenna can be selected for downlink transmission in TDD.

In FDD-CDMA transmission, a RAKE receiver is used to collect the multipath components and to obtain multipath diversity. The optimal RAKE receiver is a matched filter to the multipath channel. If the transmitter knew the multipath channel, it could apply RAKE in the transmitter (pre-RAKE). Transmission would be such that the multipath channel would act as a matched filter to the transmitted signal. In the receiver no RAKE would be needed (i.e., multipath diversity could be obtained with a one-finger receiver). It should be noted, however, that only a little multipath diversity may be available in indoor propagation environments, as shown in Chapter 9 with ITU channel models. Indoor and microcell environments are the most probable application areas for TDD communication. Therefore, antenna diversity transmission will be a more attractive diversity technique for TDD operation than the multipath diversity technique.

If such a TDD proposal, which has different solutions for the uplink and downlink (where uplink uses a single wideband carrier and downlink a multicarrier approach), is applied, channel reciprocity cannot be utilized. The effect of such a structure depends on the environment, but, as calculated in Chapter 9, the coherence bandwidth cannot be guaranteed to be so high that similar fading characteristics could be obtained for both uplink and downlink.

7.2.3 Interference Issues

7.2.3.1 Intracell and Intercell Interference Between Uplink and Downlink

In CDMA systems, the SIR may be quite low (e.g., below −15 dB) at carrier bandwidth. After despreading, the SIR is improved by the processing gain. In TDD systems, a transmitter located close to a receiver may block the front end of the receiver, because no RF filter can be used to separate uplink and downlink transmission as in FDD operation. This blocking may happen even if the transmitter and receiver are not operating in the same frequency channel but if they are operating in the same TDD band. In that case, the processing gain at baseband does not help because the signal is already blocked before baseband processing. These interference problems with TDD operation are considered in this section.

Within one TDD-CDMA cell, all users must be synchronized and have the same time division between uplink and downlink in order to avoid interference between uplink and downlink. This time division is based on the average uplink and downlink capacity need in that particular cell. Each user then applies multirate techniques to adapt its uplink and downlink capacity needs to the average need in that cell. The same time division must be applied to all carriers within one base station. If the base station transmits and receives at the same time as adjacent carriers, it would block its own reception.

Asymmetric usage of TDD slots will impact the radio resource in neighboring cells. This scenario is depicted in Figure 7.2 and the resulting signal to adjacent channel interference ratio is calculated in Table 7.1. Intercell interference problems occur in asymmetric TDD-CDMA if the asymmetry is different in adjacent cells even if the base stations are synchronized. MS2 is transmitting at full power at the cell border. Because MS1 has a different asymmetric slot allocation than MS2, its downlink slots received at the sensitivity limit are interfered with by MS1, causing blocking. On the other hand, since BS1 can have much higher effective isotropically radiated power (EIRP) than MS2, it will interfere with BS2 receiving MS2. It is difficult to adjust the asymmetry of an individual cell in a network due to interference between adjacent cells. If TDD-CDMA cells are located adjacent to each other, offering a continuous coverage, then synchronization and asymmetry coordination between these cells is required. This ensures that the near-far problems of interference between mobiles in adjacent cells can be controlled. Another scenario, where the previously described blocking effect clearly exists, is if TDD operation were also allowed in the FDD band.

Table 7.1
Adjacent Channel Interference Calculation

BTS transmission power for MS2 in downlink 1W	30 dBm
Received power for MS1	−100 dBm
Adjacent channel attenuation due to irreducible noise floor	50 to 70 dB
Signal to adjacent channel interference ratio	−60 to −80 dB

Figure 7.2   TDD interference scenario.

7.2.3.2 Interoperator Interference with Continuous Coverage

If there are several operators offering the service in the same geographical area in the TDD band, base station synchronization for the different operators is required and asymmetry flexibility between uplink and downlink becomes considerably more difficult. Asymmetric TDD-CDMA systems are therefore not well suited if several operators share the spectrum and the same area. In those cases, TDD-TDMA systems are better off because the signal is concentrated in the time domain. In TDD-TDMA, interference can be averaged with time hopping or avoided with dynamic channel allocation techniques. A difficult problem for operation in an unlicensed band is the case where different operators use different multiple access techniques in the same TDD band.

Adjacent channel interference between operators may also cause problems. As the terminals have limited dynamic range and neighboring channel filtering capability, adjacent channel interference in this kind of uncoordinated operation may prove to be very severe. The power differences between the transmission of the desired base station and the interfering mobile transmitting at the same time as adjacent carriers can block the receiver terminal's A/D converters and can also cause problems in the RF components.

7.2.3.3 Synchronization of Base Stations

Intercell and interoperator interference problems are present in a TDD-CDMA system if the base stations are not synchronized. Synchronization is therefore desirable for TDD-CDMA. Synchronization accuracy must be at the symbol level but not at the chip level. It can be achieved with, for example, GPS receivers at the base station or by distributing a common clock with extra cabling. These methods increase the cost of the infrastructure.

7.2.3.4 TDD Power Pulsing

If fast power control frequency with open loop is desired to support higher mobile speeds, then short TDD frames must be used. The short transmission time in each direction results in the problems listed below:

- Audible interference from pulsed transmission both internally in the terminal and to the other equipment. Generated pulsing frequency in the middle of voice band will cause problems to small size speech terminal design where audio and transmission circuits are relatively close to each other. Achieving the needed isolation is costly and requires design considerations. At high power levels this may not be achievable at all.
- Base station synchronization requirements are tight and more overhead must be allocated for guard times and also for power ramps as EMC requirements limit the ramping speed.

- Fast ramping times set tighter requirements to the components (e.g., to the power amplifier).

Lower pulsing frequency, say, 100 Hz (i.e., a TDD frame of 10 ms), results in less audible pulsing but limits the maximum tolerable mobile speeds. In the TDD-CDMA in [8], the uplink slot and the downlink slot are both 0.625 ms, resulting in an audible interference at 800 Hz.

7.2.3.5   *Remedies Against TDD Interference Problems*

TDD interference issues has been studied in [9–14]. Based on these studies, the following remedies have been proposed to ensure proper system operation [9]:

- Frame level synchronization of TDD base stations within an operator's network;
- Frame level synchronization of the base stations of different TDD operators if the base station are near each other;
- Dynamic channel allocation;
- Intersystem and interfrequency handovers as means to reduce interference impacts;
- Co-siting of FDD and TDD macro cell base station should be avoided totally;
- Careful overall network planning.

7.3   WCDMA TDD SPECIFICATIONS

The air interface description in the following is based on the 3GPP wideband CDMA specifications as listed in Table 7.2. The TDD physical layer is specified in TS25 series of 3GPP specifications.

7.4   WCDMA TDD SYSTEM DESCRIPTION

The physical layer of WCDMA TDD is a combination of TDMA and CDMA. As in the FDD component, each TDD frame of length 10 ms is divided into 15 time slots. Each slot is comprised of several (maximum 16) orthogonal spreading symbols. The WCDMA FDD and TDD components have been harmonized. Due to different duplex methods, however, some inherent differences remain. These are highlighted in Table 7.3 [9].

Table 7.2
3GPP TDD Specifications

Specification number	Name	Scope
TS 25.221	Physical channels and mapping of transport channels onto physical channels (TDD)	Establishes the characteristics of the layer 1 transport channels and physical channels in the TDD mode, and specifies: • transport channels; • physical channels, structure, and contents; • mapping of transport channels onto the physical channels.
TS 25.222	Multiplexing and channel coding (TDD)	Describes multiplexing, channel coding, and interleaving in the TDD physical channel segmentation and mapping mode and specifies: • coding and multiplexing of transport channels; • channel coding alternatives; • coding for layer 1 control information; • the different interleavers; • rate matching; • physical channel segmentation and mapping.
TS 25.223	Spreading and modulation (TDD)	Establishes the characteristics of the spreading and modulation in the TDD mode, and specifies: • the spreading (channelization plus scrambling); • generation of channelization and scrambling codes; • generation of synchronization codes.
TS 25.224	Physical layer procedures (TDD)	Establishes the characteristics of the physical layer procedures in the TDD mode, and specifies: • cell syncronization; • timing advance; • idle mode tasks.
TS 25.225	Physical layer measurements (TDD)	Establishes the characteristics of the physical layer measurements in the TDD mode, and specifies: • the measurements performed by layer 1; • reporting of measurements to higher layers and network; • handover measurements and idle mode measurements.

7.4.1 Transport Channels

Similar to FDD, two types of transport channels exist for TDD — common and dedicated. In addition to the transport channels defined for FDD (see Chapter 6), there are three TDD specific transport channels:

• Opportunity-driven multiple access (ODMA) dedicated transport channel (ODCH);

• ODMA random access channel (ORACH);

• Uplink shared channel (USCH).

Table 7.3
WCDMA TDD and FDD Differences

	TDD	FDD
Multiple access method	TDMA, CDMA	CDMA
Spreading factor	4–512	1–16
Multirate concept	Multislot, multicode, and variable spreading factor	Multicode and variable spreading factor
Burst types	Traffic Random access Synchronization	No burst defined
Intrafrequency handover	Hard	Soft
Power control for dedicated channels	Uplink: open loop with rate of 100 or 200 Hz Downlink: closed loop with rate of equal or less than 800 Hz	Closed loop with rate of 1500 Hz
Channel allocation	Slow and fast DCA	No DCA required
Capacity allocation between uplink and downlink	5 MHz carrier divided between uplink and downlink. Downlink / uplink capacity can be adjusted between 2–14 out of 15 slots	5 MHz for uplink 5 MHz for downlink

7.4.2 Physical Channels

A TDD physical channel is defined by frequency, timeslot, channelization code, burst type, and radio frame allocation. The TDMA frame has a duration of 10 ms and is subdivided into 15 time slots (TS). The physical content of the time slots are the bursts of corresponding length. The TDD time slots can be allocated to either the uplink or the downlink to support asymmetric services. In any configuration at least one time slot has to be allocated for the downlink and at least one time slot has to be allocated for the uplink. Figure 7.3 depicts examples for multiple and single switching point configurations as well as for symmetric and asymmetric uplink/downlink allocations.

When operating ODMA at least one common time slot has to be allocated for the ORACH. If large quantities of information have to be transferred between ODMA nodes, then it is normal to use at least one time slot for the ODCH. As Figure 7.4 shows, any time slot in the TDD frame may potentially be used by the ODCH. A common time slot indicates a carrier-time slot combination that can be used for transmission and reception by a group of mobiles operating ODMA.

Multiple-switching-point configuration (symmetric DL/UL allocation)

Multiple-switching-point configuration (asymmetric DL/UL allocation)

Single-switching-point configuration (symmetric DL/UL allocation)

Single-switching-point configuration (asymmetric DL/UL allocation)

Figure 7.3 TDD frame structure examples (DL=downlink, UL=uplink). (*Source*: [15], reproduced with permission from ETSI.)

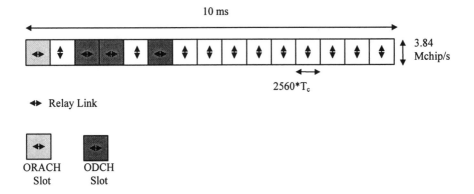

Figure 7.4 TDD frame structure example for ODMA operation. (*Source*: [15], reproduced with permission from ETSI.)

7.4.3 Dedicated Physical Channel (DPCH)

The DCH (or in the case of ODMA networks, the ODCH) are mapped onto the dedicated physical channel. Spreading is applied to the data part of the physical channels and consists of two operations.

Downlink physical channels shall use a spreading factor of 16. Multiple parallel physical channels transmitted using different channelization codes can be used to support higher data rates. Operation with a single code with spreading factor 1 is possible for the downlink physical channels.

The spreading factor of uplink physical channels can range from 16 down to 1. For multicode transmission a mobile station uses a maximum of two physical channels per time slot simultaneously.

7.4.4 Burst Types

Two types of bursts exist for dedicated physical channels: the burst type 1 and the burst type 2, as shown in Figures 7.5 and 7.6, respectively. Both bursts consist of two data symbol fields, a midamble, and a guard period. The bursts type 1 has a longer midamble of 512 chips than the burst type 2 with a midamble of 256 chips.

Because of the longer midamble, the burst type 1 is suited for the uplink, where up to 16 different channel impulse responses can be estimated. The burst type 2 can be used for the downlink and, if the bursts within a time slot are allocated to less than four users, also for the uplink.

Figure 7.5 Burst structure of the burst type 1 (GP = the guard period and CP = chip periods). (*Source*: [15], reproduced with permission from ETSI.)

Figure 7.6 Burst structure of the burst type 2. (*Source*: [15], reproduced with permission from ETSI.)

Both burst types 1 and 2 provide the possibility for transmission of transport format combination indication (TFCI) both in the uplink and the downlink. The transmission of TFCI is negotiated at call setup and can be renegotiated during the call. Additionally, for each allocated time slot it is signaled individually whether that time slot carries the TFCI or not. If a time slot contains the TFCI, then it is always transmitted using the first allocated channelization code in the time slot, according to the order in the higher layer allocation message.

Both burst types 1 and 2 for dedicated channels provide the possibility for transmission of transmit power control commands (TPC) in uplink. The transmission of TPC is negotiated at call setup and can be renegotiated during the call. If applied, transmission of TPC is done in the data parts of the traffic burst. Hence, the midamble structure and length is not changed. The TPC information is transmitted directly after the midamble.

For every user the TPC information is to be transmitted once per frame. If the TPC is applied, then it is always transmitted using the first allocated channelization code and the first allocated time slot, according to the order in the higher layer allocation message. The TPC is spread with the same spreading factor and spreading code as the data parts of the respective physical channel.

7.4.4.1 Training Sequences for Spread Bursts

Two options are being considered for the spreading. The training sequences presented here are common to both options. The training sequences (i.e., midambles) of different users active in the same time slot are time-shifted versions of one single periodic basic code. Different cells use different periodic basic codes (i.e., different midamble sets). In this way a joint channel estimation for the channel impulse responses of all active users within one time slot can be done by one single cyclic correlation. The different user-specific channel impulse response estimates are obtained sequentially in time at the output of the correlator.

7.4.5 Common Physical Channels

7.4.5.1 Primary Common Control Physical Channel (P-CCPCH)

The BCH is mapped onto the P-CCPCH. The position (time slot/code) of the P-CCPCH is known from the physical synchronization channel (PSCH). The P-CCPCH uses fixed spreading with a spreading factor of 16. The burst type 1 is used for the P-CCPCH. No TFCI is applied for the P-CCPCH. Block STTD antenna diversity can be optionally applied for the P-CCPCH.

7.4.5.2 Secondary Common Control Physical Channel (S-CCPCH)

PCH and FACH are mapped onto one or more S-CCPCH. In this way the capacity of PCH and FACH can be adapted to the different requirements. The S-CCPCH uses fixed spreading with a spreading factor of 16. The burst types 1 or 2 are used for the S-CCPCHs. TFCI may be applied for S-CCPCHs.

7.4.5.3   Physical Random Access Channel (PRACH)

The RACH (or in the case of ODMA networks, the ORACH) are mapped onto one or more uplink physical random access channels (PRACH). In this way the capacity of RACH and ORACH can be flexibly scaled depending on the operators need.

Figure 7.7 shows the structure of the PRACH burst. The PRACH uses either spreading factor of 16 or 8. The set of admissible spreading codes for use on the PRACH and the associated spreading factors are broadcast on the BCH (within the RACH configuration parameters on the BCH).

The mobile stations send the uplink access bursts randomly in the PRACH. The PRACH burst consists of two data symbol fields, a midamble, and a guard period. The second data symbol field is shorter than the first symbol data field by 96 chips in order to provide additional guard time at the end of the PRACH time slot.

Figure 7.7   PRACH burst. (*Source*: [15], reproduced with permission from ETSI.)

The training sequences, i.e., midambles, of different users active in the same time slot, are time-shifted versions of a single periodic basic code. The basic midamble codes used for PRACH bursts are the same as for burst type 1. The necessary time shifts are obtained by choosing either all $k = 1, 2, 3, ..., K'$ (for cells with small radius) or uneven $k = 1, 3, 5, ... , \leq K'$ (for cells with large radius). Different cells use different periodic basic codes (i.e., different midamble sets).

For cells with large radius additional midambles may be derived from the time-inverted basic midamble sequence. Thus, the second basic midamble code m_2 is the time inverted version of basic midamble code m_1.

In this way, a joint channel estimation for the channel impulse responses of all active users within one time slot can be performed by a maximum of two cyclic correlations (in cells with small radius, a single cyclic correlator suffices). The different user-specific channel impulse response estimates are obtained sequentially in time at the output of the cyclic correlators.

7.4.5.4   Physical Synchronization Channel (PSCH)

In TDD mode, the code group of a cell can be derived from the synchronization channel. Additional information, received from higher layers on SCH transport channel, is also transmitted to the mobile station in PSCH in case 3 from below. In order not to limit the uplink/downlink asymmetry, the PSCH is mapped on one or two downlink slots per frame only.

There are three cases of PSCH and P-CCPCH allocation as follows:

- Case 1: PSCH and P-CCPCH allocated in TS#k, $k = 0, ..., 14$.
- Case 2: PSCH allocated in two TS: TS#k and TS#k+8, $k = 0, ..., 6$; P-CCPCH allocated in TS#k.
- Case 3: PSCH allocated in two TS, TS#k and TS#k+8, $k = 0, ..., 6$, and the P-CCPCH allocated in TS#i, $i = 0, ..., 6$, pointed by PSCH. Pointing is determined via the SCH from the higher layers.

These three cases are addressed by higher layers using the SCCH in TDD mode. The position of PSCH (value of k) in frame can change on a long-term basis in any case. Because of this PSCH scheme, the position of PCCPCH is known from the PSCH. Figure 7.8 shows an example for transmission of PSCH, $k = 0$, of case 2 or case 3.

Figure 7.8 Scheme for PSCH consisting of one primary sequence C_p and $N = 3$ parallel secondary sequences in slot k and k+8. (*Source*: [15], reproduced with permission from ETSI.)

As depicted in Figure 7.8 the PSCH consists of a primary and three secondary code sequences with 256 chips length. The secondary codes are transmitted either in the I channel or the Q channel, depending on the code group.

Due to mobile interference, it is mandatory for public TDD systems to keep synchronization between base stations. As a consequence of this, a capture effect concerning PSCH can arise. The time-offset t_{offset} enables the system to overcome the capture effect.

7.4.5.5 Physical Uplink Shared Channel (PUSCH)

For PUSCH the burst structure of DPCH is used. User-specific physical layer parameters like power control, timing advance, or directive antenna settings are derived from the associated channel (FACH or DCH). PUSCH provides the possibility for transmission of TFCI in uplink.

7.4.5.6 Physical Downlink Shared Channel (PDSCH)

For PDSCH the same burst structure as for DPCH is used. User-specific physical layer parameters like power control or directive antenna settings are derived from the associated channel (FACH or DCH). PDSCH provides the possibility for transmission of TFCI in downlink. To indicate to the mobile station that there is data to decode on the DSCH, three signaling methods are available:

- Using the TFCI field of the associated channel or PDSCH;
- Using on the DSCH user-specific midamble derived from the set of midambles used for that cell;
- Using higher layer signaling.

7.4.5.7 Page Indicator Channel (PICH)

PICH is used to carry the page indicators. The PICH substitutes one or more paging subchannels that are mapped on a S-CCPCH. The page indicator indicates a paging message for one or more mobile stations that are associated with it.

7.4.6 Mapping of Transport Channels on Physical Channels

Figure 7.9 shows the mapping between transport channels and physical resources.

7.4.7 Power Control

Power control for the TDD mode is applied to limit the interference level within the system, thereby reducing the intercell interference level and the power consumption in the mobile station. Power control is performed on a frame basis (i.e., every 10 ms).

Based on the signal-to-interference ratio of the received signal, the transmitter is able to know the fast fading of the multipath channel. This assumption holds only if the frame length of 10 ms is shorter than the coherent time of the channel. The uplink power control uses an open loop power control. The mobile station can use this information and set its transmission power to the target SIR in order to achieve the suitable quality within the radio access network.

Transport channels	Physical channels
DCH	Dedicated physical channel (DPCH)
ODCH*	
BCH	Primary common control physical channel (P-CCPCH)
FACH	Secondary common control physical channel (S-CCPCH)
PCH	
RACH	Physical random access channel (PRACH)
ORACH*	
SCH	Physical synchronization channel (PSCH)
USCH	Physical uplink shared channel (PUSCH)
DSCH	Physical downlink shared channel (PDSCH)
	Page indicator channel (PICH)

* in case of ODMA networks

Figure 7.9 Mapping of transport channel to physical channel.

For the downlink closed loop power control is used so that the mobile station can compare the estimated received SIR with the target SIR. This information can then be transferred to the base station, which accordingly adapts its transmitted power for that particular mobile station.

7.4.8 Handover

In TDD mode, only hard handover is used. The handover is based on measurements, which the mobile station performs in its idle time slots. From the network the mobile station receives a list of neighboring cells to measure. The list contains the characteristics of each of these cells like frequency, codes, and midambles in order to detect and identify the cells as well as to facilitate the measurements.

The handover procedure itself and the measurements to be performed depend on the handover case. For the handover from TDD to another TDD cell, the mobile station measures the receive power of the neighboring cells' beacons. In its idle time slots it performs a correlation in order to find and measure the midamble of a neighboring cell. In general, neighboring cells of a TDD system will be time synchronized, so that the UE knows at what position to search for midambles, and the search window can be kept small. If the neighboring cells are unsynchronized, a more extensive search has to be performed. In this case, the synchronization codes can be used first to get the timing in the neighboring cell in a similar way as for initial cell search. The measured receive powers provide information about the relative distance and potential quality in the neighboring cells and can be used in the handover algorithm. For a handover to a neighboring FDD cell, the E_c/N_0 of the FDD CPICH, the common pilot channel, is measured. This gives an indication of the quality and SIR to be expected in the new cell. In case of a handover to GSM the received power level of the GSM broadcast channel is measured and in certain intervals some information on the GSM synchronization channel is decoded in order to verify the power level measurements. When the network decides that a handover is necessary, it exchanges information with the target cell and then sends a command to the terminal to perform a handover to the new cell. This command contains all information necessary for the terminal in the new cell. The terminal does not have to decode information in the new cell while camping on the old one. In order to make the handover seamless, the terminal keeps its frame numbering scheme for the uplink also in the new cell.

7.4.9 Dynamic Channel Allocation

In UTRA TDD, the TDMA component enables an effective strategy of interference avoidance, which can be implemented by dynamic channel allocation (DCA). It allows the deployment of TDD for coordinated as well as for uncoordinated operation, since interference in certain time slots, originating from neighboring cells, can be efficiently avoided. With bursty high-peak rate packet data, power control cannot converge. In this situation, DCA can be used to shift the packet data traffic to certain time slots, and hence, it does not degrade the performance of circuit switched and signaling bearers.

7.4.10 Joint Detection

It is an essential part of the TDD mode that all spreading codes that are simultaneously active during the same time slot are separated by joint detection (JD) on the uplink as well as the downlink [16]. JD technique can be used to simultaneously extract the signal of each active user. The framing structure of a TDMA component allows for synchronous (joint) detection of the CDMA signals, which can significantly enhance the system capacity. JD can make use of multiple access interference (MAI) characteristics to eliminate it. It can be used in order to avoid excessive intracell MAI and reduce the intersymbol interference introduced by the channel. Therefore, the proposed system can cope with power differences of several decibels, which relaxes the requirements on the near-far effect. Thus, only low power control accuracy is required.

REFERENCES

[1] Klein, A., R. Pirhonen, J. Sköld, and R. Suoranta, "FRAMES Multiple Access Mode 1 – Wideband TDMA with and Without Spreading," *Proceedings of Personal, Indoor and Mobile Radio Communications (PIMRC'97)*, Vol. 1, September 1997, Helsinki, Finland, pp. 37–41.

[2] Baier, P. W., T. Bing, and A. Klein, "TD-CDMA," in *Third Generation Mobile Communication Systems*, R. Prasad, W. Mohr, and W. Konhauser (eds.), Boston: Artech House, 2000, pp.25–72.

[3]   Litva, J., T., Lo, *Digital Beamforming in Wireless Communications*, Norwood, Massachusetts: Artech House, 1996.

[4]   Povey, G., "Frequency and Time Division Duplex Techniques for CDMA Cellular Radio," *Proceedings of 3rd International Symposium on Spread Spectrum Techniques & Applications (ISSSTA'94)*, Oulu, Finland, 1994, pp. 309–313.

[5]   Povey, G., "Capacity of a Cellular Time Division Duplex CDMA System," *IEE Proc.-Commun.*, Vol. 141, No. 5, October 1994, pp. 351–356.

[6]   Esmailzadeh, R., and M. Nakagawa, "Pre-RAKE Diversity Combination for Direct Sequence Spread Spectrum Mobile Communications Systems," *IEICE Transactions on Communications*, Vol. E76-B, No. 8, August 1993, pp. 1008–1015.

[7]   Esmailzadeh, R., M. Nakagawa, and E. Sourour, "Time-Division Duplex CDMA Communications," *IEEE Personal Communications*, April 1997, pp. 51–56.

[8]   Hayashi, M., K. Miya, O. Kato, and K. Homma, "CDMA/TDD Cellular Systems Utilizing a Base-Station Based Diversity Scheme," *Proceedings of 45th Vehicular Technology Conference (VTC'95)*, July 1995, Chicago, Illinois, Vol. 2, pp. 799–803.

[9]   Holma, H., and A. Toskala, (eds.), *WCDMA for UMTS*, Chichester, England: John Wiley & Sons, 2000.

[10]   Holma, H., G. Povey, and A. Toskala, "Evaluation of Interference Between Uplink and Downlink in UTRA/TDD," *Proceedings of VTC'99 Fall*, Amsterdam, Netherlands, September 19–22, 1999, pp. 2616–2620.

[12]   Heikkinen, S., "System Simulations for UMTS TDD and FDD Modes," M.Sc. Thesis, University of Oulu, Finland / University of Edinburgh, Scotland, 1999.

[13]   Haas, H., and G. Povey, "The Effect of Adjacent Channel Interference on Capacity in a Hybrid TDMA/CDMA-TDD System Using UTRA-TDD Parameters," *Proceedings of VTC'99 Fall*, Amsterdam, Netherlands, September 19–22, 1999, pp. 2616–2620.

[14]   Haas, H., S. McLaughlin, and G. Povey, "The Effects of Interference Between the TDD and FDD Mode in UMTS at the Boundary of 1920 MHz," Proceedings of ISSSTA 2000, New Jersey.

[15]   TS 25.221, "Physical Channels and Mapping of Transport Channels onto Physical Channels (TDD)," 3GPP Technical Specification, Version 3.2.0, March 2000.[1]

[16]   Klein, A., G. K. Kaleh, and P. W. Baier, "Zero Forcing and Minimum-Mean-Square-Error Equalization for Multiuser Detection in Code Division Multiple Access Channels," *IEEE Transactions on Vehicular Technology*, May 1996, Vol. 45, pp. 276–287.

---

[1] The present technical specifications are the property of 3GPP. No reproduction shall be made without ETSI authorization. The original version of these 3GPP technical specifications can be obtained from the publication office of ETSI.

Chapter 8

cdma2000

8.1   INTRODUCTION

This chapter presents the cdma2000 air interface developed by the third-generation partnership project two (3GPP2). Originally cdma2000 specification work was started in TIA TR 45.5; however, the work was moved to 3GPP2 in order to gather global expertise.

Table 8.1 lists the parameters of cdma2000. The basic chip rate of the system is 1.2288 Mcps and bandwidth 1.25 MHz. This mode of cdma2000 is referred as cdma2000 1x. The second specified mode is cdma2000 3x, also referred to as the multicarrier (MC) mode. As illustrated in Figure 8.1, in the downlink there are three consecutive 1.25-MHz carriers where each carrier has a chip rate of 1.2288 Mcps. The multicarrier approach maintains orthogonality between the cdma2000 and IS-95 carriers [1]. In the downlink this is more important because the power control cannot balance the interfering powers between different layers, as it can in the uplink. In the uplink of the 3x mode, direct spreading is used. The chip rate is 3.6864 Mcps and nominal carrier bandwith is 3.75 MHz. A hybrid 3x/1x configuration can also be used [2]. In this hybrid mode, three 1.25-MHz carriers are transmitted in the downlink and one 1.25 MHz carrier in the uplink.

In contrast to WCDMA, cdma2000 is synchronous system and thus timing of all cdma2000, as well as IS-95, networks are aligned within a few microseconds throughout the world using signals from, for example, Global Positioning System (GPS).

The frame length is 20 ms and each frame is divided into 16 slots (power control groups). cdma2000 defines also a 5-ms frame structure for signaling bursts. Spreading factors range from 256 to 4 in the downlink. For separating channels from the same source, orthogonal variable spreading factor Walsh channelization codes are used. In

225

the downlink, a single code of length 2^{15} chips is used for scrambling and different cells are separated by time shifts (PN offsets) of the same code. In the uplink, a code of length 2^{42}-1 chips is used.

For the channel coding, three options are supported: convolutional coding, turbo coding, or no channel coding. Channel coding selection is indicated by upper layers. Bit interleaving is used to randomize transmission errors. The modulation scheme is QPSK.

Table 8.1
cdma2000 Parameter Summary

	cdma2000 1x	cdma2000 3x
Timing	Synchronous (timing derived, for example, from GPS)	
Channel bandwidth	1.25 MHz	1.25 MHz
Downlink RF channel structure	Direct spread	Multicarrier
Chip rate	1.2288 Mcps (downlink and uplink)	1.2288 Mcps (downlink carriers) 3.6864 Mcps (uplink)
Roll-off factor	Similar to IS-95	
Frame length	20 ms / 5ms option for signalling	
Spreading modulation	Balanced QPSK (downlink) Dual-channel QPSK (uplink) with hybrid phase shift keying (HPSK) Complex spreading circuit	
Data modulation	QPSK (downlink) BPSK (uplink)	
Coherent detection	Pilot time multiplexed with PC and EIB (uplink) Common continuous pilot channel and auxiliary pilot (downlink)	
Channel multiplexing in uplink	Control, pilot, fundamental and supplemental code multiplexed I&Q multiplexing for data and control channels	
Multirate	Variable spreading and multicode	
Spreading factors	4 – 256	
Power control	Open loop and fast closed loop (800 Hz, higher rates under study)	
Spreading (downlink)	Variable length orthogonal Walsh sequences for channel separation, M-sequence 2^{15} (same sequence with time shift utilized in different cells, different sequence in I&Q channel) Quasi-orthogonal functions (QOF) in code limited case	
Spreading (uplink)	Variable length orthogonal Walsh sequences for channel separation, M-sequence 2^{15} (same for all users different sequences in I&Q channels), M-sequence $2^{41}-1$ for user separation (different time shifts for different users)	
Handover	Soft handover Interfrequency handover	
Transmit diversity	Orthogonal transmit diversity Space-time spreading	Demultiplexing of carriers and transmit using different antennas

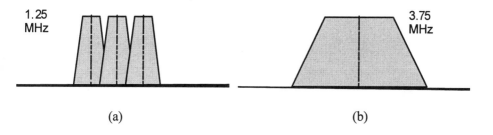

Figure 8.1   Illustration of (a) multicarrier downlink and (b) direct spread uplink.

8.2   CDMA2000 SPECIFICATIONS

The air interface description in the following is based on the 3GPP2 cdma2000 specifications as listed in Table 8.2 [2]. The corresponding specification number of TIA is also listed.

cdma2000 standards are being developed in several releases. The first release, called Release 0, was published in 1999, and it provided physical layer specification for cdma2000 1X and 3X using the IS-95 common channels. Release A was published in early 2000 and provides signaling support using a new set of common channels. Release B is planned to be published at the end of 2000 and is discussed in Section 8.4.

Table 8.2
cdma2000 Air Interface Specifications

3GPP2	TIA	Title
C.S0001	TIA/EIA/IS-2000-1	cdm2000 – Introduction
C.S0002	TIA/EIA/IS-2000-2	cdm2000 – Physical Layer
C.S0003	TIA/EIA/IS-2000-3	cdm2000 – MAC
C.S0004	TIA/EIA/IS-2000-4	cdm2000 – Layer 2 MAC
C.S0005	TIA/EIA/IS-2000-5	cdm2000 – Layer 3
C.S0006	TIA/EIA/IS-2000-6	cdm2000 – Analog
C.S00010	TIA/EIA/97	Mobile station minimum performance
C.S00011	TIA/EIA/98	Base station minimum performance

8.3 CDMA2000 SYSTEM DESCRIPTION

8.3.1 Physical Channels

Table 8.3 shows the physical channels of cdma2000. Their fuctionality is discussed in the following subsections.

Table 8.3
cdma2000 Physical Channels

Channel name	Physical channel
F/R-FCH	Forward/reverse fundamental channel
F/R-DCCH	Forward/reverse dedicated control channel
F/R-SCCH	Forward/reverse supplemental code channel
F/R-SCH	Forward/reverse supplemental channel
F-PCH	Paging channel
F-QPCH	Quick paging channel
F/R-CCCH	Forward/reverse common control channel
F/R-PICH	Forward/reverse pilot channel
F-APICH	Dedicated auxiliary pilot channel
F-TDPICH	Transmit diversity pilot channel
F-ATDPICH	Auxiliary transmit diversity pilot channel
F-SYNCH	Sync channel
F-CPCCH	Common power control channel
F-CACH	Common assignment channel
R-ACH	Reverse access channel
R-EACH	Enhanced access channel
F-BCH	Broadcast channel

8.3.1.1 Downlink Physical Channels

Downlink has six different dedicated channels and three common control channels. Similar to the uplink, the fundamental and supplemental channels carry user data and the dedicated control channel carries control messages. The dedicated control channel contains power control bits and rate information. The synchronization channel is used by the mobile stations to acquire initial time synchronization. One or more paging channels are used for paging the mobiles. The pilot channel provides a reference signal for coherent detection, cell acquisition, and handover.

In the downlink, the F-PICH is a common pilot channel, which is used as a reference signal for coherent detection when adaptive antennas are not employed. The pilot channel is similar to IS-95 (i.e., it is comprised of a long PN-code and Walsh sequence number 0). When adaptive antennas are used, auxiliary pilot is used as a reference signal for coherent detection. Code multiplexed auxiliary pilots are generated by assigning a different orthogonal code to each auxiliary pilot. This approach reduces the number of orthogonal codes available for the traffic channels. This limitation is alleviated by expanding the size of the orthogonal code set used for the auxiliary pilots. Because a pilot signal is not modulated by data, the pilot orthogonal code length can be extended, thereby yielding an increased number of available codes, which can be used as additional pilots.

Originally, there were two alternatives for the downlink modulation in the nominal 3.75-MHz bandwidth: direct spread and multicarrier. The multicarrier transmission principle is illustrated in Figure 8.2. A performance comparison of direct spread and multicarrier can be found in [3].

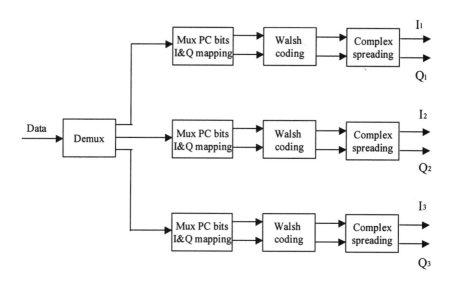

Figure 8.2   Multicarrier downlink.

8.3.1.2   Uplink Physical Channels

In the uplink, there are six different dedicated channels. The fundamental and supplemental channels carry user data. A dedicated control channel, with a frame length of 5 or 20 ms, carries control information such as measurement data; and a pilot channel is used as a reference signal for coherent detection. The pilot channel also carries time multiplexed power control symbols. Figure 8.3 illustrates the different uplink dedicated channels separated by Walsh codes.

The reverse access channel (R-ACH) and the reverse common control channel (R-CCCH) are common channels used for communication of layer 3 and MAC layer messages. The R-ACH is used for initial access, while the R-CCCH is used for fast packet access.

The fundamental channel conveys voice, signaling, and low rate data. Basically, it will operate at a low frame error rate (FER) of around 1%. The fundamental channel supports basic rates of 9.6 Kbps and 14.4 Kbps and their corresponding subrates (i.e., Rate Set 1 and 2 of IS-95). The fundamental channel will always operate in soft handover mode. The fundamental channel does not operate in a scheduled manner; thus permitting the mobile station to transmit acknowledgments or short packets without scheduling. This reduces delay and the processing load due to scheduling [4]. Its main difference compared to the IS-95 voice channel is that discontinuous transmission is implemented using repetition coding rather than gated transmission.

The supplemental channel provides high data rates. The uplink supports one or two supplemental channels. If only one supplemental channel is transmitted, then the Walsh code (+-) is used on the first supplemental channel, and if two supplemental channels are transmitted then the Walsh code (+-+-) is used. A repetition scheme is used for variable data rates on the supplemental channel.

8.3.2 Spreading

On the downlink, the cell separation for cdma2000 is performed by two M-sequences of length 2^{15}, one for the I channel and one for the Q channel, which are phase shifted by PN-offset for different cells. Thus, during the cell search process only these sequences need to be searched. Because there is only a limited number of PN-offsets, they need to

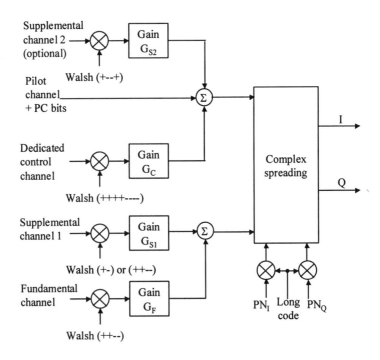

Figure 8.3   The uplink dedicated channel structure.

be planned in order to avoid PN-confusion [5]. In the uplink, user separation is performed by different phase shifts of M-sequence of length 2^{41}. The channel separation is performed using variable spreading factor Walsh sequences, which are orthogonal to each other. Fundamental and supplemental channels are transmitted with the multicode principle. The variable spreading factor scheme is used for higher data rates in the supplemental channel.

Similar to WCDMA, complex spreading is used. In the uplink, a techniques called hybrid phase shift keying (HPSK) provides one chip delay and decimation by 2 in the Q branch, thus leading to a reduction of peak-to-average statistic of the uplink signal in the order of 1 dB [2].

8.3.3 Multirate

The fundamental and supplemental channels can have different coding and interleaving schemes. In the downlink, high bit rate services with different QoS requirements are code multiplexed into supplemental channels, as illustrated in Figure 8.4. In the uplink, one or two supplemental channels can be transmitted. The user data frame length of cdma2000 is 20 ms. For the transmission of control information, 5- and 20-ms frames can be used on the fundamental channel. Also on the fundamental channel, a convolutional code with constraint length of 9 is used. On supplemental channels, a convolutional code is used up to 14.4 Kbps. For higher rates, turbo codes with constraint length 4 and rate 1/4 are preferred. Rate matching is performed by puncturing, symbol repetition, and sequence repetition.

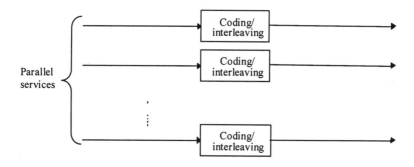

Figure 8.4   Principle of code multiplexing.

8.3.4 Packet Data

cdma2000 also uses the slotted Aloha principle for packet data transmission. Instead of fixed transmission power, however, it increases the transmission power for the random access burst after an unsuccessful access attempt. When the mobile station has been allocated a traffic channel, it can transmit without scheduling up to a predefined bit rate. If the transmission rate exceeds the defined rate, a new access request has to be made. When the mobile station stops transmitting, it releases the traffic channel but not the dedicated control channel. After a while, it also releases the dedicated control channel but maintains the link layer and network layer connections in order to shorten the channel setup time when new data needs to be transmitted. Short data bursts can be transmitted over a common traffic channel in which a simple ARQ is used to improve the error rate performance.

8.3.5 Handover

It is expected that soft handover of the fundamental channel will operate similarly to the soft handover in IS-95. In IS-95, the active set is the set of base stations transmitting to the mobile station. For the supplemental channel, the active set can be a subset of the active set for the fundamental channel. This has two advantages. First, when diversity is not needed to counter fading, it is preferable to transmit from fewer base stations. This increases the overall downlink capacity. For stationary conditions, an optimal policy is to transmit only from one base station–the base station that would radiate the smallest amount of downlink power. Second, for packet operation, the control processes can also be substantially simplified if the supplemental channel is not in soft handover. Maintaining the fundamental channel in soft handover, however, provides the ability to reliably signal the preferred base station to transmit the supplemental channel when channel conditions change [4].

8.3.6 Transmit Diversity

The downlink performance can be improved by transmit diversity. For direct spread CDMA schemes, this can be performed by splitting the data stream and spreading the two streams using orthogonal sequences. For multicarrier CDMA, the different carriers can be mapped into different antennas.

8.4 CDMA2000 1X EVOLUTION–1XEV

In further development of cdma2000, the focus is on the evolution of the 1X mode, termed 1xEV. The main enhancements relate to increase of maximum bit rate. The 1xEV comprises two phases; phase 1 is termed 1xEV-DO (data only), and phase 2 1xEV-DV (data and voice).

8.4.1 1xEV-DO

1xEV-DO is based on the proposal from Qualcomm, originally called high data rate (HDR). It is a packet data system that uses a dedicated 1.25-MHz channel to provide data-only service. It provides a peak data rate of 2.4 Mbps on the forward link and 307

Kbps on the reverse link. Average throughput on a loaded sector is an estimated 600 Kbps on the forward link and 220 Kbps on the reverse link. 1xEV-DO is a time division multiple access technology where users are allocated some number of time slots based on the required QoS and current channel conditions. Each user can access full capacity of the cell during assigned time slots. The base station transmits at full power because there is no downlink power control. A dynamically assigned data rate can be adjusted every 1.67 ms.

8.4.2 1xEV-DV

Currently, there are several proposals pending for phase 2 of 1xEV. One of them, 1XTREME, was originally proposed by Nokia and Motorola. It is a set of modulation and coding enhancements to cdma2000 1x. The main features of 1XTREME are the following:

- Increases data rates on the order of 5 Mbps;
- Supports mixed voice and data;
- Supports legacy mobiles;
- Can be deployed and an upgrade to cdma2000 1X;
- Can be deployed as a packet data-only solution.

REFERENCES

[1] Chia, S., "Will cdmaOne Be the Third Choice?" *CDMA Spectrum*, No. 2, September 1997, pp. 30–34.

[2] Willenegger, S., "Cdma2000 Physical Layer: An Overview," *Journal of Communications and Networks*, Vol. 2, No. 1, March 2000, pp. 5–17.

[3] Jalali, A., and A. Gutierrez, "Performance Comparison of Alternative Wideband CDMA Systems," *Proceedings of ICC'98*, Atlanta, Georgia, June 1998, pp. 38–42.

[4] Tiedemann, E. G., Jr., Y.-C. Jou, and J. P. Odenwalder, "The Evolution of IS-95 to a Third Generation System and to the IMT-2000 Era," *Proceedings of ACTS Summit 1997*, Aalborg, Denmark, October 1997, pp. 924–929.

[5] Yang, J., D. Bao, and M. Ali, "PN Offset Planning in IS-95 Based CDMA System" *Proceedings of VTC'97*, Vol. 3, Phoenix, Arizona, May 1997, pp. 1435–1439.

Chapter 9

PERFORMANCE ANALYSIS

9.1 INTRODUCTION

In this chapter, the performance of a wideband CDMA cellular system in different radio environments is investigated. Because absolute performance results are valid only for the used assumptions and models, the intention of this chapter is to study relative performance figures and to analyze the impact of different techniques on wideband CDMA performance.

In Section 9.2, the effect of diversity, including multipath diversity, antenna and polarization diversity, macro diversity, and time diversity, on wideband CDMA performance is investigated. In Section 9.3, the principles of the wideband CDMA link and system simulations to obtain spectrum efficiency are explained. Section 9.4 investigates the impact of fast power control on the system's performance. In particular, the effect of fast power control on the average transmission powers is studied. In addition, the impact of imperfect power control on CDMA cellular capacity is discussed.

Section 9.5 investigates spectrum efficiency for different performance enhancement techniques and radio environments. In the uplink, the impact of MUD with different efficiencies compared to a conventional receiver is simulated. Furthermore, analytical results are presented. As multipath partially destroys the orthogonality of orthogonal Walsh codes used in the downlink, the impact of multipath delay spread of different radio environments on the orthogonality factor is studied. The uplink and downlink spectrum efficiencies for service bit rates of 12 and 144 Kbps and 2 Mbps are simulated. Finally, the spectrum efficiency differences between the uplink and the downlink are analyzed.

Section 9.6 presents the principles of cell range calculation. Furthermore, the impact of system load on CDMA range is explained. Multiuser detection can increase range by canceling intracell interference. Range extension from MUD in the uplink is calculated analytically, and the impact of different MUD efficiencies is studied. If the gain from MUD is not used for range extension, it can reduce the mobile station

235

transmission power. The decrease in the transmission power of a mobile station as a function of multiuser detection efficiency is calculated analytically.

9.2 DIVERSITY

Diversity is essential in achieving good performance in a fading channel. With wideband CDMA, the means for achieving diversity are multipath, antenna, polarization, macro, and time diversity. The use of frequency hopping typical of TDMA systems, like GSM, is not employed with wideband CDMA to gain frequency diversity.

9.2.1 Gain from Multipath Diversity

The uncoded bit error rate as a function of SNR and the number of diversity branches is given as [1]

$$P_b = \left[\frac{1}{2}(1-\mu)^L \sum_{k=0}^{L-1} \binom{L-1+k}{k} \left[\frac{1}{2}(1+\mu)\right]^k\right]$$ (9.1)

where L is the number of diversity branches and

$$\mu = \sqrt{\frac{\bar{\gamma}_c}{1+\bar{\gamma}_c}}$$ (9.2)

where $\bar{\gamma}_c$ is the average SNR per diversity branch. It is assumed that all diversity branches have equal SNR on average.

The importance of diversity can be seen in Figure 9.1, where the average uncoded BER is plotted as a function of the total SNR in a Rayleigh fading channel for PSK modulation with Lth order diversity. Ideal coherent maximal ratio combining of uncorrelated diversity branches is assumed. Maximal ratio combining requires estimation of the amplitudes and phases of the diversity components. When the number of multipath components increases, the SNR per each multipath gets lower, and it becomes more difficult to perform coherent combining in a RAKE receiver. Also, when the diversity branches are correlated, the diversity gain is lowered.

9.2.2 Multipath Diversity in Different Radio Environments

One of the key factors that differentiate the third-generation wideband CDMA from the second-generation narrowband CDMA is the wider bandwidth. In addition to the ability to provide wideband services, the increased bandwidth makes it possible to resolve more multipath components in a mobile radio channel. If the transmission bandwidth is wider than the coherence bandwidth of the channel, the receiver is able to separate

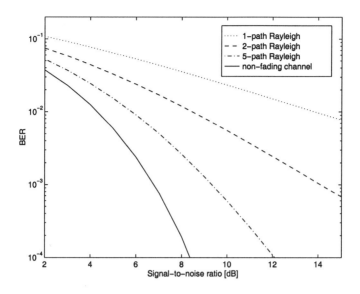

Figure 9.1   Uncoded BER with diversity assuming ideal coherent combining of diversity branches.

multipath components. This brings more diversity and higher capacity, along with power control.

We already discussed the effect of bandwidth on the diversity order in Chapter 4, where ATDMA and CODIT channel models were used. In this section, we use the ITU channel models presented in [2–3]. ITU channel models are based on the JTC (Joint Technical Committee for the U.S. PCS standardization) and ATDMA models [2]. Table 9.1 presents the coherence bandwidths for different environments. Also, the theoretical number of separable paths with chip rates of 1.2288, 4.096, and 8.192 Mcps is calculated.

The coherence bandwidth is calculated as in [4]:

$$(\Delta f)_c = \frac{1}{2\pi S} \tag{9.3}$$

where S is the rms delay spread defined as in [5]:

$$S = \sqrt{\frac{\int_0^\infty (\tau - D)^2 P(\tau)d\tau}{\int_0^\infty P(\tau)d\tau}} \tag{9.4}$$

and $P(\tau)$ is the power delay profile, and D is the average delay defined as

$$D = \frac{\int\limits_0^\infty \tau P(\tau)d\tau}{\int\limits_0^\infty P(\tau)d\tau} \qquad (9.5)$$

The maximum number of rays (paths) in the ITU channel models is six.

Table 9.1
Coherence Bandwidths of ITU Channel Models

ITU Channel model	Number of paths in channel model	Coherence bandwidth	Maximum number of separable paths		
			1.25 Mcps	4.0 Mcps	8.0 Mcps
Indoor A	6	4.3 MHz	1	1	2
Indoor B	6	1.6 MHz	1	3	6
Outdoor to indoor A	4	3.5 MHz	1	2	3
Outdoor to indoor B	6	250 kHz	5	17	33
Vehicular A	6	430 kHz	3	10	20
Vehicular B	6	53 kHz	24	78	155

Source: [6].

With a chip rate of 1.2288 Mcps, only very limited multipath diversity is available in indoor and in outdoor-to-indoor environments, for ITU channel models.

In Table 9.2, the number of useful multipath components is presented with three different environments measured in Australia [7]. The measured results clearly indicate the superiority of the wideband CDMA compared to narrowband CDMA in terms of achieving diversity gain with the RAKE receiver. With the 1.25-MHz bandwidth only one or two multipath components are available, where a single finger collects most of the energy. Very limited multipath diversity can be achieved with 1.25 MHz. With the 4-MHz bandwidth the achieved number of multipath components is in all measurements either two or four, which means a clearly achievable diversity gain.

If the autocorrelation properties of the spreading codes are not ideal, multipath components interfere with each other and degrade the multipath diversity gain. This effect will be noticeable with very low spreading factors (i.e., with high bit rates).

Table 9.2
Number of Useful Multipath Components

	1.25 MHz	4.0 MHz	8.0 MHz
Melbourne	1	2	2
Adelaide	1	2	3
Sydney	2	4	6

Source: [7].

9.2.3   Antenna Diversity

If antenna diversity branches are close to each other, they are correlated and diversity gain is smaller. Therefore, antenna diversity is best suited for base stations and for large sized mobile stations.

Receiver antenna diversity can be used to average out receiver noise in addition to providing diversity against fading and interference. At base stations, receiver antenna diversity can be used to increase cell coverage by increasing the noise limited uplink range.

Transmission antenna diversity can be used to generate multipath diversity in such environments where only limited multipath diversity is available (i.e., in indoor and macrocell). Transmission diversity can be obtained in several ways (see Section 5.11.2.1). The same signal can be transmitted from more than one antenna with a delay that the RAKE receiver can separate. Alternatively, the data can be divided between transmission antennas without any delays, thus maintaining the orthogonality of the diversity signals. Transmission antenna diversity is best suited for the downlink transmission in the base station together with the uplink diversity reception.

9.2.4   Polarization Diversity

Different polarization directions may experience different fading. Especially in indoor environments, polarization directions have been shown to be nearly uncorrelated, thus providing diversity [8]. The advantage of polarization diversity over antenna diversity is that polarization diversity does not require separation between antennas, and thus, it could be applied to small sized equipment as well.

9.2.5   Macro Diversity

With CDMA systems, the use of macro diversity (i.e., soft handover) is essential for achieving reasonable system performance because of frequency reuse 1 and fast power control. If the mobile station is not connected to the base station to which the attenuation is the lowest, unnecessary interference is generated in adjacent cells. In the uplink, the macro diversity effects are only positive, because the more base stations that try to detect the signals, the higher the probability is for at least one to succeed. In the uplink direction, the detection process itself does not utilize the information from the other base stations receiving the same signal, but the diversity is selection diversity, where the best frame is selected in the network based on the frame error indication from a CRC check. The selection can be done in the base station controller (BSC) or in some other network element, depending on the implementation.

In the downlink, macro diversity is different as the transmission now originates from several sources and diversity reception is handled by one receiver unit in the mobile station. All extra transmissions contribute to the interference. Capacity improvement is based on a similar principle as with a RAKE receiver in a multipath channel, where the received power level fluctuations tend to decrease as the number of separable paths increases. With downlink macro diversity, the RAKE receiver capability to gain from the extra diversity depends also on the number of available RAKE fingers. If a RAKE receiver is not able to collect enough energy from the transmissions from two or, in some cases, three base stations due to a limited number of RAKE fingers, the extra transmissions to the mobile can have a negative effect on the total system capacity due to increased interference. This is most likely in the macro cellular environments because the typical number of RAKE fingers considered adequate for capturing the channel energy in most cases is four. If all connections offered that amount of diversity, then the receiver has only one or two branches to allocate for each connection.

Table 9.3 illustrates the gain of soft handover over hard handover in the downlink. The results are generated with the system simulator as presented later in Section 9.3.2. The CODIT macro- and microcell environments, described in Chapter 4, were used in simulations [9].

From the macrocell results it can be seen that system downlink capacity decreases by 10% if macro diversity is not utilized. Capacity loss is not remarkable even if soft handover gain and macro diversity gain is lost. A CODIT macrocell channel gives sufficient diversity even if the mobile station was in a hard handover state. On the other hand, the limited number of RAKE fingers cannot fully exploit the attained diversity from soft handover in the macrocellular environment. On the contrary, a remarkable part of the energy generated by soft handover base stations is lost, which contributes to the interference. The situation is different for microcell, because a CODIT microcell channel provides only little diversity. Thus, macro diversity is essential for high capacity in environments with a low diversity order.

Table 9.3
Downlink Capacity with Hard and Soft Handover

[Kbps/cell/MHz]	Hard handover	Soft handover
Macro	155	169
Micro	139	222

In addition to providing gain against multipath fading, macro diversity also gives gain against shadowing. The shadow fade margin for soft handover is analyzed to be 2 to 3 dB smaller than the fade margin for hard handover [10, 11]. Macro diversity can thus be used to increase the cellular range.

Soft handovers should be used with circuit switched services because macro diversity is important for low delay services to guarantee high quality with short delays. For packet data there are no strict delay requirements, and therefore, soft handover is not that important for non-real-time packet data services that can take advantage of time diversity through retransmissions.

9.2.6 Time Diversity

Time diversity is achieved by coding, interleaving, and retransmissions. Channel coding is applied to achieve lower power levels and required signal quality in terms of

BER/FER. Interleaving and channel coding processes are used to correct the errors due to channel fades and interference peaks. The longer the interleaving, the better an error burst is spread over the interleaving period. Consequently, the individual errors can be much better handled with error correction coding than when errors appear to the decoder in bursts. In practice, it is the delay requirements that limit the allowed interleaving depth. For non-real-time services the requirements are clearly looser, and retransmissions can be employed to ensure near error-free transmission of data. The total delay requirement for real-time services is typically on the order of 20 to 100 ms.

As the interleaving period gets shorter, the requirements for fast power control get tighter. Generally with CDMA signal transmission there is a peak in the BER curve when plotting the BER as a function of mobile speed; until certain velocities, the power control eliminates the fades, and at some speed, the power control is not able to follow the fades anymore [12–13]. But as the fades occur more often and get shorter in the time domain, the interleaving is able to spread the information over several fades and thus uncorrelate the signal errors to allow them to be corrected with forward error correction such as convolutional coding. When antenna diversity is applied, however, the performance is almost flat over different mobile speeds [12].

9.3   WCDMA SIMULATORS

In this section the principles of the evaluation of the WCDMA radio network performance are analyzed. There are two approaches to simulate the overall performance of a DS-CDMA system. One is a combined approach where the link level and cellular network level simulations are combined into one simulator. Another approach is to separate the link and system level simulations to reduce the complexity of the simulators. For the performance evaluation of cellular networks a single simulator approach would be preferred. The complexity of such a simulator—including everything from transmitted waveforms to a cellular network with tens of base stations —is far too high. Therefore, separate link and system level simulators are needed.

In the link level, one communication link between a mobile station and a base station is modeled. The time resolution is typically 1 to 4 samples per chip. For 4.0 Mcps with one sample per chip, the time resolution is 0.25 μs. In the system level there are tens of base stations and all the mobiles connected to those base stations. The time resolution is coarser than in the link level, typically one power control period (0.5 ms) or longer.

The interface between the link level simulator and the system level simulator is presented in Figure 9.2 and in [14]. The link level simulator provides the system level simulator with the required input parameters. Those parameters are E_b/N_0 both for uplink and downlink, MUD efficiency for uplink, orthogonality factor for downlink, and multipath channel model including antenna diversity. Here, E_b is energy per received bit and N_0 is the interference power per bit. The orthogonality factor could also be calculated theoretically based on the multipath profile.

MUD efficiency states the percentage of intracell interference that can be cancelled with base station multiuser detection. In system level simulations it is assumed that MUD efficiency is the same for different system loads.

Figure 9.2 The interface between the link level and the system level simulator.

9.3.1 Link Level Simulation

The simulation program used in the link level performance evaluations was the Communications Simulation and System Analysis Program (COSSAP) by Synopsys Inc. The COSSAP simulation program is a stream driven simulation program, with support for asynchronous operation because no timing control between simulation elements is required. All the receiver blocks were generated with the C language for efficiency and flexibility in the configuration.

Link level simulations included a multipath channel model that was modeled as a tapped delay line with one sample per chip. The CODIT macro-, micro-, and picocell channel models described in Chapter 4 were used. Amplitudes and phases of the multipath components were estimated in the receiver, but the multipath delays were known a priori. Fast power control in the link level simulations was based on the estimated received SIR values.

9.3.2 System Level Simulation

9.3.2.1 Simulator Principle

The system level capacity is studied in an interference-limited case. One capacity simulation typically contains 10,000 local mean SIR values. One local mean SIR value is simulated over a 500-ms period for each user, consisting of random mobile placement, virtual handover, and fast power control. Figure 9.3 presents the block diagram of the system level simulator. The simulations give the outage probability for the system, where outage probability represents the number of users having a worse average SIR than required. An outage probability of 5% was selected as the target value for capacity.

The mobile stations in this simulator are not actually moving, such as in dynamic simulators with mobility. The effect of mobility is modeled by adding a fast fading process to the signal in the system level. In this semistatic approach, signal strength varies due to fading as a function of the mobile speed given as a parameter. In the system level it is important to model the effect fast fading and fast power control as discussed in Section 9.4. The simulator can also be used for multioperator simulations.

The propagation model of the system simulator consists of attenuation, shadowing, and statistically generated fast fading. Both macro- and microcell models are adopted from [15].

9.3.2.2 Path Loss Models

The distance-dependent attenuation for macro- and microcell environments is used as presented in Section 4.5 for macro- and microcell environments. The simulation environment parameters are given in Table 9.4.

The path loss between a mobile station and a macrocell base station is modeled as

$$L = 29 + 36\log(R) + 31\log(f) \tag{9.6}$$

where f is the carrier frequency in megahertz and R the distance between the transmitter and the receiver in kilometers. Resulting path loss L is given in decibels.

Path loss between a mobile station and a microcell base station is calculated with a multislope model. The microcell base stations are located on every second street intersection. The slopes are non-line-of-sight slope L_{NLOS}, line-of-sight slope L_{LOS} for small distance, and line-of-sight slope for long distances. If the connection between transmitter and receiver is line-of-sight, attenuation is calculated as

$$L_{LOS} = \begin{cases} 82 + 20\log(\dfrac{d}{300}), & \textit{if} \quad d \le 300\text{m} \\ 82 + 40\log(\dfrac{d}{300}), & \textit{if} \quad d > 300\text{m} \end{cases} \tag{9.7}$$

At a distance of 300m a breakpoint marks the separation between two line-of-sight segments.

Turning a corner causes an additional loss in (9.7). Attenuation between a transmitter and a receiver that have non-line-of-sight connections constitutes a line-of-sight segment, a non-line-of-sight segment, and an additional corner attenuation.

$$L_{NLOS} = L_{LOS}(d_{corner}) + 17 + 0.05d_{corner} + (25 + 0.2d_{corner})\log\left(\frac{d}{d_{corner}}\right) \tag{9.8}$$

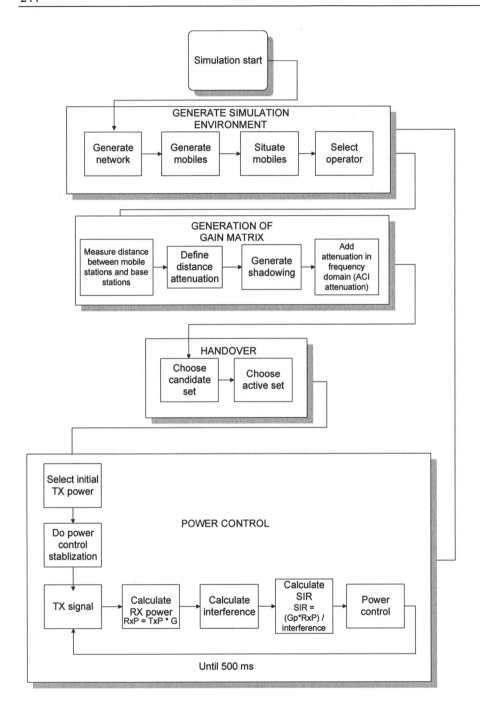

Figure 9.3 The block diagram of the system level simulator.

Line-of-sight attenuation is calculated between a corner and receiver d_{corner}, and non-line-of-sight connection between a corner and transmitter.

Table 9.4
Simulation Environment

	Micro	Macro
Shadowing	mean 0 dB, std dev 4dB	mean 0 dB, std dev 10dB
Number of base stations	128	19
Base station layout	Manhattan	hexagonal
Base station spacing	200m	800m
Building width	100m	-
Street width	30m	-
Mobile speed	3 km/h 36 km/h	5 km/h 50 km/h

Figures 9.4 and 9.5 depict, respectively, the macrocellular base station location and the base station deployment in microcellular system level simulations.

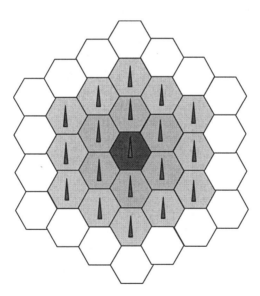

Figure 9.4   Macrocellular base station location.

Figure 9.5   Base station deployment in a microcellular system level simulator.

9.3.2.3   Fast Fading and Fast Power Control Modeling

In order to correctly model CDMA effects such as fast power control, a fast fading functionality was included in the system simulator, with the principle presented in [4]. The same kind of multipath channel with Rayleigh fading multipath components was used in the system level as in the link level simulations.

The power control algorithm is implemented as in [16]. The signal-to-interference ratio (SIR)–based power control algorithm aims to keep SIR levels of users at an appropriate level by adjusting the transmission power up or down. The received SIR level is compared to the SIR threshold. If the received SIR was lower than the threshold, a "power up" command is sent. If the SIR level was higher than the threshold, a "power down" command is sent.

In the link level simulations, realistic fast power control has been modeled. Real SIR estimation has been used in the WCDMA receiver, and errors have been added to the power control commands in the feedback channel. Also, real power control frequency and step size have been used. The effect of nonideal power control can be seen in the obtained E_b/N_0 figures.

In the system level, ideal SIR estimation has been used in the fast power control because the effect of nonideal power control is taken into account in the link level simulations. In the system level, fast fading and fast power control are still needed because we can observe how power control effects the interference in the network. Therefore, real power control frequency and step sizes are used also in the system level.

9.3.2.4   Handover Modeling

In the handover modeling, a maximum of 3 BTSs are selected in the active set. The BTSs are selected randomly with a uniform distribution from a pool of BTSs that fits in the handover margin. Such a BTS selection procedure takes the effects of nonideal handovers into account.

In the uplink, macro diversity (i.e., soft handover) is taken into account by selecting the best source (the frame with the highest average SIR) on a frame-by-frame basis. In the downlink, macro diversity can be modeled in two ways at the system level. The first approach is to simulate macro diversity in the link level so that separate E_b/N_0 levels are obtained for users in the soft handover state. There is a difference in E_b/N_0 levels because a different amount of diversity is available during soft handover and because part of the transmitted energy may not be captured by the limited number of RAKE fingers in the mobile station. Then, when we have separate E_b/N_0 values for the soft handover state in the system level, the mobile can be assumed to be able to receive all the paths directed to it in the system level.

Another way to model downlink macro diversity is to transmit the signal from several base stations to mobile stations so that the mobile station cannot receive all the paths directed to it. Paths not captured by RAKE processing contribute to the interference. Now the same E_b/N_0 threshold is used for all the mobile stations. The second method has been used in the system simulator due to straightforward modeling. The first method could be more accurate because it can also take into account the effect of increased multipath diversity on the E_b/N_0 performance.

9.3.2.5   Micro Diversity Modeling

Micro diversity (i.e., multipath diversity and antenna diversity in the BTS) is modeled by uncorrelated Rayleigh fading multipath components. The diversity combining is modeled as coherent maximal ratio combining. In this system level it is assumed that the number of RAKE fingers at the base station is sufficient and all modeled multipath components are received. The performance loss due to the limited number of RAKE fingers is modeled in the link level and can be seen in the E_b/N_0 values.

Micro diversity modeling is needed in the system level because the amount of diversity affects the intercell interference, as shown in Section 9.4.2.

9.3.2.6   Modeling Downlink Pilot and RAKE

In the downlink, the pilot signal is essential for system operation, and it is generally transmitted with a higher power level than any traffic channel. Modeling of the common pilot is not included in the link level performance figures. For this reason, its effect on downlink interference must be included in the system level modeling. In the simulator, each base station transmits user-specific signals and a common pilot signal. Powers of all the transmitted signals are summed together at the base station transmitter. Coherent maximal ratio combining at the RAKE receiver is assumed, and it is modeled in the system level by taking the sum of SNR values of paths [4].

9.3.2.7   Interference Cancellation

Intracell interference cancellation is modeled in the system simulator by removing β percentage of intracell interference in the interference calculation. The factor β is called MUD efficiency, and it is obtained from the link level simulations shown in Section 9.5.1.

In the uplink, when measuring SIR for a single user, the cancellation process includes all the users connected to the same BTS. Interference cancellation is nonideal, with the efficiency derived from the link level simulations. Thus, the total interference that the BTS experiences consists of the interference from all the mobiles in the network not having a connection with the BTS and from the partly canceled interference from the mobiles having connection to the BTS.

9.3.2.8   Sectorization and Adaptive Antennas

The effects of sectorization are not included in this study. The capacity gain from sectorization is ξN, where ξ is the sectorization efficiency, and N is the number of sectors. Coverage overlap between antenna beams causes interference spillover, which causes sectorization efficiency to be less than 1 and reduces capacity. For 3-dB overlap between antennas, $\sin(x)/x$ antenna pattern gives a sectorization efficiency of 0.88. When the beam overlap increases, the efficiency decreases. The efficiency also depends on whether softer handoff is used between overlapping sectors [17]. The use of adaptive antennas for improving capacity or coverage is not covered here, but evaluation results can be found in [18].

9.3.3   Simulation Parameters

The wideband CDMA system parameters used in the simulation are listed in Table 9.5. These system parameters are from the MUD-CDMA system concept [19], a wideband CDMA system that has been used as a basis for the FRAMES FMA2 wideband CDMA air interface [20]. Because the main system parameters are similar to the WCDMA system presented in Chapter 6, the results are representative for third-generation wideband CDMA systems.

9.4   FAST POWER CONTROL

As was discussed in Chapter 5, fast power control impacts the performance of a wideband CDMA system in three different ways. First, it compensates for the fast fading and, in a perfect case, turns the fading channel into a nonfading channel in the receiver, reducing the required E_b/N_0. The second effect is a consequence of the first: the compensation of the fading channel by power control leads to peaks in mobile station transmission power, which affect the intercell interference in the cellular network. Third, fast power control equalizes the mobile station powers in the base station and thus prevents the near-far effect. The effect of fast power control on a CDMA system can be summarized as follows:

- Lowers the E_b/N_0 requirement in the receiver;
- Introduces interference peaks in the transmitter;
- Avoids near-far problems in the uplink.

Table 9.5
Simulation Parameters

CDMA System Parameters	
Carrier spacing	6 MHz
Chip rate	Basic chip rate 5.1 Mcps
Carrier frequency	2 GHz
Spreading codes	Uplink: Extended VL-Kasami Downlink: Augmented Walsh-Hadamard
Bit rates	12 Kbps 144 Kbps 2 Mbps
Interleaving	20 ms (12 Kbps) 40 ms (144 Kbps) 100 ms (2 Mbps)
PC dynamics	Uplink 80 dB Downlink 20 dB
Link Level Parameters	
Voice activity	100%
Number of stages in base station MUD	2 (1 interference canceling)
Antenna diversity (uplink)	2 antennas, with fading correlation 0.7 and 0.0
Path combining method	Maximal ratio combining
Target BER	10^{-3}
Coding	Uplink: 1/2 rate convolutional with K = 9 Downlink: 1/3 rate convolutional with K = 9
Fast power control	2 kHz, step size 1.0 dB, 5% errors in feedback signaling
RAKE fingers	4 or 9
System Level Parameters	
Outage requirement	95% satisfied users
Active set size in soft handover	3
Handover margin in nonideal handover	3 dB
PC step size	1 dB
Pilot strength	6 dB higher than maximum traffic channel power

In practice there are imperfections in the power control because of the transmission and processing delay, command errors, limited dynamic range, and step size; the system performance is degraded because of the increased residual variation in the signal-to-noise ratio.

9.4.1 Impact of Fast Power Control in E_b/N_0

The gain in E_b/N_0 depends on the channel characteristics, the bandwidth of the system, and the mobile speed. Because a wide bandwidth gives better diversity, power control is not as critical for the wideband CDMA as for the narrowband CDMA. In a small delay spread channel, the improvement of performance from fast power control is largest. Furthermore, as discussed in Section 9.2.6, the improvement of E_b/N_0 is highest at slow mobile speeds. This is because of the ability of fast power control to follow the fast fading. As shown in Section 9.5.1 for a 2-Mbps service, fast power control is beneficial especially in micro- and picocell channels because of a limited multipath diversity. In a macrocell channel, the gain of fast power control in E_b/N_0 is not very large because of the high degree of multipath diversity. Fast power control is compared to slow power control in the downlink in [13, 21].

9.4.2 Intercell Interference with Uplink Fast Power Control

When uplink fast power control is able to follow the fast fading, the intercell interference will increase because of peaks in the transmission (TX) power of the mobile MS in Figure 9.6. The received (RX) interference level in the neighboring cells experiences peaks. In the downlink, this effect is smaller because of limited power control dynamics. Downlink power control is not considered in this subsection. The less diversity available, the higher the average transmission power is and the more intercell interference is generated. This phenomenon of increased interference is imminent with such systems with fast power control that cannot exploit the multipath diversity of the channel and do not employ antenna diversity. On the other hand, soft handovers provide more diversity and reduce intercell interference. As explained below, if a system employs fast power control and soft handovers, they should be modeled together with a relevant channel model in the system simulations to give valid capacity results.

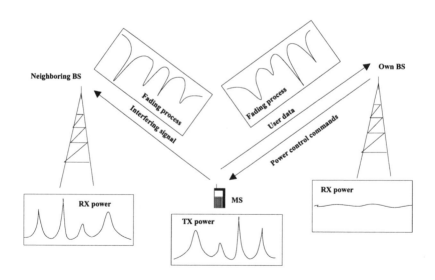

Figure 9.6 Interference to neighboring base stations in fading channel with fast power control without macro diversity.

In this section, we fix the SIR target to be the same for all channels in order to study only the impact of increased intercell interference. It should be noted that in practice this is not the case, but different channels require different SIR to achieve the required QoS.

In the following treatment, we first provide an analytical model and then the corresponding simulation results. Because of modeling difficulties, handovers are not considered in the analysis corresponding to the assumption of slow hard handovers, where a mobile is connected to the base station with the smallest long-term attenuation. This hard handover analysis leads to a worst case analysis in terms of transmission power and interference. In reality, soft handover increases diversity for mobile stations at the cell border, and thus, the interference is reduced. This is visible in the system simulations where both hard handover and soft handover are modeled.

9.4.2.1 *Average Uplink Transmission Power with Fast Power Control*

In this section an analysis on the average uplink transmission power with ideal fast power control is presented. The terminal transmission power level depends on the multipath diversity of the channel model. This result is then used in the next subsection to analyze system capacity with fast power control.

The average transmission powers with different degrees of diversity are calculated as follows. The probability density function (pdf) of the total channel power with L statistically independent Rayleigh fading multipath components, given in [1], is

$$p(\gamma) = \sum_{k=1}^{L} \frac{\pi_k}{\overline{\gamma}_k} e^{-\gamma/\overline{\gamma}_k} \quad \gamma \geq 0 \qquad (9.9)$$

where $\overline{\gamma}_k$ is the average power of the kth multipath component and

$$\pi_k = \prod_{\substack{i=1 \\ i \neq k}}^{L} \frac{\overline{\gamma}_k}{\overline{\gamma}_k - \overline{\gamma}_i} \qquad (9.10)$$

If the multipath components are equally strong, the pdf is written as

$$p(\gamma) = \frac{L^L \gamma^{L-1}}{(L-1)!} e^{-L\gamma_b} \quad \gamma \geq 0 \qquad (9.11)$$

Now we want to find out the pdf of the transmission power ρ of a given mobile station. In this section, we assume that the number of interfering users is so high that the interference at the base station can be approximated by Gaussian noise. The more users there are in the system, the better this approximation holds. With this assumption, the interference level at the base station is constant. Therefore, SIR-based power control can be replaced by power control based on received signal power only. Because we assume ideal power control, mobile transmission power $\rho = 1/\gamma$ and the pdf of ρ is [22]

$$p_\rho(\rho) = \frac{1}{\rho^2} p\left(\frac{1}{\rho}\right) \qquad \rho \geq 0 \tag{9.12}$$

The average transmission power can be obtained by

$$\overline{\rho} = \int_0^\infty \rho p_\rho(\rho) d\rho \tag{9.13}$$

In case of equally strong Rayleigh fading components and unlimited power control dynamics, the increase in transmission power can be shown to be

$$\frac{L}{L-1} \tag{9.14}$$

where L is the number of multipath components. In a 1-path Rayleigh fading channel ($L = 1$) the result would be infinity with unlimited power control dynamics. Typical total dynamics of uplink CDMA power control are 80 dB, which takes care not only of fast fading but also distance attenuation and slow fading. In these calculations only fast fading is taken into account, and therefore, the power control dynamics allowed for the fast power control is set to be 50 dB. The minimum and maximum transmission powers are chosen to be $\rho_{min} = 0.2$ and $\rho_{max} = 2.0 \times 10^4$. Now, the average transmission power can be calculated as

$$\overline{\rho} = \int_0^{\rho_{min}} \rho_{min} p_\rho(\rho) d\rho + \int_{\rho_{min}}^{\rho_{max}} \rho p_\rho(\rho) d\rho + \int_{\rho_{max}}^\infty \rho_{max} p_\rho(\rho) d\rho \tag{9.15}$$

The average transmission powers are calculated in the AWGN channel, in the 1-tap Rayleigh fading channel, in the 2-tap Rayleigh fading channel (two equally strong taps on average), in the 4-tap Rayleigh fading channel (four equally strong taps on average), and in the ATDMA macrocell channel with and without base station antenna diversity. The 2-tap and 4-tap Rayleigh fading channels, with all taps equally strong, can be considered either as multipath channels or as antenna diversity. The impulse response of the ATDMA macrocell channel model used is shown in Figure 9.7. The used bandwidth is chosen to be so high that all the multipath components in the ATDMA macrocell channel model can be resolved. The minimum bandwidth needed for separating all the paths in the ATDMA macrocell channel is approximately 3.5 MHz. Maximal ratio combining of multipath components is assumed in this analysis.

The calculated increase in average transmission power compared to nonfading channel ΔP_{tx} is shown in Table 9.6.

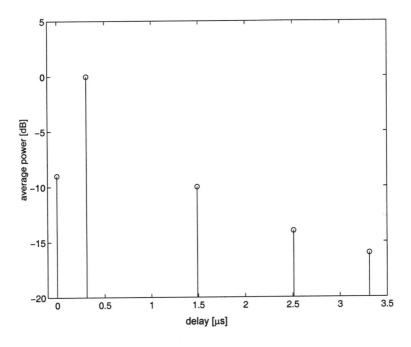

Figure 9.7 Impulse response of ATDMA macrocell channel.

Table 9.6
Average Increase of Transmission Power in a Fading Channel Compared to a Nonfading Channel with Hard Handovers

Channel	ΔP_{tx} (dB)
Nonfading channel	0
1-tap Rayleigh fading channel	10.1
2-tap Rayleigh fading channel	3.0
4-tap Rayleigh fading channel	1.2
ATDMA macro channel	2.5
ATDMA macro channel with antenna diversity	1.3

The average transmission power in a fading channel with fast power and with hard handovers is higher than in a nonfading channel. Thus, the average intercell interference from the mobile stations to the base stations in the surrounding cells is also higher.

9.4.2.2   Analysis of Impact of Fast Power Control into Capacity

The maximum number of simultaneous users N on one carrier in a cellular CDMA system can be calculated as in [23]:

$$N = F\left[G_p\left(\frac{E_b}{N_0}\right)^{-1} + 1 \right]$$

(9.16)

where G_p is the processing gain, E_b/N_0 is the received bit energy per noise density, and

$$F = \frac{I_{\text{intra}}}{I_{\text{intra}} + I_{\text{inter}}}$$

(9.17)

where I_{intra} is the intracell interference, I_{inter} is the intercell interference in a nonfading channel, and F denotes the percentage of interference originating from intracell mobiles at the base station. In this definition of F, intracell interference I_{intra} at the base station includes all the interference from the mobiles that are connected to that base station. Therefore, users in soft handover are included in intracell interference in two to three base stations.

Let this F be valid in a nonfading channel. In a fading channel the average transmission power increases, and thus, the interference from other cells I'_{inter} is higher than the intercell interference in a nonfading channel I_{inter}. Now the percentage of interference from the own cell F' is

$$F' = \frac{I_{\text{intra}}}{I_{\text{intra}} + I'_{\text{inter}}}$$

(9.18)

where I'_{inter} is the intercell interference in a fading channel.

The increase in interference is equal to the increase in the average mobile transmission power ΔP_{tx} shown in Table 9.6, where ΔP_{tx} is given by

$$\frac{I'_{\text{inter}}}{I_{\text{inter}}} = \Delta P_{tx}$$

(9.19)

Now, we can obtain F'

$$F' = \frac{1}{1 + \Delta P_{tx}\left(\frac{1}{F} - 1\right)}$$

(9.20)

F is here assumed to be 55% in a nonfading channel with hexagonal base stations. This value is obtained from the system simulator for the macrocell environment and is a typical value for this environment. The values for F' in different channels are shown in

Table 9.7. If we assume in (9.16) that the processing gain G_p and the E_b/N_0 requirement do not change between different channels, then $N = F \cdot$constant (i.e., the relative capacity):

$$\frac{N'}{N} = \frac{F'}{F} \tag{9.21}$$

The relative capacities N'/N are shown in Table 9.7.

Table 9.7
Relative Capacities Compared to a Nonfading AWGN Channel with Hard Handovers

Channel model	Percentage of interference from intracell mobiles F'	Relative capacities N'/N compared to AWGN (%)
AWGN channel	55	100
1-tap Rayleigh	11	19
2-tap Rayleigh	38	69
4-tap Rayleigh	48	87
ATDMA macro	41	74
ATDMA macro with antenna diversity	48	86

9.4.2.3   Simulated Cellular Capacity in Fading Channel

In this subsection the analytical capacity estimates are compared to the simulated values. The power control dynamic range is 80 dB, which is larger than the value used in the analytical calculations. This is because of the additional distance-dependent attenuation as well as the slow fading that must be compensated for in the simulator by the available power control dynamic range. The parameters of the power control scheme are shown in Table 9.8.

The propagation model of the system simulator consists of attenuation, shadowing, and statistically generated fast fading. The attenuation model used is presented in Section 9.3.2.

Even if the power control algorithm in the simulations has a perfect knowledge of the received SIR, the fast fading process cannot be followed perfectly because of a finite step size.

Table 9.8
Simulation Parameters for Fast Power Control Simulations

Power control frequency	1 kHz
Power control step size	1.0 dB
Power control dynamics	80 dB
Mobile speed	1 km/h
Active set size in soft handover	3
Handover margin in nonideal handover	5 dB
Log normal shadowing	mean 0 dB, std dev 6 dB
User bit rate	10 Kbps

Both hard and soft handovers are supported by the system simulation. When soft handover is used, the mobile station is able to communicate simultaneously with one or more base stations. The uplink transmission power in the mobile station is increased only if all base stations in the active set request more transmission power through fast power control signaling. Uplink macro diversity in soft handover is considered by selecting the best source (the frame with the highest average SIR) on a frame-by-frame basis. Each active base station receives the frame transmitted by the mobile station. The quality of the frame is measured by calculating the average SIR over the frame. The frame with the highest quality is utilized, while other frames are discarded.

Hard handover is modeled in simulations by connecting the mobile station to the base station to which the distance-dependent attenuation and the slow fading attenuation is the lowest. This can be done by setting the maximum active set size to one.

Simulated and calculated capacities for hard handovers can be compared in Table 9.9. The simulated capacities are lower than the calculated capacities except in the 1-tap Rayleigh fading channel. These differences are because of nonidealities in the power control in the network simulation. The power control step size was fixed and the same in different fading channels. It should be noted that the SIR target was set to be the same for all capacity simulations. In practice, however, the 1-tap Rayleigh type channel with power control imperfections would require the highest SIR target for power control to achieve comparable link quality with other cases. In other channels, the power control step size was too large, and thus, unnecessary power fluctuations increased the interference and resulted in lower capacities.

Soft handover (macro diversity) shows a considerable gain in the 1-tap Rayleigh fading channel over hard handover.

The differences in capacities between different channel models point out the importance of including fast fading, fast power control, and realistic handovers in CDMA system simulations if realistic capacity results are needed for network planning. The simulated capacity with a 1-tap Rayleigh fading channel and soft handover is only 43% of that calculated and simulated for a nonfading channel even if the same SIR target was used in the system simulations.

Table 9.9
Simulated Relative Capacities with Different Channel Models

Channel model	Hard handover (slow) (%)	Soft handover (fast) (%)
Nonfading AWGN channel	100	100
1-tap Rayleigh	25	43
2-tap Rayleigh	57	Not simulated
4-tap Rayleigh	78	Not simulated
ATDMA macro	60	Not simulated
ATDMA macro with antenna diversity	79	Not simulated

Figure 9.8   Calculated and simulated relative capacities with different channel models

The results of Table 9.9 and Figure 9.8 show capacity differences between different amounts of diversity. Therefore, if the channel does not have enough diversity, other means such as receiver antenna diversity or transmit diversity need to be used. Furthermore, in order to obtain reliable capacity results the multipath channel needs to be modeled in the same way in both link level simulations and in system level simulations.

9.4.3   Capacity Degradation Due to Imperfect Power Control

One purpose of power control is to equalize the received powers in the base station to eliminate the near-far effect. In practice, however, it cannot eliminate the near-far effect completely due to imperfections in the power control loop. Thus, the residual variation in SIR causes degradation of the system capacity [24]. The amount of residual variation, and thus capacity degradations, depends on channel response, the system chip rate, SIR estimation accuracy, power control loop delay, command rate, and power control step size. The effect of imperfect power control is included in the E_b/N_0 results of the link level simulation because a realistic power control algorithm has been used.

The effect of imperfect power control has been studied widely in connection with IS-95 uplink performance analysis. As indicated in several studies, the performance in terms of system capacity decreases by the percentages listed in Table 9.10, as the power control error increases. The power control error causes variations in the received E_b/N_0. Empirical evidence suggests that E_b/N_0 is log-normally distributed and that the standard deviation for the narrowband CDMA employing both open loop and fast closed loop power control is between 1.5 and 2.5 dB [25].

Wideband CDMA reduces the impact of imperfect power control. This is because better diversity results in less variation in residual SIR. In [24] power control feedback delay was shown to be less critical for wideband systems than narrowband systems.

Table 9.10
Capacity Reduction with Power Control Error

Power control error, standard deviation (dB)	Capacity reduction (%)
0	0
1	3
1.5	8
2.0	13
2.5	20

Source: [26].

MUD alleviates the near-far effect, and thus the degradation of performance due to imperfect power control is smaller. In practice the gain of MUD against power control errors depends on the performance of MUD (i.e., how much of the generated intracell interference a particular receiver solution is able to remove).

9.5 SPECTRUM EFFICIENCY

In this section the evaluation of WCDMA spectrum efficiency is analyzed. First, the link level performance of the uplink is simulated, taking into account multiuser detection efficiency. Data rates of 12 Kbps, 144 Kbps, and 2 Mbps are considered. Second, spectrum efficiency results are presented based on the system level simulation using the link level results as depicted in Figure 9.3. Third, to verify the simulation results, analytical spectrum efficiency calculations are presented.

9.5.1 Link Level Performance of WCDMA Uplink

As an input for the uplink capacity studies, link level performance, E_b/N_0 for the desired BER, and MUD efficiency are needed. The performance was studied in both CODIT macro- and microcell environments. The performance figures are given in Figures 9.9 and 9.10.

The BER performances in COSSAP Monte Carlo simulations are used to assess the efficiency of multiuser detection β. This efficiency denotes the percentage of intracell interference being removed by multiuser detection at the base station receiver. The efficiency of MUD is estimated from the load that can be accommodated with a specific E_b/N_0 value with a conventional RAKE receiver and with a multiuser receiver. The same efficiency of MUD is assumed for different loads in the system simulator. The additive white Gaussian noise N_0 is used to represent both thermal noise and intercell interference, while intracell interference is represented by real transmitters. The target BER is 10^{-3}. In the analysis, the number of users with a RAKE receiver is denoted by K_{RAKE} and that with a MUD receiver by K_{MUD}. The efficiency of MUD β at a given E_b/N_0 value is given by

$$K_{RAKE} = (1 - \beta)K_{MUD} \qquad (9.22)$$

This applies to the macrocell where multipath interference is significant. In a microcell, the channel model is close to a single path channel and thus, self-interference from multipath components is negligible. Therefore, we can subtract the desired user from the total number of users and calculate the MUD efficiency for the microcell environment by

$$K_{RAKE} - 1 = (1 - \beta)(K_{MUD} - 1) \qquad (9.23)$$

To keep the complexity of the receiver to a moderate level, only four RAKE fingers were used in the macrocell with which 63% of the total energy could be collected on average. In the microcell, two RAKE fingers were used with which 95% of the energy could be captured. If only part of the energy is captured by the receiver, a higher E_b/N_0 is required, resulting into lower capacity figures. Therefore, the required E_b/N_0 in the macrocell is higher than in microcell. In the CODIT macrocell channel, the capacity was increased by about 30% by using nine RAKE fingers instead of four fingers [27].

We performed the link level simulations for the MUD efficiency study with some differences to the assumptions mentioned earlier. The service studied in this case was 74 Kbps in the CODIT macro- and microcell environments with BPSK data and chip modulation, and antenna diversity with a 0.7 correlation factor. Pilot symbols for channel estimation were inserted in the data stream. It should be noted that because delay estimation was assumed to be perfect, these results are only indicative. Furthermore, the MUD efficiency varies according to system load, which has not been considered in these simulations. The multiuser detection algorithm that has been used in the link level simulations has been presented in [28]. The impact of imperfect delay estimation on the multiuser detection has been considered, for example, in [29,30].

Simulated BER curves for the microcell are shown in Figure 9.9 and for the macrocell in Figure 9.10. All the users have equal average power. The bit error probabilities shown are the mean values of all the users. The single user case without intracell interference is shown for comparison.

In the microcell we have used simulations to estimate the required E_b/N_0 when 15 users are transmitting in the multiuser detection system. The capacity of the conventional RAKE receiver with the same E_b/N_0 is found to be between five and six users yielding a MUD efficiency of 64% to 71%. In the macrocell, the corresponding figures are 10 users with MUD and three to four users without MUD. The efficiency of MUD in macrocell is then 60% to 70%. The link level simulation results and the resulting MUD efficiencies are summarized in Table 9.11. Because the calculated MUD efficiency is valid for the intracell interference only, it does not directly convert to the gain in the cellular capacity. The interference from all the cells in the network must be included in the study to obtain the capacity for the whole network see (9.27).

Figure 9.9   BER as a function of E_b/N_0 for the CODIT microcell with 36 km/h.

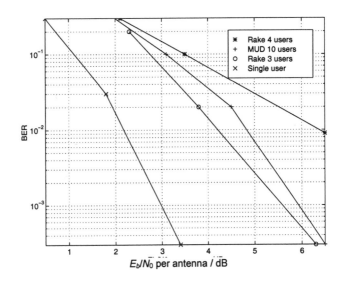

Figure 9.10   BER as a function of E_b/N_0 for the CODIT macrocell with 50 km/h.

Table 9.11
Link Level Results with and Without Base Station Multiuser Detection

Channel model, E_b/N_0 value	Number of users, no MUD, K_{RAKE}	Number of users with MUD, K_{MUD}	Efficiency of MUD β
Micro, 2.2 dB	5 – 6	15	64%–71%
Macro, 5.9 dB	3 – 4	10	60%–70%

For the spectrum efficiency comparison for the uplink, results were produced for 12- and 144-Kbps services in the macrocell environment with the channel model presented in Figure 9.7. The main difference with the downlink results is the use of uplink antenna diversity. In the simulations for macrocell, the fading on the different antenna branches has been assumed to be uncorrelated. For the studied 12- and 144-Kbps services, the performance results are given in Table 9.12 for CODIT microcell multipath profile.

Table 9.12
Link Level Results of 12- and 144-Kbps in CODIT Microcell

Studied service in the uplink	E_b/N_0
12 Kbps	5.7
144 Kbps	1.7

The difference between the uplink 12- and 144-Kbps services is because of the larger interleaving period for 144 Kbps and because more energy is available for the decision directed channel estimator, thus giving better channel estimates for 144 Kbps than for 12 Kbps.

9.5.1.1 Transmission of 2 Mbps

Link level performance simulations for the 2-Mbps user bit rate and BER of 10^{-3} were carried out using similar wideband channel models as in the CODIT project (described in Chapter 4), for urban macrocell and urban street microcell at speeds of 3 and 36 km/h and for indoor environments [9] at a speed of 3 km/h using three different chip rates of 5, 10, and 20 Mcps. The actual target BER of 10^{-6} was not simulated because of the very high computational efforts required.

The performance of uplink with 2-Mbps service is good (4 to 6 dB without antenna diversity) with all simulated bandwidths. Receiver antenna diversity was not used in the link level simulations, but diversity gain is taken into account in the range and capacity simulations in the following sections. Fast power control offered considerable gains in the microcell and indoor environments. The E_b/N_0 performance was in the best cases within 1 dB of the simulation results in nonfading channels for the coding scheme given in [1]. Uplink simulation results for BER of 10^{-3} with fast power control and without antenna diversity are shown in Table 9.13. The corresponding results without power control are presented in Table 9.14.

Table 9.13
Uplink Performance with Fast Power Control but Without Antenna Diversity (BER 10^{-3})

Bandwidth	5.12 Mcps	10.24 Mcps	20.48 Mcps
Channel	E_b/N_0	E_b/N_0	E_b/N_0
Macro 3 km/h	4.6	5.5	6.2
Micro 3 km/h	3.9	3.6	4.1
Indoor 3 km/h	N/A	3.6	3.3

Table 9.14
Uplink Performance Without Power Control but Without Antenna Diversity (BER 10^{-3})

Bandwidth	5.12 Mcps	10.24 Mcps	20.48 Mcps
Channel	E_b/N_0	E_b/N_0	E_b/N_0
Macro 3 km/h	7.0	7.0	6.5
Micro 3 km/h	10.4 (with antenna diversity)	9.9	7.5
Indoor 3 km/h	N/A	6.0	5.3

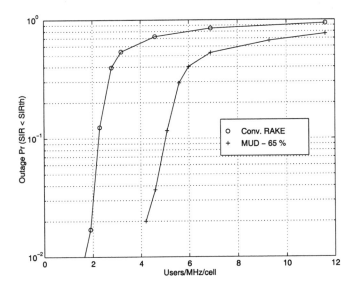

Figure 9.11 Outage probability curves for microcells with conventional receiver and with MUD receiver with $\beta = 65\%$.

9.5.2 Spectrum Efficiency of WCDMA Uplink

Figure 9.11 shows the outage figures as a function of the load in the microcell. The capacity with a conventional RAKE receiver is 2.0 users/MHz/cell, and for the system with MUD it is 4.7 users/MHz/cell, with the 74-Kbps service used in the analysis for MUD performance. The capacity is obtained with 5% outage.

In Figure 9.12 the simulated outage figures for the macrocell are presented as a function of load for different MUD efficiencies. Lines indicate performances with different MUD efficiency. It can be seen that the capacity of MUD with 65% efficiency is almost doubled compared to the conventional RAKE receiver.

The high capacity of CDMA with MUD results from the fact that most of the interference is intracell interference. Because of soft handover, users in a handover state can be included in the MUD process, which allows an extra 20% of interference to be included in the interference cancellation in the macrocellular environment. In the microcellular environment, cell separation is better, fewer users are in the soft handover state, and the amount of interference from other cells is lower than that in the macrocellular environment.

Figure 9.12 Outage probability curves as a function of MUD efficiency in macrocell. The line with 0% efficiency corresponds to the conventional RAKE receiver.

9.5.2.1 Analytical Spectrum Efficiency Calculations

To validate the simulation results, analytical spectrum efficiency calculations for the macrocell environment were carried out. The simulated performance compared with analytical results is given in Table 9.15. The reason for the conventional RAKE receiver having differences between simulated and analytical results is in the power control modeling. The analytical approach assumes perfect power control, whereas the simulator has the actual power control algorithm running with limited dynamics and accuracy. Furthermore, the analytical results assume a nonfading channel, while the simulated capacities are obtained in a fading channel.

Table 9.15
Analytical and Simulated (5% outage) Capacities in Macrocell (users/MHz/cell)

	Conventional RAKE	MUD 20%	MUD 40%	MUD 65%
Simulated	2.3	2.7	3.3	4.5
Analytical	2.8	3.3	4.0	5.3

The Gaussian approximation is generally the basis for analytical capacity calculations. The central limit theorem states that the normalized sum of independent random variables approaches a Gaussian random variable as the number of terms (here, users) increases [31]. This is highly applicable in the case of a large number of low bit rate users; but as the data rates get higher and the interference originates from only a few users, Gaussian approximation underestimates the interference. It should be noted again that a high degree of diversity allows the use of the Gaussian approximation a bit further because of the randomizing effect on the interference.

First, the capacity formula for the network without MUD in BTS is defined. The attained E_b/N_0 value is given by

$$\frac{E_b}{N_0} = \frac{S \cdot G_p}{I_{intra} + I_{inter} + N_0} \tag{9.24}$$

where S is the received signal strength, G_p is the processing gain, I_{intra} is the intracell interference, I_{inter} is the intercell interference from other cells, and N_0 is the thermal noise. In the following, thermal noise N_0 is neglected and I_{intra} is equal to $(N-1)S$.

The fraction of the intracell interference caused by the users operating in the same cell as the studied user compared to the total interference is given as

$$F = \frac{I_{intra}}{I_{intra} + I_{inter}} \tag{9.25}$$

It should be noted that in (9.25) the intracell interference I_{intra} is equal to $N \cdot S$, instead of $(N-1) \cdot S$. The simulated value for F in macrocells is 0.73. For a single cell, F is equal to 1. From the two equations above we get

$$\frac{E_b}{N_0} = \frac{G_p}{\frac{N}{F}-1} \Leftrightarrow N = F \cdot (G_p (\frac{E_b}{N_0})^{-1} + 1) \tag{9.26}$$

The value N is the number of users that are associated with the BTS. N also includes users that are connected to more than one BTS while in a soft handover state. The number of users being connected to two or three BTSs is obtained from the simulator and used to adjust the analysis correspondingly. Simulations show that typically in the simulated macrocellular environment, 80% of the users are connected to only one BTS, while 15% of the users are connected to two BTSs, and 5% are connected to three BTSs. The calculated capacity must be then scaled by 1.25, as the effect of soft handover is seen in F and the total number of connections in the system is higher than the number of mobiles in the system.

The corresponding analysis for the MUD case can be performed by

$$\frac{E_b}{N_0} = \frac{S \cdot G_p}{(1-\beta)I_{\text{intra}} + I_{\text{inter}} + N_0} \tag{9.27}$$

Intercell interference can be calculated by

$$I_{\text{inter}} = \frac{1-F}{F} I_{\text{intra}} \tag{9.28}$$

From (9.23) and (9.24) we have

$$\frac{E_b}{N_0} = \frac{S \cdot G_p}{(1-\beta)(I_{\text{intra}} - S) + \frac{1-F}{F} I_{\text{intra}}} \tag{9.29}$$

Because I_{intra} is equal to NS, we can write

$$\frac{E_b}{N_0} = \frac{S \cdot G_p}{(1-\beta)(NS - S) + \frac{1-F}{F} NS} \tag{9.30}$$

The capacity of the system is now given by

$$N = F \left(\frac{G_p (\frac{E_b}{N_0})^{-1} - (\beta - 1)}{1 - F\beta} \right) \tag{9.31}$$

If β is set to 0, then the equation above becomes the same as (9.16). The β with value 0 represents the capacity of the conventional RAKE-receiver-based system.

Capacity as a function of MUD efficiency is shown in Table 9.15. Both analytical and simulated results are shown. The simulated results compare well to the analytical results. The offset between analytical and simulated results is due to the non-ideal power control as well as real interferers in the simulator instead of Gaussian approximation.

Capacity strongly depends on the radio environment, which is defined by the path loss attenuation factor, shadowing, and the wideband channel model. The proportion of the interference coming from the own cell from the total interference F in (9.31) is greatly dependent on the environment. It ranges from the single isolated cell case with no intercell interference to the macro cell environment with up to 40% of interference having intercell origin.

Multiuser detection is shown to have the potential to increase capacity. It is more suited for the uplink, where all users need to be detected in any case, and the base station has more processing power available than the mobile station. A suboptimal multiuser receiver has been shown to offer a clearly improved link performance over the RAKE receiver by removing 60% to 70% of the intracell interference in urban micro- and macrocell environments. This interference reduction achieved in the link level with a multiuser detector was used as an input to the system level simulator, which led to a considerable increase in cellular capacity. The capacity gain depends on the ratio of intracell interference to intercell interference, and therefore, the microcellular environment offering high cell isolation gained even more from the use of MUD. Alternative solutions for capacity enhancements are, for example, the use of adaptive antennas or cell splitting, both of which require extra hardware such as antennas. With the use of MUD for capacity enhancement, only the baseband hardware needs to be modified.

9.5.3 Link Level Performance of WCDMA Downlink

As an input for the downlink system level simulations, we need link level performance results and the orthogonality factor. We first derive the orthogonality factor and then present the link level simulation results.

9.5.3.1 Derivation of the Orthogonality Factor

The interference in the downlink from the same BS originates from a single point and the parallel code channels can be synchronized. When using orthogonal spreading codes in the ideal case, the intracell interference could be completely avoided; but after the multipath channel, part of the orthogonality is lost and intracell interference exists also in the downlink.

For system level studies in the downlink, an estimate of the achievable degree of orthogonality needs to be drawn from the link level studies. An analytical capacity calculation of the orthogonality factor assuming Gaussian approximation is presented below. The attained SIR can be written as

$$\text{SIR} = \frac{\sum\limits_{i=0}^{N} g_i L_{p,l} P_{tx}}{\sum\limits_{i=0}^{N} \overline{g}_i} \frac{1}{I_{\text{total}}} \tag{9.32}$$

where N is the number of perceived paths in the channel model with the selected bandwidth, $L_{p,l}$ is the path loss between mobile stations and base station l, P_{tx} is the transmission power for the selected user, I_{total} is the total interference experienced by that user, g_i is the instantaneous path gain, and $\overline{g}_i$ the average gain for path i.

Instantaneous SIR is calculated by dividing the received signal by the interference, and multiplying by the processing gain. The downlink intracell interference is multiplied by $1-\alpha$ where α is the orthogonality factor. An orthogonality factor of 1 corresponds to perfectly orthogonal intracell users, while with the value of 0 the intracell interference is completely asynchronous. The signal-to-interference ratio is then given by

$$\text{SIR} = \frac{\sum\limits_{i=0}^{N} g_i L_{p,l} P_{tx}}{\sum\limits_{i=0}^{N} \overline{g}_i} \frac{G_p}{(1-\alpha)I_{\text{intra}} + I_{\text{inter}} + N_0} \tag{9.33}$$

where N_0 is the thermal noise, G_p is the processing gain, I_{intra} is the intracell interference, and I_{inter} is the intercell interference. Since the system is interference limited, thermal noise N_0 is assumed small and therefore neglected. I_{intra} and I_{inter} are equal to

$$I = \sum_{l=0}^{M} \left[L_{p,l} \frac{\sum\limits_{i=0}^{N} g_{i,l}}{\sum\limits_{i=0}^{N} g_i} \left(\sum_{k=0}^{R} P_{tx,k} + P_{\text{pilot},l} \right) \right] \tag{9.34}$$

where R is the number of interferers within one cell, and $P_{tx,k}$ is the transmission power to the kth user. Each path of the desired user experiences the same interference on average. M is the number of base stations causing inter- or intracell interference and $P_{\text{pilot},l}$ is the pilot power of base station l.

Orthogonality factor α is given by

$$\alpha = 1 - \frac{E_b}{I_0} \left(\frac{E_b}{N_0} \right)^{-1} \tag{9.35}$$

where N_0 is intercell interference, I_0 is intracell interference, and E_b/N_0 and E_b/I_0 are given in absolute figures, not in dB. E_b/N_0 is the performance figure with Gaussian noise, and E_b/I_0 is the corresponding figure dominating intracell interference. The values of E_b/N_0 and E_b/I_0 are obtained from link level simulations. The block diagram of downlink simulation with intracell and intercell interference is shown in Figure 9.13.

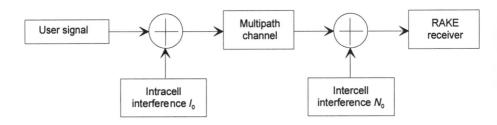

Figure 9.13 The block diagram of downlink simulation with intracell and intercell interference.

In the uplink, the signals are always asynchronous and do not have orthogonality unless the mobiles are synchronized with very high accuracy and operated in an environment where only one multipath component existed.

Figure 9.14 and Tables 9.16 and 9.17 show the results for downlink 12- and 144-Kbps link level E_b/N_0 and E_b/I_0 (in dB) in the CODIT microcell multipath channel. For 12- and for 144-Kbps results, the main differences to the uplink were the lack of antenna diversity and the use of a pilot channel for channel estimation. The pilot channel is not included in the E_b/N_0, and thus must be included in the system level modeling.

E_b/I_0 values have been calculated using the following equation:

$$E_b / I_0 \equiv \frac{W P_{\text{user}}}{R(NP + P_{\text{pilot}})} \tag{9.36}$$

where W is the chip rate of the downlink, R is the user data rate, P_{user} is the power of the wanted signal, N is the number of interfering channels, P is the power of interfering channel, and P_{pilot} the power of the pilot signal.

Table 9.16
Downlink 12-Kbps Link Level E_b/N_0 and E_b/I_0 Results in CODIT Microcell Multipath Channel

12 Kbps E_b/I_0	12 Kbps E_b/N_0	Orthogonality factor α
2.3	6.4 dB	0.61

Table 9.17
Downlink 144 Kbps Link Level E_b/N_0 and E_b/I_0 Results in CODIT Microcell Multipath Channel

144 Kbps E_b/I_0	144 Kbps E_b/N_0	Orthogonality factor α
1.9 dB	6.1 dB	0.62

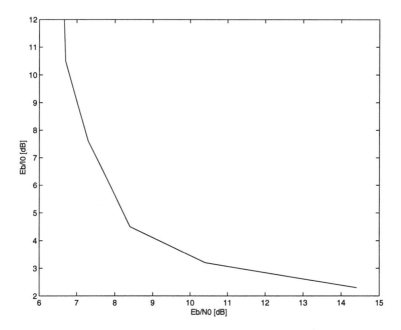

Figure 9.14 Downlink 12 Kbps with intracell interference (I_0) and with intercell interference (N_0).

9.5.3.2   Transmission of 2 Mbps

Transmission of 2-Mbps service was simulated in the CODIT macro-, micro-, and picocell channels. In each channel, enough RAKE fingers were allocated to gather all the signal energy. The maximum number of RAKE fingers was nine. The most' important parameter used in the comparison of different configurations in noise limited cases is E_b/N_0, which represents the received energy per bit versus noise power density. In the intracell interference case, energy per bit versus intracell interference power E_b/I_0 was compared.

The corresponding link performance results are presented in Tables 9.18 and 9.19. According to the downlink simulation results, the requirements of providing 2-Mbps service are achieved using reasonable E_b/N_0 values (4 to 10 dB), even with the narrowest 5-Mcps bandwidth. Only the performance in the macro channel seems to be somewhat poorer when the 5-Mcps bandwidth is used. This is due to severe interpath interference and low processing gain.

Table 9.18
Downlink Performance Without Power Control (BER 10^{-3})

Bandwidth	5.12 Mcps		10.24 Mcps		20.48 Mcps	
Channel	E_b/N_0	E_b/I_0	E_b/N_0	E_b/I_0	E_b/N_0	E_b/I_0
Macro 3 km/h	10.3	9.8	6.8	5.7	5.3	4.2
Macro 36 km/h	9.5	8.8	6.3	5.5	4.8	4.2
Micro 3 km/h	14.3	3.5	12.3	2.2	9.0	2.5
Micro 36 km/h	9.8	2.2	8.8	1.9	6.8	2.5
Indoor 3 km/h	9.8	0.4	9.3	-0.2	6.8	1.7

Table 9.19
Downlink Performance With Fast Power Control (BER 10^{-3})

Bandwidth	5.12 Mcps		10.24 Mcps		20.48 Mcps	
Channel	E_b/N_0	E_b/I_0	E_b/N_0	E_b/I_0	E_b/N_0	E_b/I_0
Macro 3 km/h	10.3	9.8	6.0	5.7	5.3	4.2
Macro 36 km/h	9.5	8.8	6.3	5.5	4.8	4.2
Micro 3 km/h	8.3	3.5	6.4	2.2	4.0	2.5
Micro 36 km/h	6.0	2.2	4.5	1.9	4.0	2.5
Indoor 3 km/h	4.7	0.4	4.2	-0.2	3.6	1.7

9.5.4 Spectrum Efficiency of WCDMA Downlink

Figure 9.15 shows the downlink spectrum efficiency of 144 Kbps for micro- and macrocells as a function of the orthogonality factor. The actual orthogonality factor was calculated to be 0.7 and 0.2 in micro- and macrocells, respectively. Therefore, multiuser detection used to remove intracell interference would still improve the spectrum efficiency, because the orthogonality is not perfect (=1). This contrast often prompted results that downlink multiuser detection was not beneficial because of orthogonality.

As shown in Figure 9.16, canceling the intercell interference clearly has a smaller effect on the total system capacity. Curves indicate the capacity improvement with a different number of intercell interferers that were included in the interference cancellation process. The curves are shown as a function of the efficiency of the interference cancellation.

Link level E_b/N_0 and E_b/I_0 values used in 12- and 144-Kbps services for spectrum efficiency comparison were shown in Tables 9.16 and 9.17.

It was noted that in the downlink macrocell environment, intracell interference is most important due to multipath propagation. In the macrocell environment, significant capacity improvements could be expected if intracell interference could be canceled in the mobile. Also, in a microcell environment, intracell interference cancellation could improve the capacity but not as much as in the macrocell environment. The difference is because there is more intracell interference in macrocell due to multipath propagation.

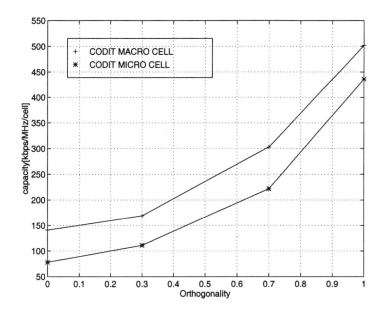

Figure 9.15 Downlink capacity as a function of intracell interference.

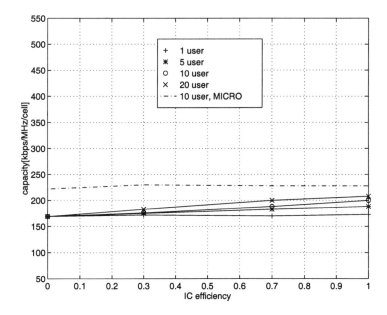

Figure 9.16 Downlink capacity as a function of intercell interference.

Additional diversity obtained from macro diversity was important in the microcell environment as there were limited number of resolvable multipath components due to large coherence bandwidth; the mobile antenna diversity would be similarly beneficial if allowed by complexity.

The downlink spectrum efficiency results were almost the same for the 12- and 144-Kbps services. This is reasonable because the downlink link level results were also very near to each other.

9.5.5 Comparison of Spectrum Efficiency for Uplink and Downlink

Table 9.20 shows capacity results for the uplink and the downlink. Downlink capacity is approximately 2 to 2.5 times lower than the corresponding uplink capacity, if both services are compared.

In addition to the base station antenna diversity, the major reason for the higher capacity of the uplink was the use of MUD, providing almost twofold capacity if compared to a conventional receiver. In the downlink, the corresponding factor for MUD gain is the orthogonality factor. There is a difference between uplink MUD gain and downlink orthogonality in the case of soft handover. In soft handover one uplink transmission is received at two base stations, and that interference can be canceled at both base stations and MUD gain can be obtained. In the downlink there are transmissions from two base stations to one mobile station, and those two signals are not orthogonal. It should also be noticed that the orthogonality factor gives only a modest capacity gain in a macrocell channel because of multipath propagation.

Table 9.20
Cellular Capacity in Kbps/MHz/cell

	Uplink	Downlink
Macro 12 Kbps	192	108
Macro 144 Kbps	388	108

9.5.5.1   Spectrum Efficiency with Circuit-Switched 2 Mbps

Table 9.21 shows the spectral efficiency in Kbps/MHz/cell for 2-Mbps communication for the downlink and for the uplink.

Table 9.21
Spectrum Efficiency in Kbps/MHz/cell

	Uplink	Downlink
Micro 5 Mcps	1040	80
Micro 10 Mcps	1080	160
Macro 5 Mcps	360	40
Macro 10 Mcps	260	100

The cellular capacity of networks offering 2-Mbps circuit-switched services seems to be quite low due to the poor downlink capacity. According to the results in Table 9.21, 15- to 20-MHz bandwidth is required to support one active 2-Mbps user in every cell. But because all the users seldom transmit simultaneously, packet reservation multiple access techniques can be exploited to alleviate the capacity problem.

The difference in spectrum efficiency between uplink and downlink is because of the following features. As in the uplink, BTS antenna diversity can be used with less complexity considerations. This easily brings a 3- or 4-dB gain to the uplink direction. This difference becomes even larger if multiuser detection is applied in the base station receiver. Even in the case of a single user, multiuser detection can be used to remove the interpath interference in multipath channels. Interpath interference is significant with low processing gain (i.e., with high bit rates). The large dynamics allowed for uplink power control also make it possible to largely compensate for the fading, especially in the microchannel case with 2 Mbps, which allows the link level results to approach the AWGN performance. In CODIT microcell, delay spread is rather narrow. As a result, the downlink receiver cannot obtain enough diversity, although the code channels remain more orthogonal. In the uplink, antenna diversity offers the required diversity.

9.6 COVERAGE

In this section, we first explain range calculation principles. Second, we consider the coverage of a DS-CDMA cellular system in both an unloaded and a loaded case. In addition, we show the impact of multiuser detection on the uplink range. If the gain from the multiuser detection is not used for coverage improvement, it can be used to reduce the mobile station transmission power. Finally, we discuss the effect of the user bit rate on the range.

9.6.1 Range Calculation

This section lists and explains the range calculation principles. The absolute range values given in this section depend on the used propagation model and cannot be directly used for network dimensioning. The values given here are relative and are to be used for comparison purposes between different services and receiver solutions. Table 9.22, which follows these principles, shows an example calculation. The parameters used in the range calculations are shown in Table 9.23 and are listed below. The link budget model is based on [3,6]. Link budget assessment is also discussed in Chapter 12.

1. *Average transmitter power per traffic channel, $P_{TX,avg}$ (dBm)*. The average transmitter power per traffic channel is defined as the mean of the total transmitted power over an entire transmission cycle with maximum transmitted power when transmitting.

2. *Cable, connector, and combiner losses at the transmitter, L_{TX} (dB)*. These are the combined losses of all transmission system components between the transmitter output and the antenna input (all losses in positive dB values).

3. *Transmitter antenna gain, G_{TX} (dBi).* Transmitter antenna gain is the maximum gain of the transmitter antenna in the horizontal plane (specified as dB relative to an isotropic radiator).

4. *Transmitter E.I.R.P. per traffic channel, P_{RX} (dBm).* This is the summation of transmitter power output per traffic channel (dBm), transmission system losses (dB), and the transmitter antenna gain (dBi), in the direction of maximum radiation:

$$P_{RX} = P_{TX,avg} - L_{TX} + G_{TX} \tag{9.37}$$

5. *Receiver antenna gain, G_{RX} (dBi).* Receiver antenna gain is the maximum gain of the receiver antenna in the horizontal plane (specified as dB relative to an isotropic radiator). The received power P_{RX} at the base station can be written as

$$P_{RX} = P_{TX,avg} - L_{TX} + G_{TX} + G_{RX} \tag{9.38}$$

6. *Cable, connector, and splitter losses at the receiver, L_{RX} (dB).* These are the combined losses of all transmission system components between the receiving antenna output and the receiver input (all losses in positive dB values).

7. *Receiver noise figure, NF (dB).* Receiver noise figure is the noise figure of the receiving system referenced to the receiver input.

8. *Thermal noise density, N_0 (dBm/Hz).* Thermal noise density is defined as the noise power per hertz at the receiver input. This can be calculated from kT, where k is the Bolzman constant and T is the temperature.

9. *Receiver interference density, I_0 (dBm/Hz).* Receiver interference density is the interference power per hertz at the receiver front end. This is the in-band interference power divided by the system bandwidth. The in-band interference power consists of both co-channel interference as well as adjacent channel interference. Thus, the receiver and transmitter spectrum masks must be taken into account. Receiver interference density for the forward link is the interference power per hertz at the mobile station receiver located at the edge of coverage, in an interior cell. In this example, an unloaded cell is assumed. The effect of interference can also be modeled under other losses calculated as presented in Chapter 12 (Figure 12.4).

10. *Total effective noise plus interference density (dBm/Hz).* Total effective noise plus interference density is the logarithmic sum of the receiver noise density and the receiver noise figure and the arithmetic sum of the receiver interference density.

$$(\text{Noise} + \text{Interference}) = 10\log\left(10^{((NF + N_0)/10)} + I_0\right) \tag{9.39}$$

11. *Information rate, $10log(R_b)$ (dB Hz).* Information rate is the channel bit rate in (dB Hz); the choice of R_b must be consistent with the E_b assumptions.

12. *Required $E_b/(N_0+I_0)$ (dB)*. The ratio between the received energy per information bit to the total effective noise and interference power density needed to satisfy the quality objectives. The translation of the threshold error performance to $E_b/(N_0+I_0)$ performance depends on the particular multipath conditions assumed. The CDMA uplink range can be calculated based on the link level E_b/N_0 value where E_b is energy per user bit and N_0 represents thermal and receiver noise. When reaching the maximum range, the mobile is transmitting at constant full power. Therefore, this E_b/N_0 value is obtained from link level simulations without fast power control, and it is higher than the value used in capacity simulations.

13. *Receiver sensitivity, $P_{RX,min}$ (dBm)*. This is the signal level needed at the receiver input that just satisfies the required $E_b/(N_0+I_0)$. The minimum reception power $P_{RX,min}$ represents a long time average power. If real receiver sensitivity is required, an activity factor must be added to this value. This is the case if voice activity is employed. In this case, continuous reception is assumed. The sensitivity can be calculated as

$$P_{RX,min} = kT + NF + 10\log(R_b) + E_b/(N_0+I_0) \qquad (9.40)$$

14. *Handover gain, G_{HO} (dB)*. This is the gain brought by handoff to maintain specified reliability at the boundary. In case of a single base station, there is probability P_{OUT} to have outage at distance r from the base station. In case of multiple base stations, at a point at distance r from, say, two base stations, the probability for outage relative to one base station is the same, but the probability to have outage from both the base stations is smaller and the resulting P_{OUT} is thus smaller than in the single cell case. In network planning, this reduces the shadowing margin needed in multiple cell case. The difference in shadowing margins is called handover gain. Depending on how the handover is done, the gain may be different. Here the handover gain G_{HO} is assumed to be 5 dB [10].

15. *Other gain, G_{other} (dB)*. An additional gain may be achieved as a result of future technologies. For instance, space diversity multiple access (SDMA) may provide an excess antenna gain.

16. *Log-normal fade margin, ξ (dB)*. The log-normal fade margin is defined at the cell boundary for isolated cells. This is the margin required to provide a specified coverage availability over the individual cells. In practice the propagation conditions vary considerably, and the path loss attenuation factor ranges from 3 to 4 [5]. The coverage requirement of 95% sets the actual shadowing margin. If 95% coverage is to be achieved over the whole cell area and we know that we have shadowing with a standard deviation σ of 6.0 dB and path loss model with $n = 3.6$ we get from [4] that the coverage probability at the area boundary is 84%. To have this, we need a shadowing margin of approximately 6.0 dB.

17. *Maximum path loss, L_{max} (dB)*. This is the maximum loss that permits the required performance at the cell boundary.

$$L_{max} = P_{TX} - P_{RX,min} + (G_{RX} - L_{RX}) + G_{HO} + G_{other}$$
$$- \xi \tag{9.41}$$

18. *Maximum range, d_{max} (km).* The maximum range is computed for each deployment scenario. Maximum range is given by the range associated with the maximum path loss. The propagation loss in decibels in a macrocell environment is calculated as in [32]:

$$PL = 123 + \alpha \times 10\log(d) + \sigma \tag{9.42}$$

where α is the path loss attenuation factor, d is distance in kilometers, and σ is a Gaussian stochastic variable that models the shadow fading. It is assumed that the average of the shadow fading equals zero and the standard deviation of the shadow fading is 6 dB. This might vary, however, between 6 and 10 dB [5]. The maximum range d_{MAX} is given by

$$d_{max} = 10^{\frac{L_{max} - 123 - \sigma}{36}} \tag{9.43}$$

Table 9.22
Link Budget Calculation Template

		Downlink	Uplink	Unit
1	Average TX power/TCH	30	24	dBm
2	Cable, connector, and combiner losses	2	0	dB
3	Transmitter antenna gain	13	0	dBi
4	TX EIRP/TCH = a – b + c	41.00	24.00	dBm
5	Receiver antenna gain	0	13	dB
6	Cable, connector, and splitter losses at the receiver	0	2	dB
7	Receiver noise figure	5	5	dB
8	Thermal noise density	−174	−174	dBm/Hz
9	Receiver interference density	−1000	−1000	dBm/Hz
9	Total noise + interference density (g + h + i)	−169	−169	dBm
10	Information rate R_b	8	8	kHz
11	$10\log(R_b)$	39.03	39.03	dBHz
12	$E_b/(N_0 + I_0)$ (link level sim. Result)	8	6.6	dB
13	Receiver sensitivity (j + j + l)	−122.0	−123.4	dBm
14	Handver gain	5	5	dB
15	Other gains	0	0	dB
16	Log normal fade margin	11.3	11.3	dB
17	Maximum path loss (d – m + (e – f) + n + o)	153.67	149.07	dB
18	Range	4.84	3.61	km

Table 9.23
Parameters for Range Calculations

Service	Medium bit rate data
User data rate	144 Kbps
BER	10^{-3}
Delay	100 ms
Environment	Macrocellular
Cell layout	Hexagonal
Path loss exponent α	3.6
Log-normal shadowing σ	6 dB
Handover gain G_{HO}	5 dB
Fractional cell loading F	70%
Transceiver parameters	
Base station antenna gain G_{BS}	6 dBi
Mobile station antenna gain G_{MS}	0 dB
Maximum mobile transmission power $P_{TX,MS}$	1W = 30 dBm
Thermal noise kT	−174 dBm
Receiver noise figure NF	7 dB
Capacity with 5% outage [33]	
Downlink capacity	169 Kbps/MHz/cell = 7.0 users at 6.0 MHz
Uplink E_b/N_0 at constant (full) transmission power	4.0 dB

9.6.2 Range in Unloaded and Loaded Networks

In cellular CDMA, the achievable range is very much dependent on cell load: If the number of users increases, the range decreases. This agrees with all cellular solutions where system capacity is considered to be mainly limited by the interference generated by the system itself. With CDMA, the effect is also seen in a single cell case, unlike in systems where intracell users are completely orthogonal, like in pure TDMA-based systems.

In this section, the decrease in cell range is calculated when network load is increased, and the effect of base station multiuser detection in increasing the range in a loaded network is also analyzed. Base station multiuser detection is shown to lower the mobile station transmission power in a loaded network, thus making uplink range and average mobile transmission power more insensitive to network load. The impact of intercell interference at the base station in different propagation environments is also studied. For actual network planning, multiuser detection and propagation environment must be taken into account.

In the case of an unloaded network, the uplink direction limits the achievable range and coverage, as the maximum transmission power of the mobile station is low compared to the maximum transmission power of the base station in the downlink. In a loaded network the downlink may limit the range if there is more load and thus more interference in the downlink than in the uplink.

The received SIR at the base station is obtained by [23]

$$\frac{E_b}{I_0 + N_0} = \frac{E_b}{I_{intra} + I_{inter} + N_0} \qquad (9.44)$$

where E_b is the received energy per bit, I_{intra} is the intracell interference from own cell mobiles, I_{inter} is the intercell interference from those mobiles not connected to this particular base station, and N_0 is the thermal noise.

In the case of an unloaded network, $I_{intra} = 0$, $I_{inter} = 0$, and the required E_b/N_0 for range calculations is simply equal to $E_b/(I_0 + N_0)$. In a loaded network, the percentage of own cell interference from the total interference (fractional cell loading) is defined as

$$F = \frac{I_{intra} + S}{I_{intra} + S + I_{inter}} \qquad (9.45)$$

where $S = E_b/G_p$ is the received signal power from one user and G_p is the processing gain.

The value for F depends on the propagation environment. The higher the pathloss attenuation factor, the higher the F. Intracell interference I_{intra} can be given in terms of intercell interference I_{inter} as

$$I_{inter} = I_{intra}\left(\frac{1}{F} - 1\right) + \frac{E_b}{G_p}\left(\frac{1}{F} - 1\right) \qquad (9.46)$$

Intracell interference is in terms of number of users N given by

$$I_{intra} = (N - 1)\frac{E_b}{G_p} \qquad (9.47)$$

and the total interference can be written as

$$I_{intra} + I_{inter} = \left(\frac{N}{F} - 1\right)\frac{E_b}{G_p} \qquad (9.48)$$

Now, setting the required $E_b/(I_0 + N_0)$ in the loaded network equal to the required E_b/N_0 in the unloaded network gives

$$\frac{E_{b,\text{loaded}}}{N_0 + I_0} = \frac{E_{b,\text{loaded}}}{\left(\dfrac{N}{F} - 1\right)\dfrac{E_b}{G_p} + N_0} = \left(\frac{E_b}{N_0}\right)_{\text{unloaded}} \tag{9.49}$$

and solving the needed E_b/N_0 in the loaded case gives

$$\left(\frac{E_b}{N_0}\right)_{\text{loaded}} = \frac{1}{\left(\dfrac{E_b}{N_0}\right)_{\text{unloaded}}^{-1} - \left(\dfrac{N}{F} - 1\right)\dfrac{1}{G_p}} \tag{9.50}$$

In Figure 9.17, range is presented as a function of uplink load for fractional cell load, $F = 70\%$. Load is shown as a percentage of maximum downlink capacity.

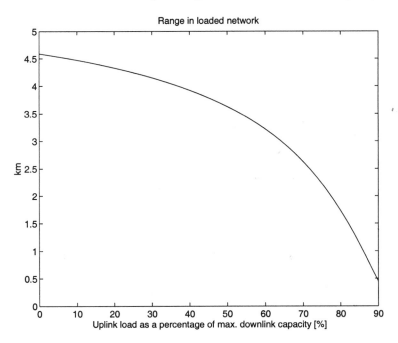

Figure 9.17 Range as a function of load. Load is shown as a percentage of maximum downlink capacity.

The maximum cell range decreases significantly when the load increases beyond about 50% of maximum downlink load. In an unloaded network the maximum cell range is 4.6 km, at 50% load range is 3.6 km, and at 90% load range is only 0.5 km. In

Figure 9.18 the E_b/N_0 requirements as a function of number of users is shown for $F = 50\%$, 70%, and 90% corresponding to different propagation environments.

For $F = 50\%$ the increase in E_b/N_0 requirements is much higher than for $F = 70\%$ since there is, in addition to intracell interference, high intercell interference.

9.6.3 Range Extension with Base Station MUD in Loaded Networks

The range decrease as a function of load can be partially avoided by using interference cancellation or multiuser detection at the base station. Base station MUD can be used to increase the uplink capacity, but it can also be used to extend the range in a loaded network. It is assumed that MUD is able to cancel part of the intracell interference at the base station.

Base station MUD can be used to make range and cellular coverage more insensitive to uplink load. In practice, MUD efficiency β will depend on the channel estimation algorithm, interference cancellation algorithm, and mobile speed. This introduces one new parameter into network planning. Also, the actual impact of MUD into cell range depends on the fractional cell loading F, which has to be predicted for each cell individually.

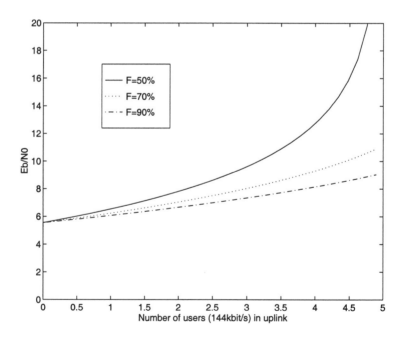

Figure 9.18 E_b/N_0 requirements in dB as a function of load in different propagation environments.

The effect of MUD can be taken into account by using the efficiency of MUD β as a measure of performance of MUD. That gives the percentage of intracell interference that is removed by MUD. With MUD the intercell interference $I_{intra,MUD}$ can be written as

$$I_{intra,MUD} = (1 - \beta)I_{intra} \tag{9.51}$$

The total interference is now

$$I_{intra,MUD} + I_{inter} = \left(\frac{N(1-\beta)+\beta}{F} - 1 \right) \frac{E_b}{G_p} \tag{9.52}$$

The required E_b/N_0 in the loaded network with MUD becomes

$$\left(\frac{E_b}{N_0} \right)_{loaded,MUD} = \frac{1}{\left(\dfrac{E_b}{N_0} \right)_{unloaded}^{-1} - \left(\dfrac{N(1-\beta)+\beta}{F} - 1 \right) \dfrac{1}{G_p}} \tag{9.53}$$

Range as a function of efficiency of base station MUD is shown in Figure 9.19. An efficiency of 0% corresponds to the base station without MUD. We can see that the range varies considerably depending on MUD efficiency.

From Section 9.5 we obtain a simulated figure for efficiency of MUD β: 70%. Now, the ranges with and without MUD are presented in Table 9.24 and in Figure 9.20.

Table 9.24
Range with and without Base Station MUD

Uplink load (%)	Range without MUD (km)	Range with MUD (km) (β=70%)
0	4.6	4.6
50	3.6	4.4
70	2.6	4.2
90	0.5	4.0

9.6.4 Mobile Transmission Power Savings with Base Station MUD

Even if the required coverage in the network can be achieved without base station MUD, advanced receiver algorithms at the base station can be used to decrease the transmission power of the mobile stations. The transmission power at maximum range $P_{TX,MS}$ can be written from (9.36)

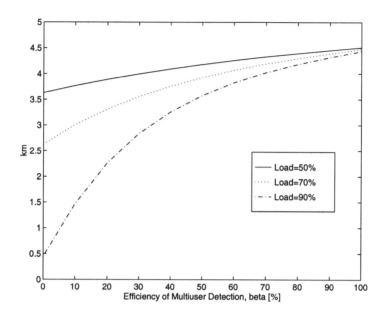

Figure 9.19 Range extension with base station MUD.

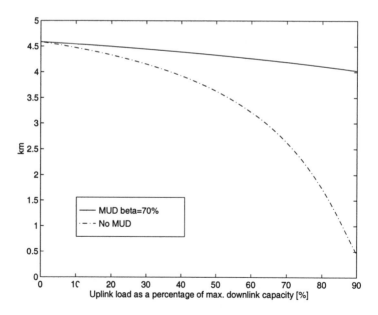

Figure 9.20 Range as a function of load with and without MUD.

$$P_{\text{TX,MS}} = P_{\text{RX,min}} + L_{\text{HO}} - G_{\text{MS}} - G_{\text{BS}} \tag{9.54}$$

and by replacing $P_{\text{RX,min}}$ with (9.35):

$$P_{\text{TX;MS}} = \frac{E_b}{N_0} + R_b + NF + kT + L_{\text{HO}} - G_{\text{MS}} - G_{\text{BS}} \tag{9.55}$$

Because all the terms in (9.50) are the same except for E_b/N_0, regardless of the base station receiver algorithm, $P_{\text{TX,MS}}$ is determined only by the E_b/N_0 requirement. The decrease in required transmission power with MUD is therefore

$$\frac{P_{\text{TX,MS}}}{P_{\text{TX,MS,MUD}}} = \frac{\left(E_b\big/N_0\right)_{\text{loaded}}}{\left(E_b\big/N_0\right)_{\text{loaded,MUD}}} \tag{9.56}$$

and the gain in E_b/N_0 can be written as

$$\frac{\left(\dfrac{E_b}{N_0}\right)_{\text{loaded}}}{\left(\dfrac{E_b}{N_0}\right)_{\text{loaded,MUD}}} = \frac{\left(\dfrac{E_b}{N_0}\right)_{\text{unloaded}}^{-1} - \left(\dfrac{N(1-\beta)+\beta}{F} - 1\right)\dfrac{1}{G_p}}{\left(\dfrac{E_b}{N_0}\right)_{\text{unloaded}}^{-1} - \left(\dfrac{N}{F} - 1\right)\dfrac{1}{G_p}} \tag{9.57}$$

In the above equations, it must be assumed that the propagation loss is lower than the maximum loss allowed for a system without MUD. If the mobile is not transmitting at full power but is able to use fast power control, then (9.52) is not exact but an approximation. In Figure 9.21 the estimated decrease in average transmission power of mobile station is shown as a function of MUD efficiency.

Assuming MUD efficiency $\beta = 70\%$ and fractional cell loading $F = 70\%$, the decrease of transmission power is 3 dB at 50% load and 7 dB at 70 % load.

9.6.5 Effect of User Bit Rate on the Range

The maximum range depends on the user bit rate. The higher the user bit rate, the shorter the range. The goal of third-generation networks is to provide full coverage for low bit rates (<144 Kbps), and for higher bit rates a limited coverage may be acceptable. In Table 9.25 the maximum ranges for different bit rates are calculated assuming the same E_b/N_0 target of 4.0 dB for all services and the parameters in Table 9.23.

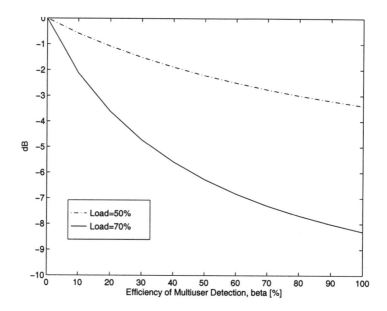

Figure 9.21   Savings in mobile transmission power with MUD.

Table 9.25
Maximum Range as a Function of User Bit Rate

User bit rate	Relative maximum range (km)
8 Kbps	10.0
14.4 Kbps	8.5
64 Kbps	5.6
144 Kbps	4.5
384 Kbps	3.4
1 Mbps	2.6
2 Mbps	2.2

9.6.6   Summary

Coverage of a loaded DS-CDMA cellular network has been considered by analyzing the uplink range. Base station multiuser detection (MUD) has been proposed as an upgrade solution to provide good coverage even with high system load after the initial deployment. MUD has also been shown to decrease mobile station transmission power in a loaded network. Therefore, a network with base station MUD can be operated with a higher percentage of maximum load if the system capacity is limited by downlink.

Good coverage need not be sacrificed in order to use high system loads. For actual network planning we need to take into account the performance of MUD to be able to predict and plan the coverage. The impact of MUD into cell range depends also on the propagation environment (i.e., the percentage of the intracell interference from the total interference). The increase in the data rate will inevitably mean reduced range in the uplink as the transmission power is limited. Thus, in the cell design the coverage area for the low rate services is likely to be different than for high rate service. This effect has not been visible in the second-generation networks because services are narrowband services with similar range performance.

REFERENCES

[1] Proakis, J. G., *Digital Communications*, 3rd ed., New York: McGraw-Hill, 1995.

[2] Xia, H. H., "Reference Models for Evaluation of Third Generation Radio Transmission Technologies," *Proceedings of ACTS Summit*, Aalborg, Denmark, October 1997, pp. 235–240.

[3] Recommendation ITU-R M.1225, "Guidelines for Evaluation of Radio Transmission Technologies for IMT-2000," *Question ITU-R 39/8*, 1997.

[4] Jakes, W. C., *Microwave Mobile Communications*, New York: Wiley, 1974.

[5] Parsons, J., *The Mobile Radio Propagation Channel*, London: Pentech Press, 1992.

[6] ETSI/SMG-5, "Selection procedures for the choice of radio transmission technologies of the UMTS," UMTS 30.03, v. 3.0.0, May 1997.

[7] Martin, G., and M. Faulkner, "1.9 GHz Measurement-based Analysis of Diversity Power versus the Number of RAKE Receiver Tines at Various System Bandwidths," *Proceedings of PIMRC'97*, September 1997, Helsinki, Finland, pp. 1069–1073.

[8] Otmani, M., and M. Lecours, "Indoor Radio Impulse Response Measurements with Polarization Diversity," *Proceedings of VTC'96*, Atlanta, Georgia, May 1996, pp. 151–154.

[9] CEC Deliv. No. R2020/TDE/PS/DS/P/040/b1, CODIT Final Propagation Model, June 1994.

[10] Rege, K., S. Nanda, C. F. Weaver, and W.-C. Peng, "Analysis of Fade Margins for Soft and Hard Handoffs," *Proceedings of IEEE 6th Personal, Indoor and Mobile Radio Communications Conference (PIMRC'95)*, Toronto, Canada, 1995, pp. 829–834.

[11] Chopra, M., K. Rohani, and J. D. Reed, "Analysis of CDMA Range Extension due to Soft Handoff," *Proceedings of VTC'95*, 1995, pp. 917–921.

[12] Andoh, H., M. Sawahashi, and F. Adachi, "Channel Estimation Using Time Multiplexed Pilot Symbols for Coherent Rake Combining for DD-CDMA Mobile Radio," *Proceedings of PIMRC'97*, Vol. 3, Helsinki, Finland, September 1997, pp. 954–958.

[13] Tiedemann, E. G., Jr., Y.-C. Jou, and J. P. Odenwalder, "The Evolution of IS-

95 to a Third-Generation System and to the IMT-2000 Era," *Proceedings of ACTS Summit*, Aalborg, Denmark, October 1997, pp. 924–929.

[14] Hämäläinen, S., P. Slanina, M. Hartman, A. Lappeteläinen, and H. Holma "A Novel Interface Between Link and System Level Simulations," *Proceedings of ACTS Mobile Communications Summit '97*, October 7–10, 1997, Aalborg, Denmark, pp. 599–604.

[15] Hata, M., "Empirical Formula for Propagation Loss in Land Mobile Radio Services," *IEEE Transactions on Vehicular Technology*, August 1980, pp. 317–325.

[16] Ariyavisitakul, S., "SIR Based Power Control in a CDMA System," GLOBECOM'92, Orlando, Florida, December 1992.

[17] Shapira, J., and R. Padovani, "Spatial Topology and Dynamics in CDMA Cellular Radio," *Proceedings in VTC'92*, Denver, Colorado, May 1992, pp. 213–216.

[18] Special Issue on Smart Antennas, *IEEE Personal Communications*, Vol. 5, No. 1, February 1998.

[19] Ojanperä, T., K. Rikkinen, H. Hakkinen, K. Pehkonen, A. Hottinen, and J. Lilleberg, "Design of a 3rd Generation Multirate CDMA System with Multiuser Detection, MUD-CDMA," *Proceedings of ISSSTA'96*, Vol. 1, Mainz, Germany, September 1996, pp. 334–338.

[20] Nikula, E., A. Toskala, E. Dahlman, L. Girard, and A. Klein, "FRAMES Multiple Access for UMTS and IMT-2000," *IEEE Personal Communications Magazine*, April 1998, pp. 16–24.

[21] Jalali, A., and P. Mermelstein, "Effects of Diversity, Power Control, and Bandwidth on the Capacity of Micro Cellular CDMA System," *IEEE Journal on Selected Areas in Communications*, Vol. 12, No. 5, June 1994, pp. 952–961.

[22] Papoulis, A., *Probability, Random Variables, and Stochastic Processes*, 3rd ed., New York: McGraw-Hill, 1991.

[23] Hämäläinen, S., H. Holma, and A. Toskala, "Capacity Evaluation of a Cellular CDMA Uplink with Multiuser Detection," *Proceedings of ISSSTA'96*, Mainz, Germany, September 1996, pp. 339–343.

[24] Gass, J. H., D. L. Noneaker, and M. B. Pursley, "Spectral Efficiency of a Power-Controlled CDMA Mobile Communication System," *IEEE Journal on Selected Areas in Communications*, Vol. 14, No. 3, April 1996, pp. 559–569.

[25] Padovani, R., "Reverse Link Performance of IS-95 Based Cellular System," *IEEE Personal Communications*, Vol. 1, No. 3, Third Quarter, 1994, pp. 28–34.

[26] Viterbi, A. M., and A. J. Viterbi, "Erlang Capacity of a Power Controlled CDMA System," *IEEE Journal on Selected Areas in Communications*, Vol. 11, No. 6, August 1993, pp. 892–899.

[27] Ojanperä, T., P. Ranta, S. Hämäläinen, and A. Lappeteläinen "Analysis of CDMA and TDMA for 3rd Generation Mobile Radio Systems," *Proceedings VTC'97*, May 1997, Phoenix, Arizona, pp. 840–844.

[28] Holma, H., A. Toskala, and A. Hottinen, "Performance of CDMA Multiuser Detection with Antenna Diversity and Closed Loop Power Control," *Proceedings of VTC'96*, Atlanta, Georgia, May 1996, pp. 362–366.

[29] Parkvall, S., E. Ström, and B. Ottersen, "The Impact of Timing Errors on the

Performance of Linear DS-CDMA Receivers," *Journal on Selected Areas in Communications*, Vol. 14, No. 8, October 1996, pp. 1660–1668.

[30] Buehrer, M., A. Kaul, S. Striglis, and B. D. Woerner, "Analysis of DS-CDMA Parallel Interference Cancellation with Phase and Timing Errors," *IEEE Journal on Selected Areas in Communications*, Vol. 14, No. 8, October 1996, pp. 1522–1535.

[31] Simon, M., J. Omura, R. Scholtz, and B. Levitt, *Spread Spectrum Communications Handbook*, New York: McGraw-Hill, 1994.

[32] Pizarroso, M., and J. Jiménez (eds.), "Common Basis for Evaluation of ATDMA and CODIT System Concepts," *MPLA/TDE/SIG5/DS/P/001/b1*, MPLA SIG 5, September 1995.

[33] Hämäläinen, S., H. Holma, A. Toskala, and M. Laukkanen, "Analysis of CDMA Downlink Capacity Enhancements," *Proceedings of PIMRC'97*, Helsinki, Finland, September 1997, pp. 241–245.

Chapter 10

HIERARCHICAL CELL STRUCTURES

10.1 INTRODUCTION

A third-generation system must be able to support a wide range of services in different radio operating environments. Different types of cells are needed for different requirements: large cells guarantee continuous coverage, while small cells are necessary to achieve good spectrum efficiency and high capacity. Small range cells are used by low mobility and high capacity terminals, while high range cells serve high mobility and low capacity terminals. In addition, different cell types should be able to operate one upon another. Hierarchical cell structure (HCS) describes a system where at least two different cell types (e.g., macrocell and microcell) operate one upon another, as shown in Figure 10.1. Microcells are small cells covering areas with a radius of a few hundred meters. Low powered base stations are typically placed at lamp post level and they serve a street block. Macrocells, or umbrella cells, cover a radius of 1 km or more. Macrocells cover rural areas, provide continuous coverage of areas that are covered by microcells, and serve rapidly moving users. Picocells cover indoor areas with a cell radius of a few dozen meters. A low power base station usually covers an office, a floor of a high building, or a residence. Satellite cells introduce global continuous coverage by covering rural areas where terrestrial systems cannot be installed. Whenever possible, traffic should be directed to the smallest available cell, so that the system's spectral efficiency is improved. Particularly in the satellite segment, it is crucial to direct traffic to the macro- or microcells whenever possible [1].

In this chapter, two approaches for HCS design in CDMA are described, and advantages and disadvantages of the two methods are discussed. In the first approach, coexisting hierarchy layers operate in the same frequency band [2, 3]. Users operating on different layers are separated by handovers and signal fading. In the second approach, different hierarchy levels are separated in the frequency domain. The aspects related to HCS are also relevant in a multioperator environment where an operator has to consider interference from the adjacent frequencies belonging to another operator. Network planning aspects such as guardbands related to multioperator environments are discussed in Chapter 12.

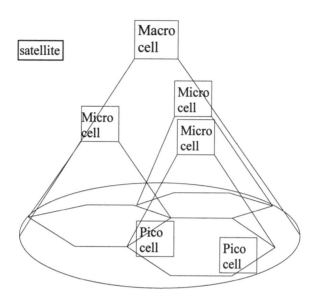

Figure 10.1 Hierarchical cell scenario.

First, nonlinear power amplifiers are studied in Section 10.2 to establish an understanding of the adjacent channel interference mechanism in a system with different cell layers at different frequencies. A HCS system with micro- and macrocells in the same frequency, and at different frequencies, are studied in Sections 10.3 and 10.4, respectively. It should be noted that aspects related to interfrequency handover are not covered here but rather in Chapters 2 and 5. Section 10.4 evaluates link and system level performance of a cellular HCS network utilizing DS-CDMA and multiuser detection.

10.2 NONLINEAR POWER AMPLIFIERS

Even if hierarchy layers have different carrier frequencies, some adjacent channel interference (ACI) is generated between adjacent channel carriers. When band limited linear modulation methods are used, spectrum leakage between adjacent channel carriers depends on the linearity of the power amplifier (PA) [4]. Spectrum leakage to adjacent channels can be controlled by backing off the PA or by using some linearization method to equalize the nonlinearities of the power amplifier. Unfortunately, both methods decrease the achievable PA efficiency when compared to modulation schemes that have constant envelope and can be amplified with power efficient but nonlinear power amplifiers [4].

Real power amplifiers are always nonlinear. Linearity requirements of a CDMA transmitter are mainly determined by the spectrum ACI attenuation requirements, rather than from link level performance losses. If low ACI attenuations (i.e., larger spectrum spreading into adjacent carriers) can be tolerated, more efficient power amplifiers can be used in mobile terminals.

For link level simulations, a model of a nonlinear power amplifier is needed. In order to get real life amplifier model, a power amplifier IC was measured. The power amplifier IC is manufactured with the Gallium Arsenide heterojunction bipolar transistor (HBT) process, and the final stage of the IC is biased to class-AB.

10.2.1 Power Amplifier Characteristics

A typical power amplifier introduces both amplitude and phase distortion into the transmitted signal, resulting in AM-AM and AM-PM conversion effects. Figure 10.2 shows the AM-AM and AM-PM characteristics of the measured amplifier. The power amplifier amplifies input power linearly when input RMS power is −15 to 0 dBm. This is the linear area of the amplifier. When the input power is 0 to 5 dBm, the amplifier operates in its conversion or saturation area. The measured power amplifier has saturated output power at the level of 34 dBm. This value corresponds to 0 dB output backoff value [4]. Output backoff is defined as the difference between output power in the saturation point and output power in the operation point. Backing off the amplifier reduces output powers in the operation point and increases output backoff.

Figure 10.2 (a) AM-AM curve and (b) AM-PM curve for measured power amplifier IC.

10.2.2 Power Amplifier Efficiency

Efficiency of the power amplifier IC is measured using a constant envelope sine wave. Efficiency compared to output backoff is shown in Figure 10.3. At saturation power (0 dB output backoff), the power amplifier has slightly less than 50% efficiency. This means that if constant envelope modulation methods, such as Gaussian minimum shift keying (GMSK), were used, up to 50% efficiencies could be achieved. Because band limited linear modulation methods have nonconstant RF-envelope, output power needs to be lowered in order to avoid spectrum spreading [4]. The accurate efficiency of the power amplifier with band limited linear modulation methods (nonconstant RF-envelope) can be predicted by using the measured efficiency curve and the simulated pdfs of the used modulation method [4].

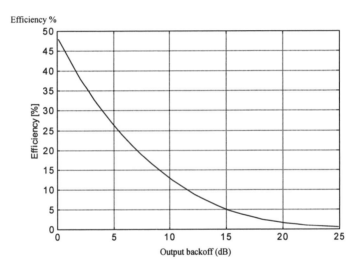

Figure 10.3   Power amplifier efficiency versus output backoff, class=AB PA.

10.3   MICRO- AND MACROCELLS AT THE SAME FREQUENCY

In the case of micro- and macrocells operating at the same frequency, the frequency reuse factor is set to one and spatial isolation is used for separating micro- and macrocell layers [2, 3]. A reuse factor of one can be offered because processing gain allows users to experience interference originating from any cell layer. Intralayer interference is controlled by power control and interlayer interference by spatial isolation. In a generic flat-Earth model, the average transmission loss follows R^{-2} until a breakpoint that marks the separation between two segments. After the breakpoint, R^{-4} is followed. The location of the breakpoint depends on receiver and transmitter heights. As the base station is installed lower, the breakpoint occurs nearer the base station.

Thus, the signal from a microcell base station attenuates faster than the signal from a macrocell base station. In Figure 10.4, spatial isolation is depicted for both hotspot case and for continuous coverage of microcells.

The problem with the presented method is that the mobile station should be instantly handed to the corresponding cell layer when it arrives to the intersection of micro- and macrocell attenuation curves. Soft handover eases the problem, but if the mobile station arrives in the microcell area, it is most often connected to both micro- and macrocell base stations. This occurs because the handover region has to be rather large in order to avoid ping-ponging between cells (or cell layers), especially if the mobile stations are fast moving. Thus, the area where mobile stations are in a soft handover state is rather wide compared to the area of microcell.

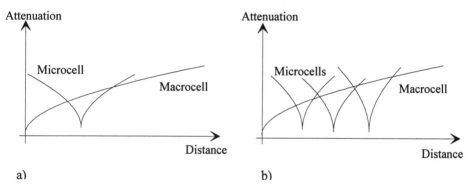

Figure 10.4 Spatial separation of umbrella cells and underlay microcells along: (a) hotspot case covered by one microcell base stations and (b) the full coverage of microcells [2].

Handover can never be very fast. If the number of candidate base stations is high, the pilot signal of a candidate base station can be measured only after some delay. In addition, this pilot has to be measured for a rather long period in order to filter fast fading. Handover can not be performed when the mobile station encounters a fading dip due to fast fading. After handover decision has been taken, signaling between the mobile station and the base station takes place. Before both the network side and the mobile side get all the handover related parameters, some delay has been experienced. The total delay due to pilot measurement, filtering, and handover signaling may be on the order of several seconds. If the mobile station runs with a 50 km/h speed, it moves almost 14m during 1 second. If the size of a microcell is 100m, only a 7-second delay on handover is needed to go to the center of the cell.

10.4   MICRO- AND MACROCELLS AT DIFFERENT FREQUENCIES

A system with different cell layers at different frequencies is easier to manage because cell layers do not interfere with each other as much as when they are at the same frequency. The main interference comes from the adjacent channel spill-over. A drawback of this approach is large spectrum requirements because each cell layer requires its own frequency. For example, as was discussed in Chapter 6 (Figure 6.1), the WCDMA scheme implements three layers within 15 MHz.

In this section, first ACI attenuation masks for the system level simulations are generated by using the characteristics of a measured power amplifier IC. The capacity for the uplink and downlink of a HCS network is then evaluated. The impact of power control, handover, and channel spacing on the system capacity is also discussed.

10.4.1 Adjacent Channel Interference and Link Level Performance

Adjacent channel interference and link level performance degradation determine the impact of nonlinear power amplifiers on overall spectrum efficiency. As shown below, the increased adjacent channel interference will dominate the degradation of spectrum efficiency due to nonlinear power amplifiers.

In order to simulate a HCS network with the system level simulator, adjacent channel interference has to be modeled. Here, adjacent channel interference is modeled so that the output power of an interfering user is reduced by *adjacent channel attenuation*. Adjacent channel attenuation describes how much the power of the interfered user is attenuated if it is received at the adjacent channel. Adjacent channel interference attenuations are simulated with OQPSK chip modulation, band limited with square-root raised cosine pulse shaping and using the described power amplifier model. Roll-off factor is selected to be 0.20 in the simulations. Because the system includes different chip rates for macro- and microcells (2.5 and 5.1 Mchip/s), two

Figure 10.5 ACI masks for system level simulations.

separate ACI masks need to be created for system level simulations. These masks are created by backing off the power amplifier by 3.5 dB and by measuring the spectrum spreading power with the receiver square-root raised cosine filter also having a roll-off factor of 0.2. By sliding the receiver filter in frequency domain, continuous ACI masks are created for system level simulations, as shown in Figure 10.5. The backoff used corresponds to about 31% power amplifier efficiency [4]. Because the measured amplifier effects are quite equal to OQPSK and QPSK modulation methods [4], the same ACI masks can be used for the downlink system level performance simulations.

Link level performance was also simulated with the same power amplifier model and with OQPSK and QPSK chip modulations, band-limited with square-root raised cosine filter having a roll-off factor of 0.2. CDMA was found to be very robust against the nonlinearity of the power amplifier. Less than 0.2 dB link level performance losses occurred (both uplink and downlink) when the same output backoff (3.5 dB) as in the ACI mask definitions was used. This means that in a CDMA system, the spectrum mask requirements will limit the use of nonlinear amplifiers rather than link level performance losses. On the other hand, this means that if moderate ACI attenuation requirements can be tolerated in CDMA, the use of highly efficient power amplifiers becomes possible.

10.4.2 System Level Simulations of an HCS Network

In Figure 10.6, a HCS case with a hotspot of traffic in the middle of the area is shown. The high density of users located at the hotspot is served by microcells. The whole system area is covered by macrocells, which are used to provide continuous coverage.

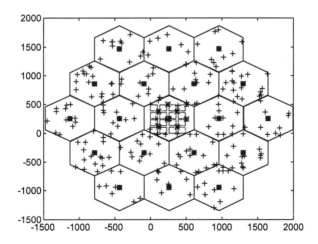

Figure 10.6 HCS environment where the macrocell base stations are on a hexagonal grid and the mobile stations (plus signs) are generated only to the streets. Eight microcell base stations are placed in the middle of the map (hotspot).

Figure 10.7 shows continuous coverage of microcell base stations. Here, both micro- and macrocell base stations provide continuous coverage. Fast moving mobile stations are handled by macrocell base stations, while high capacity terminals are served by microcell base stations.

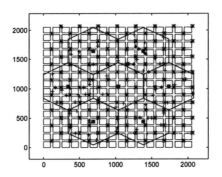

Figure 10.7   HCS environment with the full coverage of microcells.

System level simulations are performed at two HCS environments so that the hierarchy layers are separated in frequency. In the first case, the simulated system is covered by full coverage of macro- and microcell base stations. In this case, the number of macro- and microcell base stations are 7 and 128, respectively. In the second case, microcells are placed in a hotspot, while macrocells continuously cover the simulated area. The number of microcells is 8, and the number of macrocells is 19. In the simulation, macrocell base stations are placed in a hexagonal grid and the microcell base stations in a Manhattan-like grid. Microcell base stations are located in every second street intersection. All mobile stations are located in street canyons.

Instantaneous SIR is calculated by dividing the received signal by the interference and multiplying by the processing gain. In the uplink, it is assumed that a signal propagates from a mobile station to a base station via four equal strength Rayleigh faded paths. The observed base station adds together paths received from the observed user. It is supposed here that signals are combined coherently with maximal ratio combining. Maximal ratio combining is modeled by taking the sum of SNR values of each received path [5]. When simulating medium bit rate services, interference can be assumed to be Gaussian, because it is composed of transmissions to several independent users: signals from other users are just added together and seen as interference. If an interfering user was connected to the observed base station, the multiuser detection process is able to cancel the signal partially. The cancelled signal is multiplied by $(1-\beta)$, where β is *multiuser detection efficiency* and given as a parameter. SIR can be written as

$$SIR_{UL} = \frac{G_p \cdot S}{(1-\beta) \cdot I_{\text{intra}} + I_{\text{inter}} + N_0} \tag{10.1}$$

Because the system is interference limited, thermal noise N_0 is assumed to be small and neglected. External interference I_{inter} consists of intercell and interhierarchy level interference. Interhierarchy level interference is multiplied by *adjacent channel attenuation* δ. Thus, I_{inter} is

$$I_{inter} = I_{intercell} + \delta \cdot I_{interlayer} \tag{10.2}$$

In the downlink, signals from each base station to the mobile station propagate via four independent equal strength Rayleigh faded paths. Because macro diversity was considered at link level, the mobile station is able to receive all the paths from the base stations transmitting to it. Interference that propagates via the same paths than the desired signal is multiplied by an *orthogonality factor* α. This orthogonality factor states the orthogonality loss due to multipath propagation and energy losses due to limited receiver capacity. The orthogonality factor α is a simulation parameter because it varies according to radio environment, depending on the multipath delay spread (see Section 9.5.3). The transmission of the base stations that do not direct their transmission to the observed user contributes to the noise. SIR will then be

$$SIR_{DL} = \frac{G_p \cdot S}{(1 - \alpha) \cdot I_{intra} + I_{inter}} \tag{10.3}$$

In (10.3) I_{inter} consists of intercell and interhierarchy level interference, where interference from other hierarchy levels is multiplied by δ.

A performance measure for system simulations is outage percentage. A maximum of 5% of the obtained SIR values are allowed to be lower than the threshold SIR. SIR distributions are generated for both hierarchy layers and outage requirements that have to be fulfilled for both. Thus, the total outage requirement becomes even tighter than 5%. When the developed outage for one layer is 5%, the corresponding outage with the selected load will probably be less than 5% for the other layer. Simulation parameters are shown in Table 10.1.

10.4.3 Spectrum Efficiency Results

Spectrum efficiency results are shown in Table 10.2. Single layer micro- and macrocell capacities were simulated and used as references for HCS simulations. As seen previously, the capacity of the uplink with multiuser detection and antenna diversity is higher than the near orthogonal downlink capacity. In HCS simulations with full coverage of microcells, the microcell layer was loaded to 80% of a single layer microcell capacity. In case of a hotspot, macrocells were loaded to 80% of a single layer capacity. As seen from the results, the coexisting layer capacity is almost the same as the single layer capacity, even when the interfering layer was loaded to 80% of its capacity. Even if the difference between path loss of a microcell user and a macrocell user may be several dozen decibels, power control with high dynamics makes the situation easy.

Capacities shown in Table 10.2 are obtained with high dynamic power control range. If the downlink power control range is set to 10 dB, a high difference between path loss values cannot be compensated and outage becomes poor with the selected loading. Outage requirements can be fulfilled if the load is decreased or if channel spacing is increased. Very high power control dynamics may be difficult to implement in the downlink. Therefore, power control dynamics and channel spacing should be selected so that performance versus bandwidth usage is optimized and implementation requirements fulfilled. As an example, power control dynamics could be 10 dB and channel spacing 1.6 MHz higher than in the reference case. Then, on average, a 15 dB higher difference on path loss values can be compensated. Another way to increase the ACI attenuation is to raise the output backoff of the power amplifier. Unfortunately, this decreases the achievable power amplifier efficiency quite fast, as shown in Figure 10.3. If the number of microcells is low, then the intercell interference ($I_{intercell}$) becomes low and the HCS microcell capacity can be even higher than the reference capacity. This is the case if microcells are in a hotspot. The same effect occurs for macrocell, as can be seen in the case where microcells cover a large area and the number of macrocell base stations is low (seven base stations).

Simulation results shown in this chapter are different from the results shown in Chapters 5 and 9, as the selected parameter set and channel models were different.

Table 10.1
System Level Simulation Parameters

Parameter	Uplink	Downlink	Unit
Chip rate: - macrocell	2.5	2.5	MHz
- microcell	5.1	5.1	MHz
Channel separation	4.6	4.6	MHz
Handover margin	3	3	dB
Active set size	3	3	–
MUD efficiency: - 144 Kbps, micro	60	–	%
- 12 Kbps, macro	40	–	%
Orthogonality, - macro	–	68	%
- micro	–	62	%
ACI attenuation (δ): - Macro-to-micro	28	28	dB
- Micro-to-macro	32	32	dB
Power control step size	1	1	dB
Power control dynamics	80	10 and 80	dB
E_b/N_o, - 144 Kbps, micro	0.4	6.4	dB
- 12 Kbps, macro	5.7	6.4	dB

Table 10.2
Simulation Results [Kbps/MHz/cell]

	Uplink		Downlink	
	Micro	Macro	Micro	Macro
Hotspot	1138	204	317	175
Full	634	271	216	276
Reference	792	236	274	218

In [6], the case where adjacent channels overlap (totally or partially) was neglected. From the simulations, it is expected that a CDMA-based network could operate so that both hierarchy levels use the same frequency. This, however, requires very fast power control and handover, and a very wide power control dynamic range for both transmission directions. Because the simulations were performed as snapshot simulations, the mobile station did not move during the simulation, and effects due to mobility were neglected. Mobility may destroy systems where the two hierarchy levels operate with the same frequency. Consider a case where a mobile station connected to the macrocell moves rapidly to the area of a microcell. The handover process is so slow that a mobile station runs deep inside microcell area before it is handed to the microcell base station. The only solution is very fast power control. If the mobile stations connected to the microcell base station are able to adjust their uplink powers fast enough, the system will not crash. The mobile stations connected to macrocells should have such a fast downlink power control that they can adjust downlink power in a way that the connection does not drop due to increased interference power from the microcell. In addition, the power control range for downlink should be so large that mobile stations can compensate for increased interference. The conclusion from system simulations is that hierarchical cell structure networks are feasible with CDMA, but some implementation problems may occur.

REFERENCES

[1] Andermo, P. G., "System Flexibility and Its Requirements on Third Generation Mobile Systems," *Proceedings of PIMRC'92*, Boston, Massachusetts, 1992, pp. 1–3.

[2] Shapira, J., and R. Padovani, "Spatial Topology and Dynamics in CDMA Cellular Networks," *Proceedings of VTC'92*, Denver, Colorado, 1992, pp. 213–216.

[3] Shapira, J., "Microcell Engineering in CDMA Cellular Networks," *IEEE Transactions on Vehicular Technology*, Vol. 43, No. 4, November 1994, pp. 817–825.

[4] Lilja, H., "Characterizing the Effect of Nonlinear Amplifier and Pulse Shaping on the Adjacent Channel Interference with Different Data Modulations," M.Sc. Thesis, Oulu University, Finland, 1996.

[5] Proakis, J. G., *Digital Communications*, New York: McGraw-Hill, 1995, pp. 778–785.

[6] Hämäläinen, S., H. Lilja, and J. Lokio, "Performance of a CDMA Based Hierarchical Cell Structure Network," *Proceedings of PIMRC'97*, Helsinki, Finland, September 1997, pp. 863–866.

Chapter 11

IMPLEMENTATION ASPECTS

11.1   INTRODUCTION

This chapter concentrates on the implementation aspects of a wideband CDMA transceiver. The goal is to help the reader understand the basic analog and digital structure of a wideband CDMA terminal and get an idea of the problems encountered in real implementation.

Section 11.2 highlights different optimization criteria (power consumption, cost, and size) that should be taken into account while designing a CDMA transceiver. The task is certainly not easy because the number of degrees of freedom is large. The main concern is on the mobile terminal side, but in the future, power consumption, cost, and size will become important also for base stations.

In Section 11.3, the transceiver modularity concept is discussed in conjunction with wideband CDMA systems. A wide set of defined third-generation system services calls for terminals capable of adapting to different data rates and quality of services. In practice this makes the physical terminal design much more complicated because many functions have multiple sets of operating parameters.

Baseband signal processing is investigated in Section 11.4. The emphasis is on the basic building blocks, but attention is also paid to more advanced techniques like interference cancellation. First, the different processing units are considered from the functional architecture point of view. This is followed by a discussion of how to make the mapping from the functional description to a physical implementation. The problem is not, however, unambiguous, making an optimal solution difficult to find. In all cases, digital application specific integrated circuit (ASIC) or digital signal processing (DSP)-based implementation is assumed, although some baseband processing like matched filter could be performed in an analog domain. At the end of the section, different control loops for AGC, AFC, and power control are highlighted.

Section 11.5 presents well-known RF architectures suitable for a wideband CDMA transceiver. The RF section is considered to incorporate everything from radio frequency signal to baseband, independent of analog or digital implementation. Complexity of the front end is increased compared to second-generation systems, mainly because of the need for higher linearity, better selectivity, and larger dynamic range requirements. Most radio frequency processing is still done in the analog domain, while in the future the digital border is moving towards the antenna.

Section 11.6 describes the concept of software configurable radio, and Section 11.7 illustrates an example configuration of a wideband CDMA mobile terminal architecture, integrating the different parts.

Section 11.8 discusses implementation of multimode terminals. The different implementation requirements between spread and nonspread, between slotted and continuous, between circuit and packet switched, and finally between FDD- and TDD-based systems are discussed.

11.2 OPTIMIZATION CRITERIA

In the design of a wideband CDMA terminal and base station, critical factors to be optimized are power consumption, cost, and size. Each of these can be evaluated by qualitative criteria such as input power requirements, number of components, number of instructions, or ASIC gate primitives. Figure 11.1 illustrates a generic wideband CDMA transceiver partitioned into RF, IF, and baseband sections. Traditionally, RF and IF sections have been realized by analog technology, and baseband by digital technology. The interface, however, between analog and digital sections depends on the selected receiver and transmitter architectures, as discussed in Section 11.5.

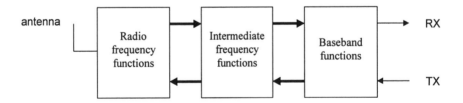

Figure 11.1   Block diagram of a CDMA transceiver.

11.2.1 Power Consumption, Cost, and Size

Because of radical evolution in the digital signal processing field in terms of DSP processors and new ASIC technologies (see Table 11.1), the analog RF section is becoming the major power consumer in normal speech terminals. On the other hand, more complex baseband functions are being required (e.g., multiuser detector), and if the digital-analog border shifts towards antenna (see Section 11.5.2), the baseband may still form a significant part of the power consumption.

Table 11.1
Development of DSP and ASIC Technologies

	1992	1994	1996	1998	2000
Gate density (gates/mm^2)	2000	5000	15,000	30,000	50,000
Processing power (MIPS)	20	50	80	120	200
Relative power consumption (1/MHz)	100	30	10	3	1
Relative power consumption (max MIPS)	100	75	40	18	10

Source: [1].

On the RF side, the dominating current consumers are the power amplifier and the frequency synthesizers. The former is seen as a critical component in third-generation systems because of their high linearity requirements, which likely cause a loss of efficiency. It should be noted, however, that because at smaller cell sizes (pico/microcells), output power is reduced, the overall power consumption can be kept acceptable. The power consumption of a synthesizer is directly related to the activity factor and is thus larger for systems such as CDMA, which transmit continuously. In addition, two synthesizers are needed in CDMA terminals because transmission and reception take place simultaneously in contrast to TDMA terminals, which need only one synthesizer.

The transceiver cost factor depends quite evenly on every required component. So far, the division between analog RF and digital baseband sections has been approximately fifty-fifty. The trend, however, leans toward the analog side because the integrated digital processing side is shrinking into a single chip.

The actual analog and digital data processing functions (encoding/decoding, filtering, mixing, equalization) are not very critical in terms of the total terminal size. This may have been the case in the past, but today the dominant components are the battery and the user/external interfaces. On the data path side, some RF section components such as the duplex filter and possibly the antenna may occupy a large amount of space.

The development of the first third-generation mobile terminals is a difficult task. They should have all the features of the second-generation systems and include new features related to higher bit rate services. The fact that consumers also expect to be able to connect to the existing systems calls for a dual/multimode engine.

11.2.2 Evaluation Methods

There are a few quantitative parameters that can be used to evaluate the transceiver power consumption, size, and cost. This section explains what should be taken into account when assessing analog and digital section complexities. Radio frequency parts are seen as analog, and baseband functions as digital. IF processing lies somewhere in between.

On the analog RF side, power consumption can be assessed by estimating the average input power requirements in both talk and idle modes. The power consumption is a function of user data rate, activity factor, required effective radiated power, and transmission spectral mask (linearity requirements). Emphasis is on deriving the power amplifier efficiency in different situations. The activity factor of a synthesizer has a large impact on the power consumption. Therefore, effective power-down/up procedures should be used to turn off the synthesizer whenever possible.

Cost and size of the RF section depends strongly on the number of individual components. The major problems relate to isolation requirements and integration of different semiconductor processes, which prevent a single chip RF front-end implementation. Increased integration, especially on the intermediate frequency side, has already been achieved (possibly joined with some baseband blocks), but still there are a few discrete components such as channel selection and duplex filters that must be implemented externally.

Evaluation on the digital processing side is a much more straightforward task because an analytical approach can be followed. Each function can be divided into instruction and/or gate primitive operations, which can further be mapped to real DSP processors and ASIC processes to give direct estimates for power consumption, cost, and size. The most problematic task here is to compare DSP- and ASIC-based implementations. A good method for doing this is to take the word length requirements into account when counting the MIPS requirements. This leads to a concept called normalized MIPS (nMIPS), which states the number of 1-bit operations needed to execute within a second. The example below compares the complexities of a RAKE receiver correlator and complex phase rotation primitives. The former is a clear ASIC-oriented function, while the latter is more ideal for a DSP processor.

Example. The despreader is for a QPSK chip modulated signal with an input sample rate of 4.096 Msps and input word length of 4 bits. The spreading ratio is 64. The same spreading code is used in both I and Q branches, so only two correlations are needed. The phase rotator takes a complex 10-bit input vector and multiplies that with another 10-bit vector. The operation rate is 64 Ksps. In this case, multiplication is assumed to take (Wlin · WLin) and addition (Wlin + WLin) primitive operations. Thus,

$$\text{nMIPS(despreader)} = 2 \cdot F_s \cdot [\text{WLin} \cdot \text{WLc} + ((\text{WLin} \cdot \text{WLc}) + \text{WLin2})] = 148 \text{ nMIPS}$$

where F_s = sampling frequency = 4.096 Msps, WLin = input sample word length = 4, WLc = reference code word length = 1, and WLin2 = integrator second input word length = 10.

nMIPS(rotator) = Fs·[4·WLin·WLin+2·(WLin+WLin)] = 28 nMIPS

where Fs = sampling frequency = 64 Ksps and WLin = input sample word length = 10.

11.3 MODULARITY CONSIDERATIONS

Third-generation wideband CDMA terminals will provide a set of different services varying from a low rate speech service to a high rate, high quality service for data. When designing RF and baseband architectures this should be taken into account. The more adaptable the architecture, the better the available processing resources can be utilized to provide this variety of operation modes.

In wideband CDMA systems, the higher data rates can be obtained by decreasing the spreading factor or by introducing more parallel code channels (see Section 5.8). The latter option, especially in the case of only a few parallel channels, results in a high peak-to-average power requirement and thus to high backoff. Therefore, it is not very good from the transmitter power amplifier point of view. It may be best to use parallel channels in the downlink to obtain good modularity and variable spreading factor in the uplink to avoid parallel transmission in the mobile station.

The baseband section chip rate elements have a fairly fixed complexity as a function of data rate. This is because the data rate is mainly defined by the correlator dumping period, not the chip/sample rate. Multiple code channels, however, create extra hardware requirements, but still many blocks (like receiver front-end, multipath delay estimation, and RAKE finger allocation, as well as complex channel estimation units) may not need to be duplicated for different codes. The critical parts from the complexity point of view are the transmitter pulse shaping filter, additional RAKE receiver fingers, and code generators. The complexity of a pulse-shaping filter depends heavily on the input word length, and in the case of several parallel code channels plus high adjacent channel attenuation requirements (long filter), implementation becomes difficult. Each parallel code channel used requires code generator(s) for spreading and despreading. With short M-sequences (simple shift registers) the complexity increase can be tolerated, while with the long Gold and Kasami codes, which require several shift registers, the silicon area consumption will grow significantly. Although a RAKE finger itself is a reasonably simple device, the total receiver complexity increases because each multipath component requires as many despreaders as there are parallel code channels.

The main effect of increasing user bit rate, however, is seen on the narrowband side because processing requirements at the symbol rate are closely tied to the user data rate. This mainly affects the channel encoding, decoding, and interleaving functions. The key aspect here is how to make the architecture as scalable as on the wideband side.

The required scalability can be achieved in two ways. Extra processing power is achieved by installing several processing units operating in parallel; or existing unit throughputs are increased by using a higher clock rate. The former seems justified from the ASIC technology evolution point of view because the available silicon area is constantly expanding. Clock rate increase can be used to save silicon but may also cause hotspots. It is not desirable, however, to have high rate system clocks at the mobile station.

Another thing to consider is the specified set of different coding schemes for several bearer services. It is obvious that data services require better BER than speech services. This must be taken into account while designing the encoder/decoder architecture in order to provide maximum resource utilization. The more common lower level processing primitives that can be found, the higher the resource sharing that can be provided. This is easier with encoders, while decoders tend to have more individual constructions. A good example is to use the same physical shift register for convolutional and Reed-Solomon encoders. Software oriented implementation would be a good solution here, but for high data rates the processing power requirements may be too high and too expensive.

11.4   BASEBAND SECTION

Baseband processing implements all the data path functions operating on a signal whose spectrum center is located at zero frequency. In wideband CDMA, the baseband section can be divided into bit/symbol and chip/sample domains, each having a different operation rate. This section describes the basic baseband functions (data path/control) required for a WCDMA terminal implementation. In all cases, the physical implementation is assumed to take place in the digital domain (not counting the converters).

11.4.1 Baseband Receiver

Figure 11.2 illustrates the generic block of a direct sequence CDMA receiver. After analog-to-digital (A/D) conversion and filtering, the wideband I/Q sample stream is fed into a correlator bank that performs signal transform (despreading) into the narrowband domain where the multipath diversity combiner collects the channel energy from different RAKE fingers. Channel delay profile estimation is required to resolve the multipaths and to set the RAKE fingers to track a specific code phase. After combining, the bits are deinterleaved and decoded to produce transmitted information bit stream. After the despreading and before the multipath combining, MUD can be used to increase system performance. MUD is primarily a base station-oriented function and can be implemented in the wideband domain as well.

11.4.1.1   Analog-Digital Conversion

In a wideband CDMA system, because of the wideband signal, a high sampling rate A/D converter is required. Power consumption and cost are not, however, radically increased because the processing gain obtained by spreading allows usage of lower dynamic range converters than in narrowband systems. The direct sequence signal as such is robust against ADC nonlinearities. For a brief discussion of the aberrations, refer to [2].

Figure 11.2   Generic receiver block diagram with optional interference cancellation stage.

As stated by Nyquist, the sampling rate must be at least twice the input signal bandwidth. Practical reasons, however, dictate that the sampling rate should be a multiple of the chip rate (if, for example, 4.096 Mchip/s takes 5-MHz band, it is better to use a rate of 4 × 4.096 Msamples/s than 2 × 5 Msamples/s). Furthermore, a higher oversampling rate means that a better multipath resolution can be obtained without utilizing interpolation techniques, which may be quite complex from the RAKE receiver point of view.

The required dynamic range of an A/D converter should be selected in such a way that the converter does not saturate with proper AGC setting and still quantization noise does not contribute to the overall link performance. The CDMA signal differs in this respect from nonspread systems because the actual information bearing signal is buried in noise. For example, with a spreading ratio of 128 and symbol-energy-to-noise density ratio (E_s/N_0) of 8 dB, the carrier-to-interference ratio (CIR) is only −13 dB. In practice, 4- to 6-bit converters have been noted to provide adequate performance [3].

11.4.1.2   Baseband Signal Filtering

An optimal receiver filter is matched to the transmitter pulse shaping filter to provide maximum SNR. A purely digital implementation is too complex; a better choice is to find a compromise between analog IF/BB and digital filtering. In addition, the latter should be designed in such a way that it is able to compensate for the possible signal distortion caused by the transmitter/receiver analog filters and other data path non-linearities.

The main problem with the receiver filter compared to the transmission filter is the higher input dynamic range (4 to 6 bits) and sampling rate requirement. This means that long filters cannot be provided practically, which is the case especially with the mobile terminal. A shorter nonoptimal filter, however, can be used because it deteriorates the signal quality only slightly. For filter implementation examples in the IS-95 system, refer to [4].

It should still be noted that filtering must be performed to both I and Q signals, which doubles the processing requirements. One option for reducing the hardware complexity is to use only high throughput filters with two delay lines, and clocking this with twice the sampling rate.

11.4.1.3 Signal Despreading

The purpose of signal despreading is to transform the received wideband signal into a narrowband signal. This can easily be implemented using the despreading circuits shown in Figure 11.3. (See Section 5.6.2 for discussion on different spreading circuits.) The number of required correlators depends on the modulation method, possible multi-code reception, and the number of multipath taps to be despread per code channel. The dumping rate of the correlators is set according to the spreading ratio. The wideband clock may be set equal to the chip clock given that the delay phase is correct. This can be achieved by utilizing an interpolator before each RAKE finger, or a higher accuracy clock on the multipath estimator side. In the latter case, the despreader wideband clock is divided from the higher rate clock, and the phase of the chip rate clock must be adjustable.

It should be noted that by using BPSK spreading, the number of correlators can be halved. The case is the same with QPSK spreading if the crosscorrelations between the I and Q branch codes are zero. Still another way of reducing the number of correlators is to use the same spreading code for both I and Q signals. This approach, however, is vulnerable to phase errors and requires more accurate channel estimation.

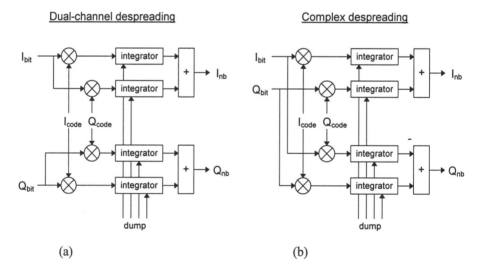

(a)   (b)

Figure 11.3   Despreader architectures for QPSK chip modulated signal: (a) dual-channel despreading and (b) complex despreading.

Especially in the downlink, flexible data rates can be implemented by using multiple parallel code channels. Extra correlators are needed for each new code, and if the number of codes increases to several tens, mobile terminal implementation becomes cumbersome. Phase and delay estimation, however, can be shared because each code channel passes through the same multipath, and hence, some alleviation can be obtained in terms of implementation.

Higher data rates implemented with the variable spreading factor (VSF) scheme result in a receiver implementation whose hardware complexity is fairly constant as a function of the data rate. Only despreaders for the multipath taps of a single code channel are needed, and the different data rates are implemented by changing the correlators' dumping periods. The integration unit must be designed according to the largest spreading ratio, which defines the worst-case register word length.

An important aspect is the number of bits required for the wideband signal presentation. The implementation complexity and cost of the receiver depend heavily on the required wideband signal dynamic range. The input signal consists of additive white Gaussian noise mainly due to the receiver front end and different user signals propagated through multipath channels. Receiver gain control is used to adjust the input signal level suitable for the A/D-converter dynamic range. By noting that quantization noise depends on the desired input signal level with respect to the converter input range, an estimate for the needed word length can be obtained. Figure 11.4 shows signal-to-noise degradation with respect to the effective quantization word length.

Figure 11.4   SNR degradation as a function of quantization word length. The four curves have different Gaussian noise variances δ_n into ± 1 A/D input (dashed: $\delta_n = 0.5$; dashdot: $\delta_n = 0.37$; dotted: $\delta_n = 0.25$; solid: $\delta_n = 0.125$).

11.4.1.4 Channel Estimation

The channel estimation block estimates multipath delays and complex tap coefficients (phase and amplitude). There are numerous different algorithms and approaches to solve the estimation problem. Some of the more frequently used ones are discussed below, emphasizing the actual implementation.

The main target of the multipath delay estimation unit is to provide accurate enough delay estimates at a fast enough rate. Traditionally, the implementation has utilized a sliding correlator that calculates one delay tap power per dumping period (see, for example, [5, 6]). The more parallel correlators available, the faster the multipath window can be scanned. A good criteria for finding the "real" taps and rejecting the false ones is to employ the so-called two-dwell approach [7]. The procedure has two parts: first, the candidates are found, then a verification is performed. A more exotic approach is to use the subspace concept in which the received signal is projected against the desired user code (vector spanning a subspace) in different delay phases [8]. Should there exist a multipath tap, the projection value will exceed a threshold.

The accuracy of the delay estimate is generally defined by the sampling rate. If, however, interpolation techniques are used, better accuracy can be obtained at the expense of increased complexity. Figure 11.5 shows an example of SNR degradation due to an erroneous delay estimate. It is interesting to note that the filtered input signal is more robust against delay errors that a nonfiltered square wave. The solid line presents correlation with raised root cosine (RRC) filtered reference sequence (input is also RRC filtered); the dotted line is with rectangular input; and the dashed line is with both rectangular input and reference signals. There are four samples per chip in each case.

Figure 11.5   SNR after despreading as a function of timing error.

A commonly adopted approach, shown in Figure 11.6, is to use a coarse delay estimation unit, which triggers delay locked loops (DLL) connected to the despreader fingers. The DLL forms a signal proportional to the code phase error, which is used to adjust the code generator(s). The time constant is made long to provide a good enough estimate without excess phase jitter. Implementation of the sliding corrector is simple, but extra complexity is introduced because of decentralized control. In real conditions, multipath taps tend to merge, locking two physical fingers to the same tap. This calls for an indication signal from each RAKE finger to the multipath estimator control, carrying the corresponding finger code phase value.

The system in Figure 11.6 operates as follows. First, the coarse multipath estimator scans the defined channel impulse response window defined by the maximum delay spread. By averaging, the decision logic selects the highest peaks and allocates available RAKE fingers to corresponding code phases. Each RAKE finger's DLL locks to the phase and starts following the drifting multipath tap. From time to time, in addition to a "synch.lost"-indication, RAKE fingers send code phase information to the allocation control, which checks that no peaks are tracked by two or more units. In addition, the multipath control must continue updating the channel impulse profile to detect possible new peaks.

Figure 11.6   Traditional DLL-based receiver.

Today's ASIC technologies are at the point that a full digital code matched filter can be implemented without great effort. In practice this means that the delay estimation unit can be made fast and accurate, suggesting that the DLLs are no longer needed. Impulse response averaging must naturally be applied in order to prevent RAKE allocation due to a noise event. The advantage is a centralized control, which determines both RAKE finger allocation and the releasing strategy that can be optimized according to changing channel conditions and available hardware (number of fingers).

Complex amplitude estimation (i.e., phase and amplitude estimation) performed for individual channel taps is required for coherent detection. Complex channel estimation is performed with the help of a pilot channel consisting of known transmitted symbols (see Section 5.2.2). The accuracy of the estimate is essential from the link performance point of view, and it depends on the pilot channel energy, the algorithm used, and environment conditions. Variable mobile speed grades require a lot from the algorithm and, in the extreme case, call for an adaptive solution.

The pilot channel can be provided in two basic ways. In the continuous pilot case, there is one physical code channel dedicated fully to constant pilot symbol transmission (code multiplexed pilot), while another option is to insert pilot symbols into the data stream (time multiplexed pilot). The former is widely used in the downlink transmission, while the latter approach is needed to implement a coherent uplink. The former argument is not, however, true if adaptive downlink beamforming is applied. Here each MS faces a different channel, which is also the case with uplink transmission.

Figure 11.7 shows one possible phase estimation architecture based on a continuous pilot channel. Each RAKE finger comprises extra correlators for despreading the pilot channel in the same code phase as the data channel tracked by the finger. The obtained instantaneous phase estimate is filtered with a suitable time constant, which can also be made adaptive to the Doppler frequency. Note that before passing the signal to the combiner, a maximal-ratio phase correction is made. The pilot channel correlator can utilize any spreading ratio regardless of the data dumping period.

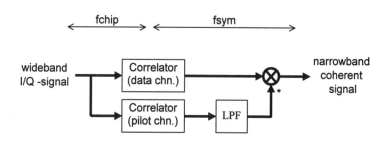

Figure 11.7   Continuous pilot signal-based phase estimation.

In case the pilot signal is transmitted as symbols in conjunction with the data channel, there is no more need for extra pilot correlators and code generators. Two well-known methods of performing the estimation for these are based on either interpolation or decision feedback techniques.

The decision feedback approach shown in Figure 11.8 performs hard decisions for the nonpilot symbols and removes the data modulation. Hence, a continuous pilot can be constructed and later filtered with a low-pass filter. Pilot symbols within the data are mainly used for taking care of phase ambiguity testing.

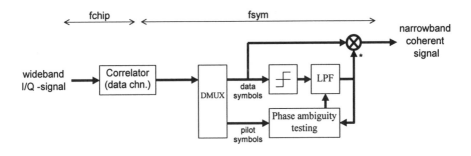

Figure 11.8   Complex phase estimation using decision feedback techniques.

If low rate coding, resulting in very small symbol energy, is used, the decision feedback performance starts to degrade because of increased symbol errors. In this case the interpolation technique may suit the decision feedback performance better because it uses only the pilot symbol energy for the estimation. The pilot energy can be increased by applying more symbols or by using a higher transmission power. The interpolating system faces difficulties with high mobile speeds because in order to get a good enough estimate, a long time constant must be used and thus, the system cannot track the channel. Another problem here is the introduced data path delay due to noncausal estimation. Figure 11.9 shows an example architecture for a RAKE finger with an interpolating estimator.

Figure 11.9   RAKE finger with interpolating phase estimator.

11.4.1.5 Multipath Combiner

A multipath combiner takes the narrowband outputs from the different RAKE fingers and combines them with each other. Maximal ratio combining produces the best performance. It weights the multipath symbols according to the SNR before combining them.

From the receiver architecture point of view, the tap weighting (maximal ratio case) can be implemented in a straightforward manner inside the RAKE fingers. This is especially the case with coherent detection where each finger incorporates a signal rotation. The rotation is done by multiplying the received signal with the complex conjugate of the phase error vector having length 1. Weighting can easily be merged into this process by setting the length of the vector according to the corresponding multipath tap SNR. In practice this means that the only task for the combiner unit is to add the coherently detected and weighted signals together.

Each RAKE finger does not always process a strong multipath component (e.g., due to fading). If the SNR of a tap is low enough, it is better to leave it out of the combining process. This is because it might have a destructive effect on the total signal quality. Some SNR level value (possibly adaptive) can be used to decide which taps are taken into the combining process.

11.4.1.6 Multiuser Detection/Interference Cancellation

The principles of multiuser detector/interference cancellation algorithms were presented in Chapter 5. A large number of different suboptimum multiuser detection approaches exists. This section concentrates on describing the best known suboptimal approaches, namely, decorrelating, MMSE, parallel interference cancellation, serial (successive) interference cancellation, decision feedback detectors, and neural network-based detectors from the implementation point of view.

Interference cancellation is a baseband processing function, which is one of the main advantages over the other capacity enhancement approaches such as adaptive antennas. State-of-the-art digital DSP and ASIC technologies already provide an opportunity for building some kind of interference cancellation unit, although optimal implementation (MLSE) is still far too complex.

Decorrelating Detector. A decorrelating detector is depicted in Figure 11.10. A matrix inversion is used for the crosscorrelation matrix, and the K signals from the traditional RAKE receiver branches are filtered with this R^{-1} operator. From the implementation point of view, the system becomes complex because the matrix inversion must be calculated every time the users' mutual time delays change. The Cholesky decomposition [9] can be used to perform the matrix inversion faster and hence ease the processing. The matrix is not always invertible (i.e., the matrix is singular), although according to [10] this should not happen in most cases. From the computation point of view, the precision required to perform the inversion is also critical.

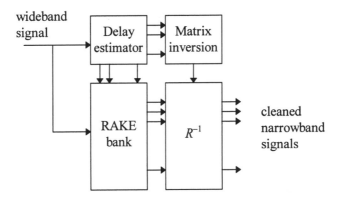

Figure 11.10   Decorrelating detector.

A decorrelating detector can also be implemented in the wideband domain. In this case, the traditional type correlator receiver is used while replacing the reference chip sequence with a signal orthogonal to all the other received users [11]. From the complexity point of view, the difference is the multilevel nature of the despreading signal and the processing element to calculate this sequence. The latter causes a radical complexity increase if operated dynamically, and hence, a suitable application would be a single user detector applied to a synchronous system.

In practice, the implementation complexity of the direct matrix inversion might be too high, especially for a large number of users and multipath components. Therefore, as was discussed in Chapter 5, iterative algorithms such as the conjugate gradient (CG) method have been proposed for implementation of the decorrelator.

Minimum Mean Square Estimate (MMSE)-Based IC. From the performance point of view, a decorrelating detector has the disadvantage of noise power enhancement in the matrix inversion process. An MMSE-criteria-based interference cancellation receiver overcomes this by taking the noise power into account. The matched filter output is fed through a linear transform derived according to MMSE criteria [12]. The linear operator is obtained by taking the inversion of the crosscorrelation matrix that is conditioned with received user signal powers and estimated noise power on the diagonal. Performance improvement comes, however, with extra detector complexity. This is because of the need to estimate individual user signals and noise power levels. The MMSE-based receiver performance is similar to that of the conventional matched filter with low SNRs, while with high SNRs, the performance approaches the decorrelating detector.

The MMSE receivers have attracted interest because of their applicability to decentralized adaptive implementation. Adaptive implementation does not require information of the interfering spreading codes, and thus, the adaptive MMSE receivers could be used in the mobile station as well. They require, however, the signal to be cyclostationary (i.e., periodic). Therefore, they can be used only with short spreading codes.

Matrix inversion turns out to be the main source of implementation complexity. Because the matrix is symmetric and still positive definite, Cholesky's method can be used to ease the calculation. It should still be noted that as in the case of the decorrelating detector, the inversion must be done every time the mutual user multipath delays change. In addition, the update must be performed every time the user or noise signal powers vary. Hence, similar to the decorrelator, iterative algorithms should be used.

Parallel Interference Cancellation. Parallel interference cancellation (PIC) cancels the interference of all users simultaneously. The multistage PIC shown in Figure 11.11 suppresses the multiple access interference in multiple consecutive steps. Stage n cancels the interference by utilizing the hard symbol decision from stage $n-1$. The cancellation process can be performed either in narrowband (NB) or wideband (WB) domains. For the corresponding detectors we use notations NB-PIC and WB-PIC. The cancellation operation itself requires knowledge of the user codes, the relative code phases, complex channel estimates for the multipath taps, and finally hard symbol decisions for each user.

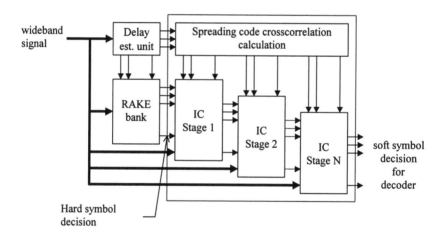

Figure 11.11   Multistage PIC detector with two cancellation stages.

Multistage PIC [13] operating in the narrowband domain performs computation with a symbol rate. Hard symbol decisions from the preceding stage are used in conjunction with complex channel estimates and mutual normalized crosscorrelations between the spreading codes to generate aggregate interference estimates for each despread multipath component of interest. Multiple access interference (MAI)-cleaned multipath signals are combined for each user, and more accurate hard symbol estimates are made from these. Most of the gain can already be obtained with two cancellation stages [13].

From an architecture point of view, implementation of the NB-PIC detector is a straightforward procedure because it can be fitted fairly directly into a traditional CDMA receiver between the correlators and the deinterleaver/decoder. Multipath combining and hard symbol decisions are inherently included in the interference cancellation stages. The actual cancellation operation is simple because it takes place at the symbol rate. The total processing requirements become high, however, for two reasons. First, the spreading code crosscorrelations must be updated every time the mutual tap delays change. In the extreme case, if long spreading codes are used, the whole matrix must be recomputed for every symbol. In addition, the aggregate interference estimates depend on the crosscorrelations and differ for each multipath. Hence, we need to compute at symbol rate $K{\times}L$ interference estimates each consisting of $(K{\times}L{-}1)$ additions and multiplications (K = number of users, L = number of multipath taps per user).

Physical implementation of the NB-PIC detector depends on the symbol rate, spreading factor, type of spreading, number of stages, and the total number of multipaths involved in the IC process. The interference estimation is likely to need an ASIC-based solution, while the actual cancellation could be done with DSPs. Multiple stages can be implemented using either serial computation with one physical instance or, if the processing rate is too high, with parallel instances in pipeline mode.

Should long spreading codes be used, the receiver complexity due to the constant need for crosscorrelation updating is increased vastly. A wideband version of PIC, also called regenerative PIC, would be a better choice in this case. Finally, it should still be perceived that the narrowband multistage detector architecture relies strongly on unified code channel symbol rate, making it more suitable for systems utilizing parallel code channels instead of variable spreading to provide variable bearer services.

A major problem with the NB-PIC detector is that each multipath component, because of different mutual crosscorrelations, sees a different interfering signal on the narrowband side. This can be prevented by conducting the cancellation in the wideband domain. WB-PIC, shown in Figure 11.12, is based on the regeneration principle, which is used to calculate the so-called residual signals. Regeneration, which requires knowledge of complex channels, hard symbols decisions, tap delay estimates, and the spreading code, is done for every multipath component taken into the cancellation process. The combined regenerated signal estimate is subtracted from the received wideband signal to get the residual signal. This is added to the regenerated individual multipath components in order to get a cleaned wideband signal for the despreaders.

WB-PIC has the advantage that it can be utilized with both short- and long-code-based systems. In addition, the residual signal provides an interference-resistant input for multipath profile estimator. Although the actual interference cancellation operation requires a higher operation rate than in the NB-PIC case, the lack of crosscorrelation calculation makes the total processing/resource requirements smaller. The complexity of the regeneration part depends heavily on the pulse shaping filtering scheme used and hence on the sampling frequency in conjunction with the word length.

The type and length of the spreading codes also have a noticeable effect on the total system complexity.

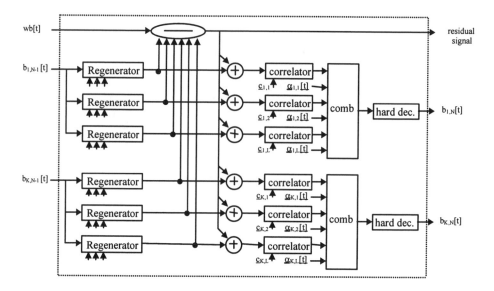

Figure 11.12   Wideband domain parallel interference cancellation (WB-PIC) stage.

Multiple simple parallel operations in WB-PIC functional architecture suggest clearly an ASIC implementation. A fully parallel architecture with multiple physical stage instances provides the lowest latency, while one WB-PIC operated sequentially would require less gate area. It would also be easy to perform serial userwise, or even pathwise, processing to further minimize the area. WB-PIC indeed suits many different processing schemes, which makes it a potential candidate for a real application.

In the previous text concerning multistage detectors it has been assumed that the interference cancellation comprised all the users with their multipath components. In practice the receiver could be made adaptive in such a way that only the strongest users/multipaths are canceled. This would require extra system/control resources for decision making but otherwise would reduce the actual PIC complexity. The system performance, however, may degrade slightly due to the selective cancellation, but in some cases it may even increase. The latter is the case when weak users are left out from the IC process. The above idea seems like a good one, but we must remember that in DS-CDMA the power control tries to keep the receiver signal levels alike. On the other hand, this is not actually the case when the VSF scheme is used for higher bit rates or when signal power control is used to adjust the quality of service.

Successive Interference Cancellation [14]. Successive interference cancellation (SIC), shown in Figure 11.13, ranks the users according to their power levels. First, the user with the strongest signal is detected, and the MAI estimate for the following users is updated with this knowledge. Next, the second strongest user is processed continuing to the weakest one. From the implementation, and also from the performance point of view, the process can be stopped after cancellation of a subset of users (the most interfering ones). The cancellation can be done on the wideband or narrowband side. The former uses the canceled user symbol hard decision, signature waveform, and amplitude and phase estimates to regenerate the wideband signal. This is then subtracted from the received wideband input stream, which is later despread by the rest of the users. Another method is to use crosscorrelations between the spreading codes of the strong user and of the "cleaned" users, and do the cancellation after the despreading operation. It should be noted that the cleaned wideband canceled signal can be used in the multipath channel estimator, thus making it MAI resistant.

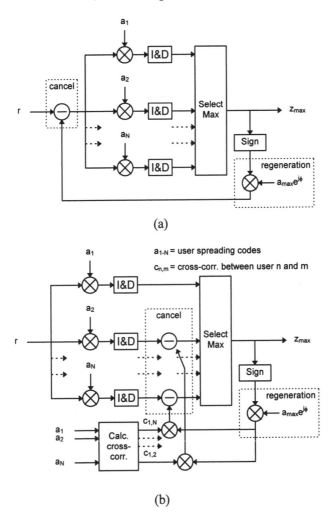

Figure 11.13   Wideband- and narrowband-based successive cancellation receivers. (*From:* [15].)

Decision Feedback (DF) Detectors. DF detectors combine linear and successive interference cancellation. They take hard symbol decisions for all, or a subset of, active users and feed this value back to estimate the MAI. The receiver generally consists of a feed-forward cancellation part relying on the previous stage symbol decisions, and a feedback part utilizing the following stage "cleaned" hard symbols in estimating the MAI [12]. The times of signal arrivals play an important role here for deriving the order of cancellation. A conventional detector is used as the first stage. The performance, however, can be improved by introducing an extra feedback part already to the first stage [16]. In case of severe near-far reception, a good approach is to rank the users according to the power levels and perform the intrastage cancellation starting from the strongest user [17].

The decision feedback detectors are very similar to the already-discussed multi-stage ones. With respect to the former, the main difference is that within a single stage, the users are detected sequentially in time (according to power levels, or time of arrival), and hence the hard decisions based on readily "cleaned" user symbols can be utilized with subsequent detection of the same stage users (see Figure 11.14). This could also be called a serial-parallel interference cancellation.

Figure 11.14   Decision feedback detector with three synchronous users.

Neural Network-Based Detectors. Neural networks (NNs) provide a means of solving optimization problems with computational elements called nodes that are connected to each other with weighted (possibly adapted) interconnections. The effectiveness of neural networks is due to parallel computation taking place in numerous simple processing elements. In multiuser detection, neural nets can be utilized to find the most likely transmitted bit sequence that produced the matched filter output. There are many different neural network architectures [18]; the most commonly used for MUD realization are the normal feed-forward multilayer perceptron [19] and a simpler Hopfield network shown in Figure 11.15 [20].

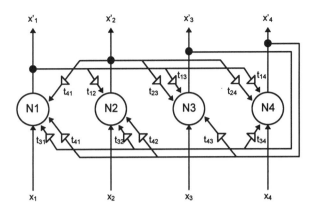

Figure 11.15   Hopfield neural network with four nodes.

A major disadvantage of the feed-forward structure is the prerequisite for training. In addition, the number of nodes increases exponentially with respect to users to be detected. Because of these drawbacks of the feed-forward structure, the Hopfield network is seen as a better neural network choice for multiuser detection problems. Its complexity expands with the square of the number of users, which will also lead to a large system in the case of many active users. Training is not needed, while the processing burden comes from the iteration rounds (depends on the number of users). Therefore, neural networks cannot be used to process the whole set of received users, and the number of inputs should be reduced. Figure 11.16 shows an approach utilizing a reduced detector (RD) in conjunction with Hopfield network. The reduced detector implemented an iterative algorithm focusing only on users with nonconfident parameters (reduced search space) [21]. If the iteration turned out to be irreducible, a Hopfield network was used to perform a more extensive search.

From the implementation point of view, the neural networks are naturally ASIC oriented (many parallel elements). In the past, analog techniques were applied to provide topologies with up to 50 neurons. Digital approaches tend to be problematic for larger nets with nonbinary inputs and interconnections, although this is about to change because of even higher integration levels. Multiplication, addition, and some nonlinear functions are the primitive operations to be provided. The number of multiplications depends on the interconnections, and when binary inputs are used, these are fairly simple. Additions, on the other hand, cause major problems here because the required adder trees call for primitive adders with nonbinary inputs.

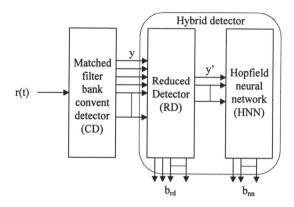

Figure 11.16   Hybrid matched filter, reduced receiver, and Hopfield neural network detector. (*From*: [21].)

11.4.1.7 Decoding Architectures

Forward error correction (FEC) and detection coding are used to protect the transmitted information bits and provide the specified quality of service. The most frequently used FEC schemes are based on convolutional and block coding. The Viterbi algorithm is used to perform convolutional decoding, while simple shift registers can be applied for block codes. Third-generation cellular systems create many new challenges for the terminal decoders. These are due to the increased number of different bearer services with different data rates, BERs, and delay requirements.

The different operation modes and coding schemes call for modular and flexible decoder architectures. For pure speech terminals the physical implementation can rely on DSPs, while the higher data rates certainly need an ASIC accelerator. The key goal here is to realize which primitive operations are common to the different coding schemes and how these can be effectively utilized to provides variable data rates.

Block coding schemes in conjunction with error detection coding (CRC) can easily make use of the same physical processing resources. This is because the decoding can be implemented using shift register chains, and the different schemes can be supported with variable polynomial lengths and changeable polynomial connections. Constant processing clock rate with alternating activity periods provides a good means for adapting the decoder to different data rates. The modest complexity of the shift register depends naturally on the polynomial degree.

The Viterbi decoder is built on top of a trellis tree consisting of stages and transitions. The basic operation consists of partial metric calculation, trellis updating based on path selection, and back-tracing. The partial metric processing involves calculation of 2^K values (K = constraint length) for each received bit. The total number of the stages is $2^{(K-1) \times L}$ where L is the so-called truncation length. After L bits have been updated to the tree, back-tracing is performed providing the decoded bits. From the complexity point of view, K determines the required processing rate, and, incorporated with L, the total decoder memory requirements can be derived. The truncation length also defines the decoding delay. Based on simulation results, L equal to $5 \times K$ is an adequate selection [22]. So far, constraint lengths less than or equal to 7 have been widely used. It is likely, however, that the third-generation cellular standards include specification for larger constraint lengths (e.g., 9). This, in conjunction with high bit rates, means that decoder processing requirements increase remarkably compared to second-generation systems.

From the physical architecture point of view, a key primitive with the Viterbi decoder is the so-called add-compare-select (ACS) primitive that is used to select the partial metrics that have lower values. For fast decoding requirements an ASIC implementation is a necessity, while low rate systems may incorporate a software solution (assuming that the selected DSP can provide an ACS command). The best Viterbi decoders at the moment are capable of providing throughputs of 1 Gbps [23]. Suboptimal Viterbi decoders are mainly based on reducing the number of states in the tree. M- and t-algorithms are examples of these [24]. The latter could also be classified as an adaptive decoder because the number of states is variable. For a comprehensive description of the Viterbi algorithm see, for example, [25, 26].

Finally, it should be noted that new decoding approaches based on iterative algorithms have emerged. These are referred to as turbo decoding, and they have claimed to possess better performance compared to convolutional schemes. On the other hand, their main disadvantage is an increased decoder complexity.

11.4.2 Baseband Transmitter

The transmitter baseband section is mainly responsible for channel coding, including both error detection and protection functions. In the case of third-generation systems, several different schemes are required to fulfill many service qualities and user data rates. Encoding functions are considered as bit level procedures.

Signal spreading specific to CDMA systems takes place on the baseband side. This operation causes the input bandwidth to widen according to a defined spreading factor or processing gain. Signal spreading is purely chip-level processing.

Digital filtering is commonly used to shape the transmitted signal spectrum. The filtering must be performed for a wideband signal and becomes complex because of higher clock rates. There exist, however, good architectures providing implementation-efficient solutions.

11.4.2.1 Channel Encoding

Error correction algorithms can be divided into block- and tree-based schemes, convolutional coding being the best known from the latter category [27]. Recursive convolutional codes, also called turbo codes, have been under intense research recently.

Traditionally, the implementation of channel encoders has concentrated on finding good solutions for some constant set of parameters like input data rate or coding parameters. This is not the case anymore because of diverse requirements and the new emphasis on flexibility. The encoders for the third-generation systems must be capable of operating with different data rates and coding parameters. In the far extreme, the same physical encoder hardware should even support several coding algorithms. Additional problems still arise when the coding scheme needs to be switched on the fly, a requirement for packet access and link adaptation.

In the second-generation systems, to attain flexibility channel encoding has been normally implemented in software mainly because the data rates have been reasonably low. Third-generation systems will place much tougher requirements on the encoders because of radically increased user data rates up to 2 Mbps. Fortunately, the well-known encoding algorithms can be efficiently implemented in ASICs. ASIC implementation, however, has the problem of not providing high flexibility, but we should remember that most of the encoding algorithms can be implemented using simple shift register chains. This is also the case with turbo encoders [5].

Hardware encoder throughput can be adjusted by three different approaches (see Figure 11.17). The first option is to provide parallel hardware and activate necessary resources when needed. This suits well systems with data rate expansion using parallel "low-rate" code channels. In the other options the same physical encoder core is used, but the operating cycle, or the clock rate, is made variable. The latter may be problematic from a mobile terminal point of view.

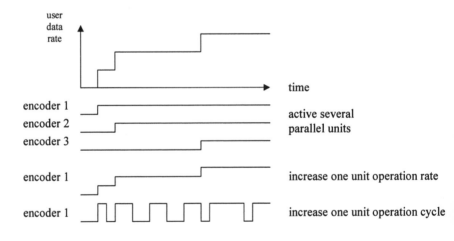

Figure 11.17   Different options to increase the encoder throughput.

11.4.2.2 Data Modulation and Spreading

After channel coding, data modulation is performed. Usually coherent phase shift keying (BPSK or QPSK) data modulation, which provides good spectral efficiency, is applied. The physical implementation of the bits-to-symbol mapping is a very simple operation.

A signal with a wideband spectrum is obtained by multiplying the transmitted symbols by a faster rate chip stream. The operation is highly hardware oriented because of its simplicity (binary multiplication) and its high frequency. Depending on the modulation method, the spreader may include only in-phase or both quadrature branches. In addition, the spreading method defines how the chip sequence(s) is multiplied with the symbol stream. As shown in Figure 11.18, for an I/Q-signal this can be done individually for both branches (dual-channel spreading) or by using a complex multiplication. The latter option means an increase in processing requirements due to extra multipliers and adders.

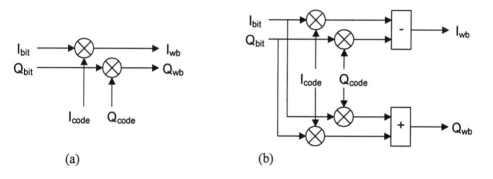

Figure 11.18   (a) Dual-channel and (b) complex spreader architectures.

From the overall transceiver point of view, the spreader procedure has only a minor effect on the total complexity. Some increase, however, may be seen in the baseband power consumption and silicon area, mainly because of fairly long code generators.

11.4.2.3 Baseband Pulse Shaping Filtering

Due to bandwidth-limited output signals, the output chip stream from the spreading modulator is filtered using either a digital or analog filter (pulse shaping filter). A digital filter with linear phase can be designed more accurately but will require a larger dynamic range from the DACs following it. Still, the implementation of long filters (close to 100 taps) may turn out to be too complex. Square RRC-type impulse response is generally used because of the Nyquist no-ISI-sampling criterion. The roll-off factor defines the sharpness of the spectrum.

For digital implementation, the filter length may have to be truncated in order to minimize complexity. Radical truncation effects on the spectral shape can be alleviated to a certain degree by using a windowing function [28]. Hamming, Hanning, Bartlett, and Kaiser functions are the best known. The filtered output signal must have a high enough word length to keep the quantization noise due to D/A conversion below spectral mask requirements. As a rule of thumb, one extra bit increases the signal-to-quantization noise ratio by 6 dB [28].

The oversampling factor, which is also an important design parameter, defines the sample-and-hold (S/H) boxcar "filter" and, more importantly, the reconstruction filter requirements. The higher the oversampling factor is, the further apart the copied spectral components appear, which relaxes the post-DAC filter specification. Analog filters, like Bessel or Chebyshev, with two to five taps would be suitable in many cases. Care must be taken, however, to prevent the filter phase response from distorting the signal. Mixed-mode simulations must be performed if tight transmitter specifications are to be fulfilled.

The two plots in Figure 11.19 show the effect of D/A conversion following pulse shaping. The RRC filter with roll-off 0.35 was employed. The signal spectrum is copied into frequencies, which are multiples of the sampling rate.

(a)   (b)

Figure 11.19 Effect of D/A conversion boxcar filter on the signal spectrum: (a) signal spectrum after RRC filter and (b) signal spectrum after D/A conversion.

11.4.2.4 AFC, AGC, Power Control

In order to make the transceiver work, there are three important functions that still need to be performed. Automatic frequency control (AFC) is needed in a mobile terminal to synchronize its local oscillator to the base station oscillator. Automatic gain control (AGC) on the receiver side is responsible for preventing the ADC from saturating, while at the same time trying to provide a maximum available dynamic range. The transmitter side still incorporates a power control (PC) adjustment, which is particularly important in CDMA systems, as discussed in Sections 5.11.6 and 9.3. Figure 11.20 depicts the control loops. The dotted lines represent optional connections. For example, in the case of AFC, the carrier frequency can be changed by adjusting the reference oscillator or possibly the digital synthesizers on both the TX and RX sides.

Figure 11.20   Mobile station internal control loops (optional ones with dotted lines).

Automatic frequency control aims at bringing the mobile up/down-conversion frequencies as close as possible to the base station ones. This can be done either by adjusting the MS reference oscillator or, if utilized, the digital frequency down-conversion. Generally, the system digital clocks are also tied to the reference clock, and by performing AFC with the reference, the side effect also provides more synchronized sample/chip/bit clocks. The AFC loop can be made very slow because the oscillator drifting in time is also a slow time varying process. Frequency control could also be performed to compensate for the Doppler effect. This is, however, a slightly different problem compared to slow AFC because on the receiver side the compensation must be done in the opposite direction, which requires a fast absolute frequency error estimate.

The received signal envelope has a large dynamic range. Gain control is required to adjust the amplitude suitable for the A/D converter input range, which tries to provide the highest dynamic range. Finding the optimal operation point is not an easy task because it depends on many aspects like modulation, fading, multipath channel, and, in DS-CDMA systems, on other user signals. AGC loop adjustment is based on the received signal power (RSSI), which can be measured from both wideband and despread narrowband signals. The wideband power can be measured either directly from the RF unit or after the receiver A/D converter.

CDMA systems are vulnerable to erroneous transmission power control because any error will introduce extra cell interference and hence reduce the capacity. The power control loop must be fast enough to react to abrupt signal changes and accurate enough not to generate undesired power on the channel. The criterion for adjusting the power level should be based on needed service quality; SIR is a good option for this purpose. In the open loop approach, the transmitted power is adjusted according to received signal; while in the closed loop power control, the commands are transmitted as part of the signaling information (e.g., every 5 ms with 1 dB up/down command). In the latter case, the time between the RX power measurement (by BS) and the actual TX power adjustment (by MS) must be kept as brief as possible. In a real implementation the power amplifier's varying characteristics in time should be taken into account. From time to time some power measurement value from the PA output should be used to calibrate the transmission.

11.5   RF SECTION

This section reviews a few of the best known RF transmitter and receiver architectures. In a wideband CDMA transceiver the RF must be designed to suit a wideband low power spectral density signal. Furthermore, transmission is continuous unlike in the case of time division systems. Requirements for high dynamic range and accurate fast power control loop cause most of the problems compared to second-generation systems. In addition, the linear modulation and multicode transmission (multilevel signal) place challenges on the power amplifier linearity and efficiency design.

11.5.1 Linearity and Power Consumption Considerations

Third-generation systems place much more difficult linearity and efficiency requirements for the RF front end. The linearity constraint is due to tighter output spectral mask specification, higher signal envelope variations (linear modulation), and, in the case of the PA, the need to keep the operation level near the compression point in order to achieve a high enough efficiency. In addition, when multicode transmission is applied, more backoff is needed causing a loss of efficiency. Adjacent channel interference (ACI) levels of –30 dB are likely to be specified, while with large umbrella cells and flexible channel allocation the requirements may increase to –60 dB.

The transmitted signal spectrum at the antenna defines how much energy is emitted to the adjacent and alternate channels (by alternate channels we mean those after the adjacent ones). Interference is increased because of the spill-over of the signal into adjacent channels, and the network capacity decreases accordingly. As was discussed in Chapter 10, systems incorporating hierarchical cell layers become very difficult because without highly orthogonal spectra, one umbrella-cell user may entirely block microcell traffic because of its radically higher transmission power. Output transmission spectrum depends mainly on the modulation method, multirate solution (multicode versus variable spreading), transmission filtering, and RF nonlinearities. In the last case, the power amplifier has the largest contribution to the signal distortion.

It should be noted that the overall transmitter linearity is an optimization problem between PA gain characteristics and transmission filtering. The former makes the spectrum narrower but at the same time increases the signal crest factor (= peak-to-mean signal ratio). The higher the crest factor, the more the signal is distorted by the power amplifier. It should be noted, however, that examining only the crest factor may be misleading. From the generated adjacent channel interference point of view, a high crest factor is tolerated if the occurrence of peak amplitudes is low. Hence, instead of crest factor, the power amplifier input signal power histogram should be investigated, as seen in Figure 11.21.

Figure 11.21   Constellation and histogram plots for RRC-filtered O-QPSK signal (roll-off = 0.35).

Power amplifier linearity depends on the operating class. The class can be selected by using different biasing levels. A-class amplifiers are linear at the expense of low power efficiency because they consume power regardless of the input signal level.

C-class amplifiers, on the other hand, are power efficient with the introduction of higher nonlinear characteristics. PA linearity can be defined by input amplitude versus output amplitude (AM-AM), and input amplitude versus output phase (AM-PM) relationships. The most nonlinear operation takes place within low/high input signal regions. These characteristics still vary within time because of aging and supply voltage, output load, and temperature change.

The PA can be made linear by applying higher biasing (to A/AB-class), increasing input signal backoff, or using linearization techniques. The two former solutions are not very suitable for hand-held terminals because the efficiency is decreased accordingly. Linearization techniques are thus seen as the key solution to overcome the tightened spectral mask requirements in conjunction with acceptable amplifier efficiency. It should be noted here that the linearization techniques do not ease the multicode transmission, which will in any case call for extra backoff.

Linearization techniques can be divided into four main categories: (1) feed-forward, (2) feedback, (3) envelope elimination and restoration, and (4) predistortion [29, 30]. Each of these have got a set of variants providing different implementation complexity, ACI improvements, and bandwidth/convergence rates. Figure 11.22 shows an example of PA architecture linearized with the complex gain predistortion technique. This approach provides a fairly simple implementation with a high bandwidth/convergence rate [29, 31]. The input signal power is used to address the predistortion vector stored in RAM. The predistortion vector is updated according to changing environment parameters. Because the number of entries is fairly small, the adaptation is fast. It is likely, however, that the high dynamic range complex multiplier turns out to be the most critical component in the system.

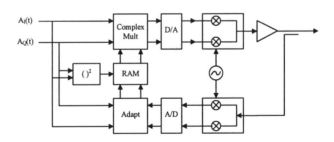

Figure 11.22   Power amplifier linearization with complex gain predistortion.

11.5.2 Receiver Architectures

While the transmitter is facing optimization between linearity and efficiency, the receiver is concerned with providing high selectivity with low noise figure (NF). The nonlinearities must be handled in order not to cause excessive signal distortion. Noise figure tells how much the receiver deteriorates the incoming SNR. While more optimal technologies and components have emerged, lower noise figure RX chains can be implemented. Good receivers provide NFs as low as 5 to 7 dB. High selectivity is desired in order to prevent adjacent/alternate channel energy from getting to the input of the ADCs, and decreasing the dynamic range requirements. The former is critical especially when operating in a hierarchical cell system where the nearby channel power may be several tens of decibels higher. A low sensitivity receiver would cause signal saturation in the A/D input.

Linearity of the signal path on the receiver side becomes more important in the case of linear, multilevel modulation or multicode transmission schemes. Still, if the system is operating close to narrowband systems with high power spectral densities, the intermodulation products may distort or even block the whole reception.

The desired channel signal strength may vary for several reasons. The receiver must ensure that if the signal goes up rapidly the ADC is not saturated, or if there is fast power level degradation the signal quality does not pass below an acceptable level. Fast and high dynamic range (up to 80 dB) AGC is responsible for adjusting the receiver variable gain amplifiers and attenuators in order to feed the ADC with as optimal a signal level as possible. In CDMA systems the adjustment can be done every symbol time without regard to slot boundaries. This is because the spread bits inherently include a training sequence, and channel estimation/equalization takes place on the symbol level.

11.5.2.1   Super-Heterodyne Receiver

The super-heterodyne architecture shown in Figure 11.23 is the most well-known receiver to implement good selectivity. The RF signal is filtered, amplified, and converted to an intermediate frequency in which the channel filtering is performed. The IF frequency is kept constant, while channel selection is done by changing the RF mixer frequency.

Several stages with different IF frequencies are needed to get the desired performance at the expense of higher receiver cost. A single conversion receiver is the simplest option and is commonly utilized.

From the total receiver noise figure point of view, the low noise amplifier (LNA) preceding the RF mixer is the key component. A typical LNA has a noise figure below 4 dB, while the best offer an NF around 1 dB with power gains up to 30 dB.

Good channel selectivity with feasible linearity can be achieved by using surface acoustic wave (SAW) filters at the IF frequency. The size of the filter depends on the channel bandwidth and IF frequency and hence can be implemented smaller for wideband CDMA systems. The problems with the SAW filter are its high insertion loss and relative cost. As a result, solutions to get rid of it have been searched for (direct conversion and digital receivers being such examples).

Figure 11.23 One-stage super-heterodyne receiver architecture.

As was the case with the IF-based transmitter, digital direct synthesis (DDS) can be utilized in super-heterodyne receivers to perform I/Q separation (see Figure 11.24). By adjusting the digital down-conversion frequency, DDS can also be applied in the AFC procedure. It should be remembered, however, that the analog channel selection filter defines the adjustment margin. Thus, it might be useful to define its bandwidth wider than the final desired channel BW and perform the final selection digitally. If, on the other hand, a wide IF filter with high dynamic range ADC is used, the DDS can make the actual frequency channel selection. This may not be as useful with wideband systems as with narrowband ones and would clearly be a base station-oriented solution.

Figure 11.24 One-stage super-heterodyne receiver with digital I/Q splitting.

11.5.2.2 Direct Down-Conversion

The direct down-conversion receiver shown in Figure 11.25 mixes the receiver RF signal straight into I/Q baseband components for which the channel filtering is performed using simple, cheap low-pass filters. The benefit of this architecture is the reduced complexity and performance, especially in terms of the SAW filter (high insertion loss), and may in the end lead to a single chip transceiver including the transmitter section.

Figure 11.25   Direct down-conversion receiver with baseband channel filtering.

There are, however, a few problems related to the direct conversion receiver technique. One aspect is the DC component, which may appear because of self-mixing the local frequency and/or nearby strong users. A CDMA signal inherently is resistant against narrowband interferers, but high level DC offset could saturate the input ADCs and destroy the reception.

The key component here is the quadrature local oscillator, which should provide good phase and gain matching in order to provide pure I/Q-branch sine waves with equal amplitude and 90-degree phase shift. Still, the isolation from the mixer input must be high enough to avoid the above-mentioned self-mixing. The mixer itself should possess good linearity to suppress intermodulation products caused by adjacent/alternate channel transmission.

Finally, it should be noted that the direct conversion receiver can be fit into a single chip by using, for example, the silicon bipolar process (in the future, a CMOS process could also be applied). The low pass filters are excluded from this and are put into the baseband side.

11.5.2.3   Harmonic IF Sampling

So far it has been assumed that the receiver I/Q baseband signal is converted into the digital domain using a sampling frequency twice as high as the highest spectral component in the input. The Nyquist criteria states, however, that it is enough to use a sampling frequency twice the desired signal bandwidth, which suggests that it is fully possible to utilize the actual signal alias term. In the frequency domain, the desired band is translated to another phase according to the selected sampling rate. Hence, the sub-sampling can also be considered as a down-conversion process (see Figure 11.26). Before feeding the signal to an A/D converter, bandpass channel selection filtering must be done for the desired signal to prevent other channels from causing their aliases within the desired frequency band.

The ADC can be used as such if the sampling rate suits the signal central frequency. If not, a separate track/hold function with a low pass filter should be applied to relax the ADC requirements. Maximum input signal frequency to the sampling device depends on the track and hold device bandwidth. In practice, several hundreds of megahertz can already be handled.

From the CDMA point of view, a subsampling receiver does not give any benefits over traditional narrowband systems. From a implementation point of view, however, only one ADC with fast T/H instead of two ADCs is required. Complexity on the digital side is increased due to I/Q separation that must be done at a high sampling rate.

Figure 11.26 Signal spectrum after performing subsampling.

11.5.3 Transmitter Architectures

This section presents the basic RF transmitter architectures suitable also for wideband CDMA transceiver implementations. Critical aspects are highlighted and, in conjunction with each approach, advantages and disadvantages are revised.

The baseband signal to be transmitted is converted to the final carrier either directly or through some intermediate frequency. The latter approach has been more common in the past, while present RF technology is capable of following the former approach.

In direct sequence spreading systems, phase and/or amplitude shift keying chip-level modulation is generally utilized. A single sideband signal with suppressed carrier is generated using an I/Q mixer with 90-degree phase difference in the frequency between in-phase and quadrature local inputs.

11.5.3.1 Traditional IF-Based Up-Conversion

Transmitted signal conversion to intermediate frequency before the final carrier provides a few advantages. First, different stages are easier from a component matching point of view. An I/Q modulator, for example, operating with low frequency can offer better phase and amplitude balance. One important aspect is that channel filtering can also be done using IF passband filters and consequently relax the processing power requirements of the digital parts.

The main sources of noise come from the nonideal local frequencies. This may be problematic if there are several IF stages because each local oscillator (LO) adds to the total transmitter noise contribution.

CDMA systems demand tough requirements for the dynamic range, accuracy, and speed of the transmission power control. In practice, the amplification/attenuation is done in many phases to fulfill these requirements, and an IF-based transmitter has an intrinsic feature to support gain control in several stages.

A functional block diagram of a one-stage IF transmitter is shown in Figure 11.27. The I/Q signal is up-converted to a predefined first IF frequency that is likely to be constant. This is followed by passband filtering and an IF amplifier. Finally, the signal is translated to carrier frequency, amplified by the PA, and input to the transmission RF filter. A duplex filter is required in case of simultaneous transmission and reception. The highest up-conversion frequency is traditionally made variable and is used in the transmission channel selection.

DDS techniques can be utilized with an IF-based transmitter. In the simplest form, DDS replaces the I/Q mixer or even the analog RF synthesizer. The disadvantage of this approach is that the generated frequencies are not as clean as the analog counterparts. The advantage is the flexibility in terms of frequency and settling speed.

By moving the digital-analog border closer to the antenna, the arrangement shown in Figure 11.28 is achieved. Here the I/Q modulation is done digitally before converting to analog format. Good amplitude and phase balance can be obtained, but the disadvantages noted before still hold. The digital-to-analog converter (DAC) also causes distortion due to the S/H-function, which must be taken into account when designing the reconstruction filter.

Figure 11.27   Typical IF transmitter architecture.

Figure 11.28   IF-based transmitter with digital I/Q modulation.

11.5.3.2 Direct Up-Conversion

The main disadvantage of a several-stage transmitter is the number of different components required, which increases the cost of a portable terminal. In a direct conversion type transmitter (Figure 11.29), the signal is transferred directly from baseband to RF band. There is a need only for one synthesizer whose output is taken to the I/Q modulator. Care must be taken to provide good enough 90-degree phase offset and gain match between the quadrature branches in order not to create too low an unwanted sideband suppression.

Channel filtering in the case of a direct up-conversion transmitter must be performed on the baseband before I/Q mixing. Digital filtering with linear phase and hence constant group delay are a good option for the purpose. The analog lowpass filters following the DAC could also be used in the process.

Power control requirements are slightly more difficult to fulfill in case of direct up-conversion. Some alleviation, however, can be provided by implementing higher dynamic range digital parts, especially the D/A converter. From the implementation point of view, higher dynamic range digital parts, and especially the D/A converter, are needed.

Figure 11.29   Direct conversion transmitter.

11.6   SOFTWARE CONFIGURABLE RADIO

The most flexible implementation for a communication terminal would incorporate a software configuration. Such an approach could provide an access to several different wireless systems with a single hardware unit simply by downloading a new configuration internally or from a network. The idea of a software radio has originally been connected to military applications. Such a device, however, would provide flexible roaming between today's multitude of commercial cellular systems. Good articles about software and multimode radios can be found in [32–34].

Figure 11.30 illustrates the concept of the software radio. There are a few basic features that relate to the software radio. The RF section must support transmission and reception of differently modulated signals within several frequency bands. This calls for a linear power amplifier operating over a wide frequency range with acceptable efficiency. Similarly, antennas must provide low loss and uniform gain across the same range. After down-conversion to IF or baseband initial channel, selection prior to the A/D conversion must be performed. This is necessary in order to restrict the signal dynamic range to within acceptable limits. If a large dynamic range A/D converter and a wide enough filter are used, the actual channel filtering can be made digitally. This may not be enough, however, if the nearby channel powers are radically larger than the desired channel. One solution here is to introduce a filter bank and select one bandwidth according to the operations mode.

The baseband section of a software radio consists of programmable hardware and a powerful DSP processor(s). In the former case, the configuration can be done by employing field programmable gate array (FPGA) technology and well-defined hardware description languages like VHDL (Very High Speed Integrated Circuit Hardware Description Language) or VERILOG. Depending on the access method, different baseband functions need to be supported. The main concern here is in the channel equalization and forward error-correction coding/decoding. The baseband architecture must be designed in such a way that both burst-type (TDMA) and symbol-type (DS-CDMA) equalization approaches can be supported in a straightforward manner. When it comes to channel encoding/decoding there are numerous methods available, each providing certain coding rates and service qualities. Software implementation is fine with low effective bit rate systems, but high rate systems call for hardware accelerators.

Figure 11.30   Mobile station software radio top-level architecture.

The application side of a mobile terminal has so far been mostly speech-oriented and several complex source codecs have been implemented into the handsets. New applications related mainly to video coding and other data processing are emerging, and hence, the requirements for the software radio application support become greater.

An important issue with the software radio is related to flexible usage of the available processing resources. The system clock rate cannot be increased easily in a mobile terminal, and hence, one of the key solutions is to utilize configurable silicon. Higher processing capabilities are obtained by creating and activating extra processing units on a need basis. This could be interpreted as a "breathing" parallel processor that expands as needed. Power consumption is occasionally increased and must be taken into account while designing the device power management.

11.7 TYPICAL WIDEBAND CDMA MOBILE TERMINAL ARCHITECTURE

This section presents an architecture for a wideband CDMA mobile terminal integrating the different functions presented earlier. For the RF section, a traditional down-conversion receiver architecture with two IFs was selected, and on the transmitter side, direct up-conversion will be employed. The baseband side consists of all the bit- and chip-level processing primitives needed to provide all bearer service quality classes, band spreading/despreading, coherent multipath diversity detection, and some extra functions for link maintenance.

Figure 11.31 shows the mobile transmitter section. On the transmitter side, the information-bearing user data is obtained from outside the radio modem part. The stream is coded and interleaved according to selected QoS. Two basic encoding schemes, namely convolutional and Reed-Solomon, are available. The rate of the user data is matched to the air interface by choosing some of the available spreading ratios and unequal symbol repetition. Frame control header information about the frame data rate and coding scheme is encoded and interleaved to the quadrature branch. In addition, the power control and the required amount of reference symbols are placed into the transmitted stream. Before D/A conversion, the physical in-phase and quadrature branches are spread with two PN codes and passed through a pulse-shaping filter to obtain the specified transmission spectrum.

The receiver side down-converts the incoming signal to the baseband before A/D conversion. Figure 11.32 depicts the mobile terminal receiver section. Channel filtering and AGC have been implemented on the analog RF side. In addition, a loop back from the baseband channel estimator provides an error signal to adjust the reference oscillator for clock synchronization purposes. The I/Q signal is converted to the digital domain using a 6-bit, four times oversampling ADC. Digital filtering following the ADC is matched to the transmitter filter, thus improving the SNR. The digital filter can also be used to compensate for nonlinearities caused by the analog side filters. Wideband power measurement is performed for the post-ADC signal and is used in the receiver AGC loop.

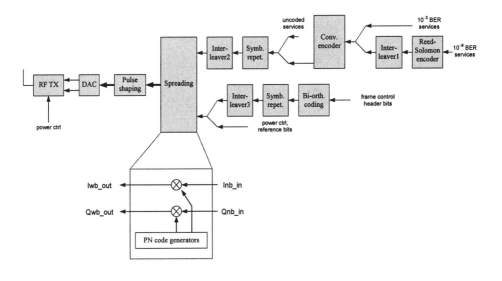

Figure 11.31   Mobile transmitter section.

Figure 11.32   Mobile terminal receiver baseband section.

The multipath channel estimator utilizes a full matched filter to find multipath components quickly and to perform accurate delay estimation without any DLL. RAKE fingers consisting of only two physical correlators perform the signal despreading for each multipath component and parallel 20 ksymbol/s channel. A maximal ratio combiner estimates complex channel taps from the despread pilot channel, rotates/weights the RAKE output signals, and adds them together. In addition, the combiner outputs signals for frequency error, narrowband signal power, and SIR. These are used in AGC, AFC, and PC-loop, as described earlier. The selector picks up the code channels (I/Q branches) owned by the specific user and routes them to de-interleavers/channel decoders.

11.8 MULTIMODE TERMINALS

Because at the beginning of wideband CDMA system deployment there will still be several second-generation systems in operation, dual-mode or multimode terminals can be used to provide seamless service for the users. From the transceiver point of view, however, this may be cumbersome because the same physical terminal must provide technical solutions to support potentially dissimilar radio technologies.

The main second-generation cellular systems are GSM, IS-136, IS-95, and PDC. Furthermore, PHS and DECT are second-generation low-tier wireless systems. Third-generation wideband CDMA systems will be implemented together with some of these second-generation systems. The most likely combinations are GSM/WCDMA, PDC/WCDMA, and IS-95/cdma2000.

Different systems place different requirements on different parts of the transceiver. On the other hand, the right selection of transceiver architecture can ease the implementation of dual-mode terminals. In the RF section, the duplexer is impacted by the fact that second- and third-generation systems have different frequency bands. RF and IF filter bandwidths are different because of the different bandwidths of the systems. Intermodulation and phase noise requirements are also most likely different. Synthesizer settling times vary, as well as A/D and D/A converter requirements. Because different systems have different symbol rates and other clocking requirements, possibly two reference oscillators are required. This could be avoided by proper selection of system parameters for the new system. In the baseband, receiver algorithms are somewhat different. For example, a TDMA-based system requires an equalizer and a CDMA-based RAKE receiver.

Because there are so many different dual-mode combinations, it is impossible to analyze each of them in detail. Instead, several commonalties can be observed, and implementation of dual-mode terminals can be analyzed in general. The main difference of the above-listed systems is the spreading used in CDMA systems and the continuous transmission compared to nonspread, slotted signal of TDMA-based systems. Therefore, the text below assumes separation between nonspread and spread systems, between slotted and continuous systems, and according to a duplex scheme.

11.8.1 Nonspread Versus Spread Systems

The difference in the implementation of nonspread and spread systems is certainly not straightforward. The major parameter is the channel bandwidth, and if these are similar for both the systems, the actual transceivers seem fairly unified. The major differences are on the baseband transmitter and receiver front ends. In the former case, the spread system faces a multilevel signal due to parallel code channels, which increases transmission filter and D/A converter requirements. On the receiver side, the nonspread equalizer (DFE or MLSE) is also replaced with a RAKE receiver, or even a multiuser detection unit.

If the bandwidths are dissimilar, like with wideband CDMA and GSM, the situation is different. First, the IF/RF section will require new filters on both TX/RX sides. For a dual-mode terminal, there should be several filters from which the correct one is selected. Another choice is to implement an adjustable filter, which has the disadvantage that it cannot always be adjusted the most optimal way.

Channel raster is directly related to the bandwidth used. If the specified rasters are not factors of each other and analog synthesizers are used, there is a problem with the reference frequency. In order to provide simple and fast PLLs, the reference frequency should equal the channel raster. For a multimode terminal this would mean a compromise between complexity (several possible references) and speed (the lowest common factor frequency).

On the baseband side the processing requirements between spread and non-spread systems can be considered fairly similar. A wideband signal calls for a higher sampling rate, while at the same time it can be presented with less bits/sample. In favor of nonspread systems, signal spreading/despreading adds some extra complexity to the terminal. For long channel delay spreads, a RAKE receiver tends to be simpler than an optimal MLSE equalizer.

11.8.2 Slotted Versus Continuous Systems

Continuous systems better suit ASIC implementation because most of the functions operate constantly. Selecting a software approach for a continuous system would cause high control overhead per received symbol because each one would be processed independently by generating an interrupt for a DSP. This could be overcome by buffering some number of symbols prior to transmission and reception. The problem here, however, is the increased data path latency.

Slotted systems, on the other hand, suit software processing fairly well. This is because symbol detection may start only when the whole/partial burst has been received. This is especially the case when the channel estimation is performed from the burst midamble sequence. Overhead due to control is minimized, and the number of instructions per processed symbol becomes feasible.

From a slotted systems' RF point of view, the power amplifier sees discontinuous transmission with high peak power. With continuous systems, the PA is active constantly, and if the average transmission power is assumed the same, the peak power is reduced respectively. Continuous systems, however, may act as discontinuous systems if DTX mode is applied. This is common especially with speech services.

11.8.3 FDD Versus TDD Systems

Continuous systems generally use frequency duplex division for providing up- and downlink channels. From an implementation point of view, this requires synthesizers on both the transmitter and receiver sides because of simultaneous operation. In addition, a duplex filters must be applied to prevent the TX signal from leaking to the RX side.

In slotted systems, another possible method is to utilize time division duplex separation. The idea is to transmit up- and downlink channels with the same frequency but at different times. The main benefit here is that the terminal is not transmitting/receiving simultaneously, and hence, only one synthesizer and no expensive duplex filter are needed.

On the baseband processing side there are few physical resources that could be shared between the transmitter and receiver. DS-CDMA code generators are such an example (polynomials, and states reloaded upon TX/RX switching). In addition, the multipath channel estimation could be made easier because the channel is reciprocal (i.e., the channel estimate obtained for one direction is valid within coherence time for the other direction).

REFERENCES

[1] Neuvo, Y., "Wireless Will Catch Up Wireline and More," *Proceedings PIMRC'97*, Helsinki, Finland, September 1997.

[2] Bernel, D. A., and T. C. Hofner, "Digital Communications Require Dynamic Performance," *Electronic Engineering*, September 1997, pp. 67–71.

[3] Gaudenzi, R. D., "The Influence of Signal Quantization on the Performance of Digital Receivers for CDMA Radio Networks," *ETT*, Vol. 8, No. 1, January–February 1997, pp. 89–97.

[4] Do, G. L., "Efficient Filter Design for IS-95 CDMA Systems," *IEEE Transactions on Consumer Electronics*, Vol. 42, No. 4, 1996, pp. 1011–1019.

[5] Dixon, R. C., *Spread Spectrum Systems with Commercial Applications*, New York: John Wiley & Sons Inc., 1994.

[6] Pickholz, R. L., D. L. Schilling, and L. B. Milstein, "Theory of Spread-Spectrum Communications—A Tutorial," *IEEE Transactions on Communications*, Vol. 30, No. 5, May 1982, pp. 855–884.

[7] Polydoros, A., and C. L. Weber, "A Unified Approach to Serial Search Spread-Spectrum Code Acquisition—Part 2: General Theory," *IEEE Transactions on Communications*, Vol. 32, No. 5, May 1984, pp. 550–560.

[8] Bensley, S. E., and B. Aazhang, "Subspace-Based Channel Estimation for Code Division Multiple Access Communication Systems," *IEEE Transactions on Communications*, December 1994, pp. 1009–1020.

[9] Kreyszig, E., *Advanced Engineering Mathematics*, 7th ed., New York: John Wiley & Sons Inc., 1993.

[10] Lupas, R., and S. Verdú, "Linear Multiuser Detectors for Synchronous Code-Division Multiple Access Channels," *IEEE Transactions on Information Theory*, Vol. 35, No. 1, January 1989, pp. 123–136.

[11] Paris, B. P., "Finite Precision Decorrelating Receivers for Multiuser CDMA Communication Systems," *IEEE Transactions on Communications*, Vol. 44, No. 4, April 1996, pp. 496–506.

[12] Xie, Z., R. T. Short, and C. K. Rushforth, "A Family of Suboptimum Detectors for Coherent Multiuser Communications," *IEEE Journal on Selected Areas in Communications*, Vol. 8, No. 4, May 1990, pp. 683–690.

[13] Varanasi, M. K., and B. Aazhang, "Multistage Detection in Asynchronous Code Division Multiple-Access Communications," *IEEE Transactions on Communications*, Vol. 38, No. 4, April 1990, pp. 509–519.

[14] Holzman, J. M., "DS/CDMA Successive Interference Cancellation," *Proceedings of ISSSTA'94*, Oulu, Finland, July 1994, pp. 69–78.

[15] Patel, P., and J. Holtzman, "Analysis of a Simple Succesive Interference Cancellation Scheme in a DS/CDMA System," *IEEE Journal on Selected Areas in Communications*, Vol. 12, No. 5, June 1994, pp. 796–807.

[16] Giallorenzi, T. R., and S. G. Wilson, "Decision Feedback Multiuser Receivers for Asynchronous CDMA Systems," *Proceedings of Globecom'93*, Houston, Texas, November 1993, pp. 1677–1682.

[17] Duel-Hallen, A., "A Family of Multiuser Decision-Feedback Detectors for Asynchronous Code-Division Multiple Access Channels, " *IEEE Transactions on Communications*, Vol. 43, No. 2/3/4, February/March/April 1995, pp. 421–434.

[18] Lippmann, R. P., "An Introduction to Computing with Neural Nets," *IEEE ASSP Magazine*, April 1987, pp. 4–19.

[19] Paris, B. P., B. Aazhang, and G. Orsak, "Neural Networks for Multi-User Detection in CDMA Communications," *IEEE Transactions on Communications*, Vol. 40, July 1992, pp. 1212–1222.

[20] Kechiotis, G., and E. S. Manolakis, "Hopfield Neural Network Implementation of the Optical CDMA Multiuser Detector," *IEEE Transactions on Neural Networks*, January 1996, pp. 131–141.

[21] Kechiotis, G., and E. S. Manolakis, "A Hybrid Digital Signal Processing-Neural Network CDMA Multiuser Detection Scheme," *IEEE Trans. on Circuits and Systems-II*, Vol. 43, No. 2, February 1996, pp. 96–104.

[22] Proakis, J. G., *Digital Communications*, 3rd ed., New York: McGraw-Hill, 1995.

[23] Dawid, H., G. Fettweis, and H. Meyr, "A CMOS IC for Gb/s Viterbi Decoding: System Design and VLSI Implementation," *IEEE Transactions on VLSI Systems*, Vol. 4, No. 1, March 1996, pp. 17–31.

[24] Anderson, J. B., and E. Offer, "Reduced-State Sequence Detection with Convolutional Codes," *IEEE Transactions on Information Theory*, Vol. 40, May 1994, pp. 944–955.

[25] Forney, G. D., "The Viterbi Algorithm," *Proc. of IEEE*, Vol. 61, March 1973, pp. 268–278.

[26] Viterbi, A. J., and J. K. Omura, *Principle of Digital Communication and Coding*, New York: McGraw-Hill, 1979.

[27] Blahut, R. E., *Theory and Practice of Error Control Codes*, Reading, Massachusetts: Addison-Wesley Publishing Company, 1993.

[28] Oppenheim, A. V., and R. W. Schafer, *Discrete-Time Signal Processing*, Englewood Cliffs, New Jersey: Prentice Hall International Inc., 1989.

[29] Sundström, L., "RF Amplifier Linearisation Unsing Digital Predistortion," M.Sc.Thesis, Lund University, Sweden, June 1993, pp. 8–17.

[30] Johansson, M., and M. Faulkner, "Linearization of Wideband RF Power Amplifiers," *Proceedings of Nordic Radio Symposium*, April 24–27, 1995.

[31] Cavers, J. K., "A Linearising Predistorter with Fast Adaptation," *Proceedings of 40th IEEE Vehicular Technology Conference*, May 1990, pp. 41–47.

[32] Special issue on software radios, *IEEE Communications Magazine*, Vol. 33, No. 5, May 1995.

[33] Kuisma, E., "Technology Options for Multi-Mode Terminals," *Mobile Communications International*, February 1997, pp. 71–74.

[34] Edmond, P., "Towards a Multi-Mode Mobile Telephone Design," *Electronics Engineering*, October 1996, pp. 67–70.

Chapter 12

NETWORK PLANNING

12.1 INTRODUCTION

When using a cellular phone, you have most likely experienced several undesired effects such as dropped calls when you move from one location to another, a busy network when attempting to make a call, or just poor speech quality. The grade of service wireless subscribers experience is dependent on the quality of the radio network planning. Thus, network planning is one of the key competence areas for any operator in order to satisfy customers and thus make a profit.

Network planning covers two major areas: radio network planning and network dimensioning. Radio network planning includes the calculation of link budget and capacities, and thus the required number of cell sites. Furthermore, radio network planning includes detailed coverage and parameter planning for individual sites. In network dimensioning, the required number of channel elements in the base station, the capacities of transmission lines, the number of base station controllers, switches, and other network elements are calculated. This chapter presents the main concepts of third-generation network planning; for details of IS-95 network planning, refer to [1].

A large number of different bit rates and the diversification of services complicate the network planning process for third-generation systems, as compared to the second-generation systems. It is more difficult to predict traffic and usage patterns of different services. Which unit to use to measure traffic density is our first topic in this chapter. Next, we introduce performance measures, such as outage and blocking, which are used to measure the network quality. High bit rates services can be offered either with uniform coverage over a cell or with smaller data rates at the cell edge. Furthermore, either continuous coverage over a wide area or only hotspot coverage could be provided. A careful assessment is required to understand the impact of different deployment strategies on the demand of services.

The radio network planning process can be divided in three phases: preparation, estimation of cell count, and network optimization. Each of these is discussed in depth.

345

In the network optimization phase, the following aspects are discussed: detailed characterization of the radio environment for individual cell planning purposes, CDMA control channel power planning, pilot pollution, planning of soft handover parameters, interfrequency handover, iterative network coverage analysis, and radio network testing. The impact on the network planning of microcell and indoor planning is discussed, as are sectorization and beamforming.

The main emphasis in the section on network dimensioning is on the calculation of the number of channel elements in the base station.

The coexistence of wideband CDMA air interface with already deployed air interfaces will be an important factor in the beginning of wideband CDMA system deployments. The last sections discuss intermodulation interference, guardband and zones, and transition considerations.

12.2 TRAFFIC INTENSITY

The starting point for network planning is the assessment of individual user traffic and traffic intensity (offered traffic) in a given area. To quantify the traffic intensity, different units can be used. Most often, traffic intensity is measured in *Erlang*, where 1 Erlang is equivalent to one circuit in use for 1 hour (3600 seconds).

Traffic intensity in Erlang =

$$\frac{\text{Number of calls per hour} \times \text{average call holding time (seconds)}}{3600} \quad (12.1)$$

For example, if a user makes one call per hour and the average call duration of calls is 120 seconds, it produces $120/3600 = 33$ mErlang of traffic. Given the user density in a geographical area, the traffic density can be calculated and is expressed in Erlang per square-kilometer (Erl/km^2).

Because third-generation systems will have a very large variety of services, a single traffic measure might not be suitable to all cases. The *equivalent telephony Erlang* (ETE) defines traffic with relation to basic telephone calls [2]. This definition, however, is dependent on the transmission rate of the basic telephone service. For data services, traffic measured in $Mbps/km^2$ will better characterize the traffic density.

12.3 PERFORMANCE MEASURES

The performance measures for the grade of service are *area coverage probability* and *blocking*. The area coverage probability is related to the quality of the radio planning and the radio network capacity. Blocking is related to the available hardware, for example, to the base station channel cards. The term *soft blocking* is used when a user is blocked because the cell runs out of capacity (i.e., a user is denied access to the network in order to ensure the quality of already admitted users). First, the network is designed for the desired area coverage probability. Within this area coverage probability, then, certain blocking requirements, usually 2%, need to be fulfilled.

Area coverage probability can also be defined by *outage*. Outage is the probability that the radio network cannot fulfill the specified quality of service target. Several definitions exist for the measure of outage. In general, it is defined as a drop in the required quality below a prespecified target value. It needs to be defined further for how long the quality must be below the target value to create an outage. In addition, quality criteria for outage must be defined. We can differentiate between a *signal outage*, when the SNR drops below the target value, and an *interference outage,* when the SIR drops below the target value. The outage is impacted by several factors, such as shadowing, path loss model, outside cell interference, handover type, and power control dynamics, and is usually evaluated by system level simulations.

If the system is *coverage limited*, outage can then be defined as the probability that path loss and shadowing exceed the difference between the maximum transmitted power and the required received signal level. Because shadowing is log-normally defined, very low outage probability leads to very high power margin, and thus, small cells and an expensive network. This is illustrated in Figure 12.1, which shows a plot of a log-normal distribution with standard deviation of 8 dB. The area of the tail of the distribution gives the outage probability. In order to achieve 10% outage probability (i.e., 90% *availability* or coverage probability), a 10.3-dB shadowing margin is required. For 5% outage target, a 13.2-dB margin is required, and for 1% outage target, an 18.6-dB margin. Thus, we can see that increasing the outage target leads to very high margins and thus reduced range. We also plot shadowing with the standard deviation of 10 dB. The increased shadowing requires higher margins since it results in higher probability of large shadowing values.

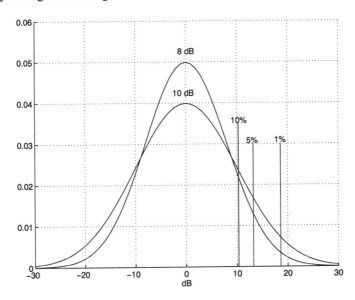

Figure 12.1 Log-normal distribution and shadowing margin.

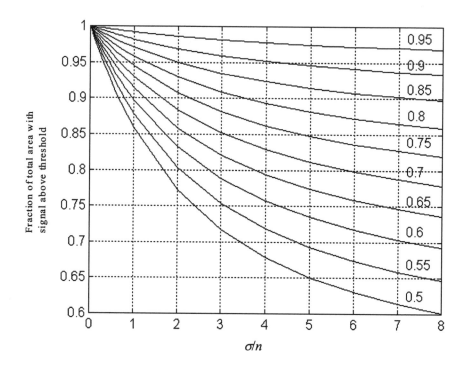

Figure 12.2   Curves relating fraction of total area with signal above threshold as a fuction of probability of signal above threshold on the cell boundary.

The coverage probability can be defined as a cell edge probability or a cell area probability. The cell edge probability refers to a probability that the mobile would receive a signal above a threshold at the cell edge, for example, RF signal strength. The threshold value chosen need not be the receiver noise threshold but may be any value that provides an acceptable signal under Rayleigh fading conditions [3]. The cell area probability means the percentage of locations within a circle of radius R in which the received signal from a radiating base station antenna exceeds a particular threshold value. The relation between area and cell edge probability has been derived in [3] (pp. 126–127) and is illustrated in Figure 12.2, where n is the path loss exponent and σ is the standard deviation of the shadowing given in decibels. In a macrocell environment, a typical value for the path loss exponent is 3.6 and for standard deviation is 8 dB. Thus, σ/n is 2.22, and a boundary coverage probability of 85% would give 95% area coverage.

Blocking is defined as the probability of a blocked call on a first attempt. A blocked call occurs when there is a lack of system resources (i.e., if all network resources are busy, a user will be denied access to the system). The reasons for blocking can be shortage of channel elements in the base station or shortage of fixed network resources such as transmission lines, BSC, or switching capacity. Thus, each network element needs to be dimensioned appropriately for the traffic demands. Usually, a blocking probability of 2% is used when dimensioning cellular systems.

Let N be the total number of channels and T be the offered traffic in Erlang. Then, the probability of all channels being busy is obtained from a Poisson distribution:

$$P(N;T) = \frac{T^N e^{-T}}{N!} \qquad (12.2)$$

where $P(N;T)$ is the blocking rate. Thus, if the traffic density (offered traffic) and blocking probability are known, the required number of network resources (e.g., channel elements in base station or capacity of the transmission lines) can be calculated using the Erlang tables. The Poisson distribution has the effect of achieving better trunking efficiency as the number of channels is increased. Trunking efficiency, also called channel utilization efficiency, is defined as

$$\text{Efficiency(\%)} = \frac{\text{Traffic in Erlang}}{\text{Number of channels}} \times 100\% \qquad (12.3)$$

Given the blocking probability, (12.2) can be used to calculate the traffic in Erlang for different numbers of channels. Thereafter, (12.3) gives the trunking efficiency. In Figure 12.3, we have plotted the trunking efficiency for a blocking probability of 2%. We can see clearly that the efficiency increases as the number of channels increases.

12.4 RADIO NETWORK PLANNING PROCEDURE

The prerequisite for good radio network planning is the know-how of the radio environment. Rough network planning and deployment plans can be made based on general radio channel models. Because the radio environment is highly variable, even within the area of one cell, detailed measurements and optimization need to be performed for each individual cell.

The radio network planning process can be divided into three phases:

- Preparation;
- Estimation of cell count;
- Detailed network planning.

12.4.1   Preparation Phase

In the preparation phase, coverage and capacity objectives are established, network planning strategy defined, and initial design and operating parameters determined.

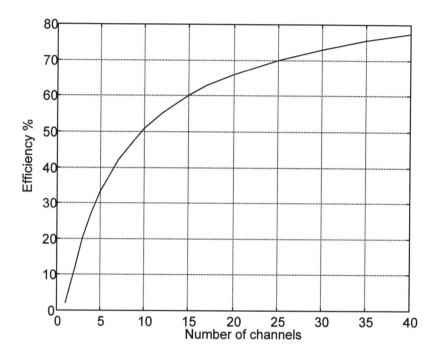

Figure 12.3   Trunking efficiency.

12.4.1.1   Coverage and Capacity Objectives

Coverage and capacity objectives are a tradeoff between desired quality and overall network cost. A smaller signal outage probability means smaller cells and thus higher overall network costs; and smaller interference outage probability means smaller capacity and thus also higher cost. A typical outage probability target is 5% to 10%, corresponding to 90% to 95% availability/coverage probability. The coverage probability could be different for different services as discussed below.

12.4.1.2   Network Planning Strategy

The network planning strategy includes issues like microcell deployment, provision for indoor and high bit rate coverage, and migration from second-generation systems. Several factors need to be considered for the most feasible network planning approach. These include cost of fixed line transmission, how easily cell sites can be acquired, and at what cost cell sites can be acquired. Furthermore, environmental issues such as cell tower appearance will impact where a base station can be deployed. Traffic distribution will of course impact the deployment strategy.

One deployment strategy could be to use macrocells for outdoor coverage and picocells for indoor coverage in office buildings. In addition, macrocells would be used to fill the gaps in the indoor coverage. This is because extensive indoor coverage is most likely required in any case. Therefore, it might be wiser to build additional capacity by increasing the number of indoor cells rather than trying to provide indoor coverage from outdoor cells and thereby be forced to introduce microcells earlier because of capacity restrictions. If coverage for indoors is provided by outdoor base stations, the building penetration margin, typically 10 to 20 dB, needs to be taken into account in the link budget calculations.

Another approach is to use microcells extensively from the beginning and to provide indoor coverage from them. This might be feasible in dense urban areas.

High bit rates can be provided uniformly over the cell area or the data rate at the cell border could be smaller than when close to the base station to allow a larger cell range. This depends on the nature of the high bit rate services. For services that utilize available bit rates (i.e., services that do not require any quality of service guarantees), nonuniform coverage might be acceptable, but for applications that require maximum bit rate such as video transmission, uniform coverage is required.

If an operator has deployed a second-generation network, migration aspects need to be considered in the network planning strategy. These include reuse of existing cell sites, handovers between the new and old systems, and coexistence requirements (see Section 12.9).

12.4.2   Estimation of Cell Count

The number of wireless users in a given area is obtained by multiplying the population by the penetration. The number of users and offered traffic per user determine the overall offered traffic. When cell capacity and range are known, a rough number of cells can be determined. Figure 12.4 illustrates the procedure for the estimation of cell count.

12.4.2.1   Offered Traffic

In the calculation of the number of users that need to be served, it is necessary to account for at least the following factors:

- Population living in given area;

- Population working in given area;
- Vehicle traffic;
- Special events and use of recreational areas.

All the above aspects contribute their share to the total number of users that might use the cellular network. When the total number of users is known, we need to estimate what percentage of the total number of users has a wireless terminal (i.e., what is the penetration?).

The offered traffic per user depends on what services are used and how often they are used. The prediction of service usage is especially difficult for third-generation systems, for which many new services will be introduced. Different services produce different bit rates. For data services, we need to estimate the average requirement of megabit per second per user. The average calling time for circuit switched users depends on the service (e.g., speech, video), customer types (business, residential), and tariffing. Typical values for the average calling time for speech service range from 120 to 180 seconds. For other services, the average calling time is more difficult to define since there is less experience. Because the average calling time has a major impact on the generated network traffic, it is very important to be able to predict it with reasonable accuracy.

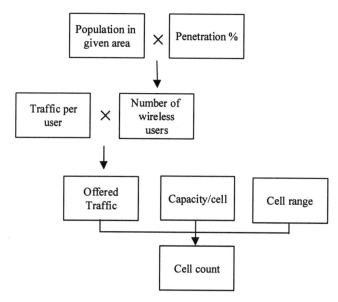

Figure 12.4 Estimation of cell count.

Because the demand for services also varies according to the time of the day, the network capacity must be dimensioned according to busy hour traffic (i.e., when the traffic demand is highest). Typically, busy hour is during business hours. New user types (e.g., young people), however, and a more widespread use of entertainment services through wireless connections might change this.

Example. The total user density is 10,000 user/km², 50% of users are using speech and 10% data. Thus, the number of users requiring speech services is 5000 user/km² and data services 1000 user/km². A speech user generates 30 mErlang traffic and a data user 100 mErlang. Thus, the total offered traffic for speech is 150 Erlang and for data 100 Erlang. If a data call requires on average 100 Kbps, the total offered data traffic is 10 Mbps.

12.4.2.2 Estimate Cell Capacity

As described in Chapter 9, cell capacity can be estimated based on simulations or analytical formulas. User data rate, traffic characteristics (variability, burstiness), quality of service requirements (delay, BER/FER), and outage probability are the main factors that determine the supported capacity. The higher the bit rate, the less users can be supported. The smaller the interference outage probability (i.e., the better the network quality), the smaller the provided capacity. It can be easily understood that provision for better quality radio connections requires more radio resources, and thus fewer users can be supported within a fixed amount of spectrum.

A possible impact of sectorization or adaptive beamforming, also called spatial division multiple access (SDMA), on spectrum efficiency needs to be estimated. It depends on specific radio environment and antenna equipment. For example, in microcells the signals tend to propagate along the street corridors no matter in what direction they were originally transmitted. Thus, the maximum sectorization in the downlink gain is less than in a more open space environment.

A CDMA radio network cannot be operated at a so-called *pole capacity*. Pole capacity can be defined as the theoretical maximum capacity. The *load factor* defines how close to the maximum capacity the network can operate. When determining a suitable load factor, a number of factors need to be taken into account. These include traffic characteristics, radio resource management algorithms, and even the planning capabilities of the network operator. Loads over 75% have been found to cause instability in the system [4].

Example. Assume 64-Kbps data service. Based on the network simulations performed in Chapter 9, we conclude that with full load the spectrum efficiency is 100 Kbps/MHz/cell (i.e., 500 Kbps/cell given 5-MHz bandwidth). Assuming 50% load, the actual spectrum efficiency is 250 Kbps/cell (i.e., slightly under 4 Erlang).

Example. Assume 8-Kbps speech users. Based on the network simulations performed in Chapter 9, we get spectrum efficiency of 108 Kbps/MHz/cell (i.e., 33.75 Erlang with 5-MHz bandwidth and 75% load). Assuming each user generates 30 mErlang traffic, one cell can support 1125 users.

12.4.2.3 Estimate Maximum Cell Coverage

The link budget is calculated according to principles presented in Chapter 9. In addition to the basic assumptions such as data rate and E_b/N_0 performance, the equipment-specific factors such as cable losses, antenna gain, and receiver noise figure need be taken into account.

Soft handover gain has a large impact on link budget. The soft handover gain is the gain brought by handoff at the boundary between two or more cells, where there is equal average loss to each of the cells. The soft handover gain can be obtained by first calculating the log-normal fade margin (i.e., the margin required to provide the specified coverage probability at the border of a single isolated cell, and then the corresponding margin required at the boundary between two or more cells). The soft handover gain is given by the difference (in decibels) between the two different margins. The soft handover gain depends on the shadowing correlation and coverage probability. The larger the coverage probability requirement (i.e., smaller outage probability), the larger the required margin. In addition to the gain brought by cell selection, soft handover brings the macro diversity gain through increased diversity (see Section 9.2.5). The actual gain depends on the radio environment and the number of RAKE fingers.

Because each radio environment has its own characteristics, for more detailed coverage prediction some correction factors for the path loss models presented in Chapter 4 are required. Field measurements can also be used.

For the uplink, the impact of load factor η in the link budget, the interference margin I_m (in decibels), can be determined from

$$I_m = 10\log\left(\frac{1}{1-\eta}\right) \tag{12.4}$$

and is illustrated in Figure 12.5. As can be seen, the interference margin increases, and thus, the range would decrease with increasing load factor. As was discussed in Chapter 9, multiuser detection can reduce the impact of load factor on the cell range.

Asymmetric traffic has to be considered in the link budget calculations. CDMA can trade the uplink system load for coverage. This is useful because usually the mobile station transmission power limits the maximum cell range. If there is three times more downlink than uplink traffic, the uplink may only be loaded to 25%, when the downlink is loaded to 75%. Consequently, the uplink coverage will improve by 4.8 dB compared to the downlink. The downlink coverage, however, can be compensated for by increasing the transmission power of the base station.

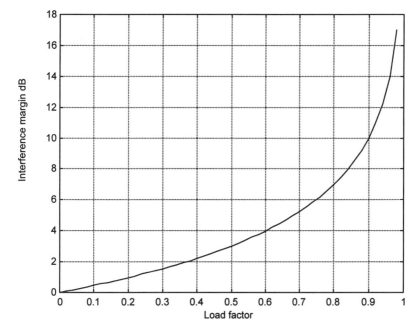

Figure 12.5   Interference margin as a function of load factor.

In Table 12.1, an example of link budget calculation is given. See Chapter 9 for a detailed explanation how to calculate the link budget. Note that in this calculation the carrier frequency and antenna height have been taken into account. We see that the uplink and downlink have different link budgets and thus different ranges. In the last column, we have different bit rates (144 and 28.8 Kbps) and unequal loading in the uplink and downlink. In practice, the downlink link budget is matched to the uplink by adjusting the base station transmission power. This has been done in the last case by increasing the downlink transmission power.

12.4.2.4 Estimate Cell Count

The number of cells required to cover a given area can be calculated based on the capacity and link budget. A network can either be *coverage* or *capacity limited*. Capacity limited means that the maximum cell radius cannot support the total offered traffic. In this case, the cell count is obtained as follows. By dividing the number of users, one cell can support by the number of user/km^2, we obtain the cell area in square kilometer. The total number of cells is obtained by dividing the total area by the cell area. Coverage limited means that there is enough capacity in a cell to support all traffic (i.e., the cell size could be larger from a capacity point of view, but the maximum cell range limits it). In this case, the maximum cell area is used to find out the required number of base stations.

Example. The area to be covered is 100 km^2 and the number of speech users is 5000/km^2. The cell area 23.32 km^2 is obtained from Table 12.1. One cell can support 1041 users with a 50% load, corresponding to 3-dB interference margin. Thus, the cell area is 1041/5000 = 0.2082 km^2 and altogether 100/0.2082 ≈ 481 cells are required to cover the area. If the user density was 30 user/km^2, then the network would be coverage limited because one cell could cover 37.4 km^2, from the capacity point of view. This, however, exceeds the maximum cell area of 23.32 km^2. The required number of cell sites in this case would be 100/23.32 ≈ 5.

12.4.3 Detailed Network Planning

After the cell count has been obtained, detailed radio network planning, taking into account the exact radio environment where each cell is deployed, can start. Because of cost reasons, zoning laws, building restrictions, or some other impediment, it is not possible to obtain optimum cell sites in a real network. This will impact the initial coverage plan. For detailed network planning, a network planning tool is used. The previously described process of obtaining a rough cell count could also be included in a network planning tool. A network planning tool has a digital map of the area to be planned. Building heights and antenna patterns are also modeled. Because each vendor has its own tools with slightly different capabilities, we try to reflect here the general principles and not the detailed procedures used in real assessment. The optimization process of the radio network coverage includes:

- Detailed characterization of the radio environment;
- Control channel power planning;
- Soft handover parameter planning;
- Interfrequency handover planning;
- Iterative network coverage analysis;
- Radio network testing.

12.4.3.1 Detailed Characterization of the Radio Environment

For optimization purposes, the radio environment has to be characterized in detail by actual field measurements or other techniques such as ray tracing [2]. Characteristics such as path loss, shadowing, and delay spread are studied. Furthermore, an RF survey can be performed using spectrum analysis. RF measurements reveal competing cellular operators and background interference and possible intermodulation products.

12.4.3.2 Control Channel Power Planning

The transmission of control channels reduces the overall capacity of the network. The total transmission power depends on how the power is divided between pilot, synchronization, paging, broadcast, and traffic channels. Power allocation principles are related to coverage.

Table 12.1
Link Budget Calculation

Service	Speech		Data		Data		
	Downlink	Uplink	Downlink	Uplink	Downlink	Uplink	Unit
Average TX power/TCH	30	24	30	24	30	24	dBm
Number of code channels	1	1	1	1	1	1	
TX power/code channel	30.00	24.00	30.00	24.00	37.14	24.00	dBm
Max TX power/TCH	30.00	24.00	30.00	24.00	37.14	24.00	dBm
Max total TX power	30.00	24.00	30.00	24.00	37.14	24.00	dBm
Cable etc losses	2	0	2	0	2	0	dB
Antenna gain	13	0	13	0	13	0	dBi
TX EIRP/TCH	41.00	24.00	41.00	24.00	48.14	24.00	dBm
Total TX EIRP	41.00	24.00	41.00	24.00	48.14	24.00	dBm
RX antenna gain	0	13	0	13	0	13	dB
Cable and connector losses	0	2	0	2	0	2	dB
Receiver noise figure	5	5	5	5	5	5	dB
Thermal noise density	-174	-174	-174	-174	-174	-174	dBm/Hz
Interference margin	3	3	3	3	6	1.25	dB
Tot. noise + interf. density, No+Io	-166	-166	-166	-166	-163	-167.75	dBm/Hz
Information rate	8	8	144	144	144	28.8	kHz
10log(Information rate Rb)	39.03	39.03	51.58	51.58	51.58	44.59	dBHz
Eb/No (incl. macro div. gain)	8	6.6	8	6.6	8	6.6	dB
Receiver sensitivity	-119.0	-120.4	-106.4	-107.8	-103.4	-116.6	dBm
Handoff gain (selection gain)	5	5	5	5	5	5	dB
Expl. diversity gain	0	0	0	0	0	0	dB
Other gains / losses	0	0	0	0	0	0	dB
Body loss	3	3	3	3	3	3	dB
Log normal fade margin	11.3	11.3	11.3	11.3	11.3	11.3	dB
Maximum path loss	150.67	146.07	138.12	133.52	142.26	142.26	dB
Carrier frequency f	2000.00	2000.00	2000.00	2000.00	2000.00	2000.00	MHz
Base station antenna height, hB	15.0	15.0	15.0	15.0	15.0	15.0	m
Range	3.97	3.00	1.84	1.39	2.37	2.37	km
Cell area	40.96	23.32	8.80	5.01	14.62	14.62	km²

In WCDMA and cdma2000, a common pilot channel is used for handover measurements and coherent detection (see Chapter 6). The coverage of these channels must be larger, when compared to traffic channels, in order for the mobile station to be able to decode other base stations before entering the soft/softer handover zone. Roughly 5% of total base station power will be allocated to the synchronization channel. Because the broadcast channel including the cell information has to be decoded before the mobile station enters the coverage area of a cell, for example, 3 dB more power than for a 8-Kbps speech traffic channel should be allocated. The paging channel(s) can manage with about the same power as a 8-Kbps speech traffic channel.

12.4.3.3 Pilot Pollution

By pilot pollution it is meant that there are a number of pilot signals but none of them are dominant enough to allow the mobile station to start a call. The symptoms of such a situation are typically good mobile received power, poor E_c/I_0, and poor forward BER. To prevent pilot pollution, network planning needs to create a cell plan where a dominant pilot exists. This can be done by scaling pilot powers, down-tilting antennas, or increasing coverage of certain sectors or cells.

12.4.3.4 Soft Handover Parameter Planning

Soft handover performance impacts the required fade margin against shadowing and the number of users in soft handover. As was shown earlier, the fade margin is one factor in the link budget and thus will influence the number of base stations that have to be deployed. Users in soft handover require additional channel elements and backhaul connections (see Section 12.8). The network operator's goal is to minimize the fade margin and number of users in soft handover, while maintaining a satisfactory quality of service.

The number of users depends on handover thresholds and service type. It is possible that packet services do not use soft handover at all. Also, antenna tilt and orientation impact the percentage of cell coverage in soft handover and should be considered in the handover parameter and region planning. The optimum active set size depends on the handover thresholds, the number of available RAKE fingers, and the radio environment. Typically, more than three base stations in the active set do not give significant soft handover gain increase.

In the following we review the soft handover planning based on the IS-95 soft handover algorithm. The main parameters for soft handover are handover thresholds, handover timers, and pilot search window at the mobile station. The parameters used in the active set updating are the following:

- T_{ADD}: This threshold indicates the point when a pilot should be added to the candidate set. The mobile station is measuring the pilot E_c/I_0.
- T_{DROP}: This threshold indicates when a pilot should be dropped from the active set if pilot strength has dropped below T_{DROP} for T_{TDROP} seconds.
- T_{COMP}: The mobile station reports that a candidate set pilot is stronger than an active set pilot only if the difference between their respective strengths is at least $T_{COMP} \times 0.5$ dB.

A typical value of T_{ADD} is between -12 and -16 dB [5]. Increasing the T_{ADD} threshold reduces the number of users in the soft handover and thus reduces the required equipment overhead. Reducing it too much, however, increases interference and thus reduces performance [5]. Because the time required to detect a new pilot should be minimized, filtering in the searcher for pilot in neighboring sets should be minimized. The T_{ADD} threshold should be high enough to prevent false alarms due to large noise in pilot measurements.

T_{DROP} should be low enough to prevent the loss of a good pilot. Typical values for T_{DROP} are between -17 and -20 dB. If the T_{TDROP} timer is too short, unnecessary handovers might occur, so T_{TDROP} should be in the order of the time required to establish handover. If T_{ADD} and T_{DROP} are too close to each other and the T_{TDROP} timer is very short, a handover ping-pong effect (i.e., adding and dropping the same pilot consecutively) might occur.

T_{COMP} should be set to a value that would prevent the mobile station from continuously sending active set update messages as a consequence of small changes in the strengths of pilots in the active set and the candidate set. Too large values, however, would introduce too long a delay before a pilot strength measurement message is issued, delaying the handover procedure [6].

The impact of T_{COMP}, path loss slope and standard deviation of shadowing (log-normal fading), and different radio environments have been studied in [7]. The following results were reported: the higher the log-normal standard deviation, the higher the shadow margin required to combat fading. Higher path loss slope causes better isolation between adjacent cells and hence a lower margin is required. T_{COMP} did not impact performance very much in [7]. The required fade margin increases with delay in establishing handover. A typical handover delay in today's systems is 2 to 5 seconds [8].

In the future, other criteria such as uplink interference might be required in the handover process. Furthermore, different services with different quality of service requirements might require different criteria for handover and even different active set sizes. This increases the network planning complexity.

Because soft handover gain depends on the correlation between the signal from different cell sites and on the correlation length of the shadow fading from the same cell site, field measurements are required to quantify the exact gain and thus the network capacity. Lower correlation between sites leads to higher call reliability. Furthermore, a large correlation length of the log-normal fading increases the fading margin [8]. The impact of correlation length depends on the mobile speed: the faster the mobile moves, the smaller the impact, because the correlation coefficient for the log-normal fading is smaller.

The searcher window determines the number of chip delays for which a mobile is instructed to look for pilot. A large window helps to collect all the multipath energy but leads to a slow measurement process, which might be detrimental for high-speed mobiles. A too narrow window leaves some of the significant multipaths unnoticed and thus degrades performance. Obviously, the window size is a function of delay spread and should be optimized for each cell, taking the expected mobile speed distribution into account.

12.4.3.5   Interfrequency Handover

The compressed mode is one alternative for seamless interfrequency handover (see Chapter 5). The main parameter of interfrequency handover is the frequency at which the base station commands the mobile station to measure neighboring carriers. The stronger the serving cell, the more seldom the mobile would check the neighboring channels. At cell edge, the mobile should measure pilot channels more often in order to be prepared for interfrequency handover.

The use of slotted frames leads to performance degradation. This degradation depends on the number of slotted frames. Assuming that the performance of a slotted frame is 2 dB worse compared to a continuous frame, the overall degradation depends on the number of slotted frames. For example, if every tenth frame is slotted, the degradation would be 0.2 dB.

If there is both micro- and macrocell coverage, then the criteria for which each layer is selected have to be decided. In general, if a fast moving mobile is detected in microcells it should be handed over to macrocells. Therefore, the speed of mobile stations needs to estimated, for example, from the fading characteristics. On the other hand, a slow moving mobile should be handed over to microcells. Wrong decisions should be avoided; for example, a mobile station in a car that has stopped at traffic lights should be kept in macrocells because it will speed up in the next moment.

12.4.3.6 PN-Offset Planning

In IS-95 and cdma2000, pilot signals belonging to different base stations and sectors are distinguished by the use of different phase offsets of the same pilot pseudo-noise (PN) sequences. If a propagation delay between two base stations exceeds the pilot signal offset, so-called PN confusion occurs. PN confusion might cause dropped calls because of high interference from another base station or a handover to a wrong target cell. Therefore, PN-offsets need to be planned [9, 10].

The distance d a pilot signal travels can be converted to a distance in chips using the following formula [10]:

$$T_d \text{ (chips)} = \frac{B}{c} \times d \tag{12.5}$$

where B is the chip rate of the system and c is the speed of light (3×10^8 km/s). If the chip rate is 1.2288 Mcps (as in IS-95) and the PN-offset index is 64 chips, one ON-offset index corresponds to a 15.6-km propagation distance. For cdma2000, one PN-offset index would correspond to 5.6 km. Figure 12.6 presents a scenario where the propagation distance difference of cell I and cell J is equal to PN-offset, therefore causing PN confusion.

12.4.3.7 Iterative Network Coverage Analysis

Once the radio network environment has been characterized, control channel power allocation decided, and handover parameters planned, a detailed coverage analysis can be carried out. The ratio of in-cell to total interference is also referred to as the *frequency reuse coefficient* (see Chapter 9). The frequency reuse coefficient is unique for each cell. Starting from the initial configuration based on the preceding steps in the planning process (cell count, detailed characterization of the radio environment, control channel power planning, soft handover parameter planning, interfrequency handover planning), network quality is analyzed and new frequency reuse coefficients evaluated [5]. These are then used again to predict the coverage within different cells. This iterative process is repeated until convergence is achieved. A network planning tool can be used to automate the process. With this type of tool, quality plots can be created and gaps or holes in the coverage can be detected.

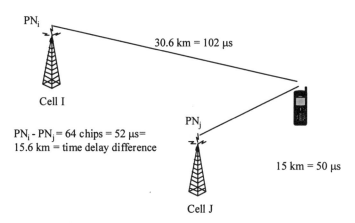

Figure 12.6   PN confusion. (*After*: [10].)

12.4.3.8   Nonuniform Traffic

In general, nonuniform traffic degrades the overall performance of the system. On one hand, the quality becomes poor because of increased interference in dense traffic zones; on the other hand, the quality becomes excessive. This dispersion of the communication quality restricts the system capacity [11]. The system efficiency can be improved by adaptive control of cell radius, antenna directivity, and uplink received power threshold. The cell radius is controlled by adapting the pilot transmission power. If the observed SIR is higher than required, the cell radius can be extended; and if it is lower, the cell radius is reduced. The desired received power threshold in the uplink is increased and reduced, respectively, to balance the cell radii of uplink and downlink. In a sector cell configuration, the central angle of each sector (i.e., the antenna directivity) is changed to equalize the communication quality in every sector belonging to the BS [11].

Of course, dynamic control of cell radius and antenna directivity requires careful planning and stable control system to prevent undesired effects. Also, interaction with handovers should be considered.

12.4.3.9   Radio Network Testing

After the network has been deployed, it is tested by drive tests to find out the network quality in practice. Furthermore, drive tests are used to collect data for optimization of the network performance. Before measuring, the optimization scope is determined so that correct measurements can be performed. Typical topics for measurements are coverage limitations, drop-call rate, quality, and pilot sets for soft handover.

Once the network has been set up, the work does not end. Testing continues as an essential part of the network operations. Therefore, proper operation and maintenance facilities are required to perform measurements and to collect data about network behavior. The measurement data are used to monitor the network quality and to locate congestion and quality gaps.

12.5 MICROCELL NETWORK PLANNING IN CDMA

Because of the propagation environment and network topology, microcell deployment differs significantly from the macrocell deployment (see Chapter 4). Microcell base stations are installed at lamp post level, and therefore, signals tend to propagate along street canyons. Two critical aspects for microcell network planning are corner effect and soft handover design.

12.5.1 Corner Effect

Figure 12.7 illustrates the so-called *corner effect*. When a mobile moves around a corner, changes in received signal level at the mobile station happen very rapidly. In case there is a new base station behind the corner, the signal strength received by the mobile station rises very fast. If the mobile station could not acquire the new base station fast enough, the increased interference leads to a dropped call. On the other hand, because the new BTS cannot regulate the power of the mobile station, the high transmission power of the mobile station can block all users in the new cell. To reduce the impact of corner effect, a fast forward handover can be performed if the old base station is dropped. Corner effect can also be avoided by proper planning of cell locations and handover thresholds. Handover thresholds can be specified in such a way that the mobile station is in soft handover before and after a street corner (i.e., overlapping cells).

12.5.1.1 Soft Handover Design in Microcell Environment

Because microcells are small, frequent handovers occur as users move along the streets. The danger is that if traditional cell planning is used, then fast-moving mobiles cause so much interference due to too slow handovers that capacity in microcells will drastically decrease. Furthermore, the signaling load increases considerably because of a large number of base stations. Thus, it is advantageous to deploy intelligence to the base stations.

The need for soft handovers in microcellular CDMA can be reduced by using distributed antenna and sectored cells. With distributed antennas, as shown in Figure 12.8, large coverage areas are achieved without the use of many small micro- or picocells. The forward link capacity is not good, because the signal is transmitted to the whole coverage area even though the mobile is only in one location.

Figure 12.7 Corner effect.

Figure 12.8 Street microcell with distributed antennas.

12.5.2 Micro/Macrocells in the Same Frequency

Placing micro- and macrocells in the same frequency is possible with low antenna installation and antenna tilting [12]. The danger is that the whole cell is in soft handover and no extra capacity is gained. In this solution, a fast soft handover from macrocell to microcell is required.

12.6   INDOOR PLANNING

The building types, sizes, construction, and materials exhibit a very large variation. Thus, no general propagation model that would be valid everywhere exists. Attenuation because of walls and floors depends on the construction of the building. Furthermore, propagation loss between floors depends also on structures like stairs and elevators. Therefore, field measurements or prediction of propagation by ray-tracing are required to characterize individual buildings.

Because of short propagation distances, the values of indoor delay spreads are very small. Thus, the coherence bandwidth is large and consequently diversity order is low. Therefore, extra diversity may be added with delayed transmission or reception. One possibility is to use distributed antennas. By introducing time delay elements between the antennas in the distributed antenna system, deliberate multipaths are created that can be processed by the CDMA RAKE receiver [13]. Creating too many multipaths, however, actually degrades the performance due to uncaptured energy.

The deployment of indoor cells into the same frequency as outdoor cells presents some challenges for the system operation. Because mobiles have to be connected to a cell where they use minimum transmission power, outdoor users can occasionally get the lowest power with picocells, for example, through windows. This situation most likely will not last very long because it begs the question: How is the handover working in this situation? One way to avoid this kind of situation is to deploy indoor cells to a different frequency band. Another possibility is careful network planning to avoid this. This means low transmission powers indoors and antennas with their backs towards windows.

12.7   SECTORIZATION AND SMART ANTENNAS

Sectorization increases the capacity of a CDMA system. In an ideal case, perfect isolation between sectors is achieved and the capacity will increase in direct proportion to the number of sectors. Of course, in practice, isolation is not perfect and the capacity increase is smaller. Furthermore, the number of softer handovers increases with the number of sectors. Several factors impact the sectorization gain. These include interference distribution and antenna location. Obstacles and structures near the antenna tend to change the antenna's impedance and radiation pattern (see also Section 9.3.2.8). This causes the coverage to deviate from the planned coverage. Another impact is that coupling between diversity antennas might increase, and consequently, correlation increases, degrading performance. Usually it is assumed that three sectors result to a capacity increase of 2.5.

Smart antennas (also called SDMA) are defined as multibeam or adaptive array antennas without handover between beams [14]. Multibeam antenna uses multiple fixed beams in a sector, while in an adaptive array the received signals by the multiple antennas are weighted and combined to maximize the SNR. The advantage of antenna arrays compared to fixed beam antennas is that in addition to the M-fold antenna gain they provide M-fold diversity gain. They require, however, a receiver for each antenna, and tracking the antenna weights at the rate of the fading.

The M-fold antenna gain will increase the range by a factor of $M^{1/\gamma}$ where γ is the path loss exponent, and reduce the number of base stations to cover a given area by $M^{2/\gamma}$ [14]. A multibeam antenna with M beams can increase the capacity by a factor of M by reducing the number of interferers. Adaptive arrays can provide some additional gain by suppressing interferers further. However, since there are so many interferers, the additional gain might not be worth the complexity.

12.8   NETWORK DIMENSIONING

12.8.1   BTS Channel Element Planning

The number of channel elements can be calculated when the offered traffic in Erlang, control channel requirements, and number of users in soft handover are known. Calculations are based on Erlang tables that give the number of required channels for a given blocking probability and offered traffic.

In addition to the traffic channels that can be determined from capacity calculations, common control channels in the downlink and access channels in the uplink need channel elements. Each wideband CDMA carrier requires one pilot and at least one common control channel, which consists of the paging channel, synchronization channel, and broadcast channel. In the uplink, the number of access channels depends on the expected access load and synchronization time requirements. In case an umbrella cell is used to overlay microcells, the microcells do not necessarily need that many common control channels and access channels if the initial access traffic is handled mainly through macrocell. For packet access, however, common control channels are required also in microcells.

The additional number of channel elements because of soft handover depends on the soft handover type. Softer handover does not need an additional channel element because the same channel element can handle signals from different sectors within the base station. Two-way soft handover needs two channel elements, and three-way soft handover needs three channel elements. As was discussed previously, the number of users in soft handover depends on the radio environment, handover thresholds, and antenna configuration and it can be found out from simulations or field measurements.

Because of the trunking gain effect, it is beneficial to pool channel elements between sectors and different carriers in the same cell site. This also applies to the channel elements due to soft handover. Therefore, the actual number of additional channel elements is less than the percentage of users in soft handover due to improved Erlang efficiency, as will be illustrated by the following example.

Example. Table 12.2 shows a calculation for the number of required BTS channel elements. We consider a reference BTS implementation, which has one CDMA frequency and three sectors and is realized as one cluster with a common pool of "floating" channel elements. The total carried traffic per sector is 39 Erlang including soft handover overhead traffic of 9 Erlang. Thus, the total traffic for the cell site is 3×39 = 117 Erlang. For a blocking rate of 2%, the total amount of needed channel elements for the cluster is 130, or 43.3 per sector (using the Erlang-B formula applied for 117 Erlang). In addition, for the downlink one channel element for pilot channel and for common control channel is required. In the uplink, one channel element for access channel is assumed. Thus, the overall number of channel elements per sector is 46.

Without additional soft handover traffic, the required number of channel elements for traffic channels would be 34.3. The additional number of channel elements because of soft handover is 26.2%, which is less than the 30% overhead traffic and is because of the Erlang effect.

If there was no pooling between sectors, the required number of channel elements for traffic channels would be 49 (Erlang-B formula applied for 39 Erlang). Thus, the pooling reduces the required number of channel elements by 12%.

For high bit rate services, we have 8 Erlang traffic. Assuming 30% overhead traffic because of soft handover, we have 10.4 Erlang per sector and 31.2 Erlang per site, 41 channel elements = 41/3 = 13.7. The additional soft handover traffic 33/3=11 is 8 channel elements (i.e., 24.2%).

Table 12.2
Channel Element Calculation

	Speech 8 Kbps	Data 144 Kbps
Erlang/sector	30 Erl	8 Erl
Soft handover overhead	30%	30%
Total traffic (3 sectors)	117 Erl	31.2 Erl
Number of channel elements/sector from Erlang B table, pooling taken into account	44	14
Channel elements for pilot and common control channels	2	2
Total number of channel elements	46	16

12.8.2 Number of BSCs and Switches, HLR and VLR Signaling Traffic

The number of BSCs is determined based on the maximum configuration. Different vendors support different traffic loads through the BSC. Similar to BSC, switches from different vendors support different traffic loads, which also determines the number of required switches.

In order to determine signaling related to mobility management (i.e., between home location register, HLR, and visitor location register, VLR), mobility models are required. The mobility models describe the mobility of wireless users (see Section 4.7).

12.8.3 Transmission Capacity

The required transmission capacity depends on the user and signaling data that needs to be transmitted between network elements. Connections include at least BTS-BSC, BSC-switch, and possibly BSC-BSC connections for handover purposes. Transmission capacity depends on the transmission technology. ATM, for example, provides good multiplexing gain. Soft handover traffic increases the required transmission capacity up to the network element, most likely to the BSC, where soft handover is terminated. In addition, soft handover between cells belonging to different BSCs over switches might require special arrangements. It also has to be considered whether additional blocking is allowed for transmission.

12.8.4 Transmission Network Optimization

When the rough network dimensioning has been obtained, an optimization phase is carried out. Depending on the costs of transmission, different network configurations can produce optimum overall cost. Sometimes it is beneficial to have more BSCs than required to tailor the transmission network for the most optimum configuration. If a BTS is situated in the middle of two BSCs, then routing of that BTS needs to be determined based on transmission costs that might be different for the two cells.

12.9 COEXISTENCE

When a wideband CDMA network is deployed over a given area, other networks will nevertheless continue to operate. Most likely the WCDMA network will coexist with GSM, IS-95, and PDC systems. Two main scenarios for the deployment of wideband CDMA are then predictable:

- Wideband CDMA introduced in third-generation frequency band;
- Wideband CDMA introduced into the second-generation frequency bands.

An existing cellular operator will most likely start UMTS with local coverage, relying on the second-generation system for low bit rate wide area coverage. Therefore, it is wise to ease the dual-mode terminal design by properly selected air interface parameters. In the longer term, the replacement of a second-generation system by wideband CDMA depends on the gained advantages, such as higher bit rates or increased spectrum efficiency. Furthermore, licensing and availability of spectrum also impact the decision. If the available bands of an operator are congested, the higher spectrum efficiency of wideband CDMA might allow the operator to pack existing users into less spectrum and to create room for higher rate services. The introduction, however, of a wideband carrier into congested frequency bands is fairly difficult. First, the operator needs to preload the network with dual-mode terminals with new wideband capabilities. Next, he needs to release the spectrum from old carriers and switch the new carrier on almost simultaneously to avoid interruption in service.

A Greenfield third-generation operator in a country with an existing cellular network also needs dual-mode terminals. Only if the data market emerges very fast and justifies the investment in a new nationwide network can an operator rely on single-mode terminals.

The transition to third-generation systems depends on technical possibilities such as increased spectrum efficiency and data rates but most of all on market needs and regulatory conditions. In this section, we highlight some of technical aspects related to network planning facilitating smooth transition and coexistence. The transition from the first-generation technology AMPS to the second-generation digital technologies in the United States gives us some background for the transition to the third generation as well. To avoid problems with the coexistence of different air interfaces, the following aspects need to be considered from the network planning perspective:

- Intermodulation;
- Deployment scenarios;
- Guard zones and bands;
- Transition aspects;
- Handover between systems.

12.9.1   Intermodulation (IM)

A power amplifier is nonlinear near the saturation point. In this nonlinear area, the signal is distorted and produces harmonics. If two or more different frequencies are present at the input of an amplifier, intermodulation products are generated [15]. Figure 12.9 illustrates intermodulation power and its relation to the power of the original signal. As can be seen, the power of the third-order intermodulation product increases three times faster than the power of the original signal. Equally, the power of the fifth-order intermodulation product increases five times faster.

Figure 12.9 Intermodulation power. (*After*: [15].)

In mobile radio systems, there are two types of intermodulation products: transmit IM and mobile-generated IM [16]. The mobile station, generated IM is more significant than transmit IM from the base station. Transmit IM is generated at the base station transmitter. Several carriers are passed through the same power amplifier creating intermodulation products that might fall in the wideband CDMA band. Mobile-generated IM is caused by the nonlinear effects in the active stages of the receiver front end; these stages typically consist of an RF low noise amplifier (LNA), mixer, and IF amplifier. As was discussed previously, the power of the third-order intermodulation products, considered the primary source for IM interference, is increasing very fast as the RF gain increases. Thus, when the mobile is close to the interfering base station, the third-order products become severe in relation to the desired CDMA signal, possibly causing a dropped call. One proposed solution to avoid this is to switch out the LNA to eliminate strong IM signals from occurring at the mobile station front end prior to mixing [17]. If the desired signal is already weak, however, this might lead to a dropped call as well.

If wideband CDMA is deployed within the same frequency band as an existing network, there are two possibilities to map CDMA cells and existing cells. One alternative is *one-to-one mapping* of the cells of the wideband CDMA and the other system. Another alternative is that one wideband CDMA cell can cover more than one cell of the existing system (1-to-N mapping), because, in the beginning, wideband CDMA deployment load is still low and the network can be built to be coverage limited. Furthermore, the existing system has already reached quite small cell sizes because of large capacity demand. Later, when wideband CDMA capacity demand increases, it is likely that it will have either the same cell size as the other systems or an even smaller cell size.

From the deployment point of view, one-to-one mapping is better with respect to intermodulation product avoidance. There should be no intermodulation interference problems because wideband CDMA has similar power than the interfering system. Of course, this will lead to increased network infrastructure costs in the beginning. In 1-to-N mapping, the CDMA mobiles with small signal power might receive high intermodulation products. When a wideband CDMA mobile at the CDMA cell edge is at the sensitivity limit, then high power intermodulation products might block the mobile. This kind of deployment requires more planning.

12.9.2   Guard Bands and Zones

A second issue is frequency coexistence. Because of spectrum regrowth, proper guard bands need to be allocated in order to not block the system by adjacent spectrum or to prevent the other system from blocking the wideband CDMA system.

A guard zone around the CDMA coverage area is needed to avoid co-channel interference. An example of an uncoordinated co-channel interference situation is when different technologies in the same frequency band are not separated well enough by a geographical barrier. Such a situation has been experienced in the United States between PCS market areas. The Code of Federal Regulations (CFR) Title 47 Part 24 addresses issues related to interference caused by the system. It specifies, for example, attenuation rules outside the operators' own spectrum (Section 24.238) and the allowed field strength at the border of an operator's licensed service area (Section 24.236). The National Spectrum Managers Association (NSMA) has prepared a document titled "Inter PCS Co-block Coordination Procedures" addressing the intersystem coordination rules for PCS systems. It specifies coordination distance, allowed degradation because of external interference, interference calculation procedures, and other coordination practices to prevent harmful interference from one system to another.

The guard zone depends on the interference that the system can tolerate from other systems and vice versa.

Example. For GSM the required carrier-to-interference ratio is

$$\frac{C_{\text{GSM}}}{I} = 9 \, \text{dB} \tag{12.6}$$

We allow 1 dB degradation in C/I due to interference from CDMA:

$$\frac{C_{\text{GSM}}}{I + I_{\text{CDMA}}} = 8 \, \text{dB} \tag{12.7}$$

Thus,

$$\frac{C_{\text{GSM}}}{I_{\text{CDMA}}} > 15\,\text{dB} \tag{12.8}$$

From this we can now calculate the minimum separation between GSM and CDMA base stations using the same frequency. The ratio between cell size R and distance D can be calculated from [18]

$$q = \frac{D}{R} = \left(6\frac{C}{I}\right)^{1/\lambda} \tag{12.9}$$

Deployment within existing footprint reduces deployment efficiency. If coverage-limited deployment is desired, this might not be feasible if existing sites and their antennas are used. Furthermore, guard zones decrease the capacity of existing systems and depend on the deployment scenario.

Guard zones between IS-95 and AMPS have been studied in [19]. For a coverage-limited system, they are larger in order to maintain the quality of the IS-95 network. Within a guard zone, the CDMA carrier plus guard bands cannot be used. So, even if IS-95 had larger capacity and coverage and it could be deployed using a lesser number of sites than AMPS, it becomes necessary because of guard zone reasons to convert enough sites to IS-95 to increase the total traffic in the network [19].

12.10 FREQUENCY SHARING

Frequency sharing means sharing the allocated frequency spectrum in the same geographical area between two or more operators. It is one potential technology to increase the overall capacity of a set of mobile radio networks in the same area. By sharing the frequencies, the trunking loss due to the subdivided total frequency band between operators is avoided. The frequency sharing is possible if the co-channel interference power of the interfering systems is less than the co-channel interference power from the own system. Because CDMA systems are due to have a DCA scheme, however, frequency sharing might be possible only if the operators use the same cell sites.

REFERENCES

[1]   Yang, S., *CDMA RF System Engineering*, Norwood, Massachusetts: Artech House, 1998.

[2]   Cheung, J. C. S., M. A. Beach, and J. P. McGeehan, "Network Planning for Third Generation Mobile Radio Systems," *IEEE Communications Magazine*, Vol. 32, No. 11, November 1994, pp. 54–59.

[3]   Jakes, W. C., *Microwave Mobile Communications*, New York: John Wiley & Sons, 1974.

[4]   Stellakis, H., and A. Girdano, "CDMA Radio Planning and Network Simulation," *IEEE International Symposium on Personal, Indoor and Mobile Radio Communications PIMRC'96*, Taiwan, 1996, pp. 1160–1162.

[5]   Wallace, M., and R. Walton, "CDMA Radio Network Planning," *Proceedings of VTC'94*, Stockholm, Sweden, 1994, pp. 62–67.

[6]   Qualcomm, *CDMA Network Engineering Handbook*, 1993.

[7]   Chopra, M., K. Rohani, and J. D. Reed, "Analysis of CDMA Range Extension due to Soft Handoff," *Proceedings of VTC'95*, Chicago, Illinois, May 1995, pp. 917–921.

[8]   Rege, K., S. Nanda, C. F. Weaver, and W.-C. Peng, "Analysis of Fade Margins for Soft and Hard Handoffs," *Proceedings of IEEE 6th Personal, Indoor and Mobile Radio Communications Conference (PIMRC'95)*, Toronto, Canada, 1995, pp. 829–834.

[9]   Chang, C. R., J. Z. Van, and M. F. Yee, "PN Offset Planning Strategies for Non-Uniform CDMA Networks," *Proceedings of VTC'97*, Vol. 3, Phoenix, Arizona, May 4–7, 1997, pp. 1543–1547.

[10]   Yang, J., D. Bao, and M. Ali, "PN Offset Planning in IS-95 Based CDMA Systems," *Proceedings of VTC'97*, Vol. 3, Phoenix, Arizona, May 1997, pp. 1435–1439.

[11]   Sato, S., and Y. Amezawa, "A Study on Dynamic Zone Control for CDMA Mobile Radio Communications," *Proceedings of ICC'97*, Vol. 2, Montreal, Quebec, Canada, June 8–12, 1997, pp. 984–988.

[12]   Shapira, J., "Microcell Engineering in CDMA Cellular Networks," *IEEE Transactions on Vehicular Technology*, Vol. 43, No. 4, November 1994, pp. 817–825.

[13]   Xia, H. H., A. B. Herrera, S. Kim, and F. S. Rico, "A CDMA-Distributed Antenna System for In-Building Personal Communications Services," *IEEE Journal on Selected Areas in Communications*, Vol. 14, No. 4, May 1996, pp. 644–650.

[14]   Special Issue on Smart Antennas, *IEEE Personal Communications*, Vol. 5, No. 1, February 1998.

[15]   Faruque, S., *Cellular Mobile Systems Engineering*, Norwood, Massachusetts: Artech House, 1996.

[16]   Hamied, K., and G. Labedz, "AMPS Cell Transmitter Interference to CDMA Mobile Receiver," *Proceedings of VTC'96*, Vol. 3, Atlanta, Georgia, April 1996, pp. 1467–1471.

[17]   Gray, S. D., and T. Kenney, "A Technique for Detection AMPS Intermodulation Distortion in an IS-95 CDMA Mobile," *Proceedings of PIMRC'97*, Vol. 2, Helsinki, Finland, September 1–4, 1997, pp. 361–365.

[18]   Lee, W. C. Y., *Mobile Cellular Telecommunications Analog and Digital Systems*, 2nd ed., New York: McGraw-Hill, Inc., 1995.

[19]   Ganesh, R., and V. O'Byrne, "Improving System Capacity of a Dual-Mode CDMA Network," *International Conference on Personal Wireless Communications*, Mumbai, India, December 1997, pp. 424–428.

Chapter 13

BASIC PRINCIPLES OF IP

13.1 INTRODUCTION

The Transmission Control Protocol/Internet Protocol (TCP/IP) suite has evolved over a period of some 30 years and is showing a large growth in recent years. Starting with services to exchange files and e-mail messages between research institutes, nowadays the Internet can be used for more advanced services as well, these include streaming audio, http, remote banking, and e-commerce. In the wireless domain a growing interest in IP technology can be observed as well; with the arrival of mobile, packet-based systems such as GPRS, it will be possible to access the Internet not only by using a fixed connection, but also by using a mobile terminal. This will only grow further with the arrival of the third-generation communication systems and especially with the all-IP–based core networks currently under standardization. All-IP networks are discussed in Chapter 14.

In this chapter the basic principles of IP protocols are described. Section 13.2 describes the basics of IP. Congestion control and routing protocols are introduced in Sections 13.3 and 13.4, respectively. The two versions of IP protocols, IP version 4 and IP version 6, are compared in Section 13.5. Provision of quality of service in IP networks is discussed in Section 13.6, and IP enhancements for mobile systems are described in Section 13.7. Telephony over Internet and its security aspects are described in Sections 13.8 and 13.9, respectively. Finally, in Section 13.10 the IP standardization in IP Engineering Task Force (IETF) is introduced.

13.2 BASICS OF IP

TCP is the connection-oriented transport layer protocol operating on top of the datagram network layer (IP). TCP applies a Selective Repeat Protocol (SRP) with positive acknowledgements and timeouts to provide reliable end-to-end byte transport. If reliable transportation is not required, the User Datagram Protocol (UDP) can be applied (a typical application of this protocol is streaming audio). UDP offers connectionless transport on top of the IP network layer. The TCP/IP protocol can be modeled in layers, as can be seen in the protocol stack, depicted in Table 13.1. Beside the OSI model, the Internet is usually modeled in a four-layer model as well. Functional comparison based on these stacks, however, is difficult, because there are basic differences between the models [1].

Table 13.1
Protocol Stack of the TCP/IP Protocol

OSI	TCP/IP Suite
Application	Application layer
Presentation	
Session	
Transport	Tranport layer (host-to-host)
Network	Internetworking layer
Data link	Link layer
Physical	

Application layer. User applications run in the application layer and may use network connections provided by TCP/IP. The interface between application and transport layer is provided in the form of sockets and ports. Multiple applications can be supported simultaneously. Examples of applications are Telnet for remote sessions, file transfer protocol (FTP) for file transfer, and SMTP for e-mail.

Transport layer. The transport layer offers a reliable means for exchange of information. Two protocols are available at this level, TCP and UDP. TCP provides connection-oriented services, with an end-to-end flow control mechanism, UDP provides connectionless services. In UDP the flow control has to be handled at the application level. Typically, UDP is used for applications requiring a fast transport with low delay, where occasional packet loss is less important. An example of such an application is streaming audio.

Internetworking layer. The internetworking layer or Internet layer provides a virtual image of the network. Here IP is the most important protocol, providing a connectionless bearer, not assuming reliable protocols at lower network layers. The main task of IP is to provide the routing function, ensuring that messages are sent to the proper destination. The IP packet is the basic information unit in TCP/IP networks. Other internetwork layer protocols are ICMP, IGMP, ARP, and RARP.

Link layer. The link layer–also named data link layer, network interface layer, and host-to-host layer–interfaces to the actual network hardware. Here the packets are physically transmitted, using techniques like IEEE802.2, asynchronous transfer mode (ATM), and Ethernet.

Each Internet interface is defined by an IP address; in IPv4 this is a 32-bit address, and in IPv6 it is 128 bits. Table 13.2 defines the different address classes as existing in IPv4. Three address classes were defined in the early days of the Internet, corresponding to three addressing formats: Class A, B, and C. Later, for multicast purposes, an extra class was added (D). The remaining address class, class E, is reserved for experimental purposes.

Table 13.2
Different Address Classes as Defined in Ipv4

Class	High order bits	Format (bits)
A	0	7 net; 24 host
B	10	14 net; 16 host
C	110	21 net; 8 host
D	1110	28 multicast group number
E	1111	Reserved for experimental purposes

Because within the Internet, addresses are not assigned to hosts, but to interfaces, a host with several interfaces will have assigned as many addresses as interfaces.

The early Internet configuration only allowed for the use of the IP address itself to contact other hosts. Although these numbers are easily handled by computers, it is much more difficult for human beings. Therefore, symbolic names for the hosts were introduced (for example, www.artechhouse.com). Databases have to be maintained, which contain the mapping between symbolic names and IP addresses. Because the Internet is growing quickly, the maintenance of one central database became too burdensome and was therefore replaced by the Domain Name Server (DNS) mechanism. The DNS system allows a program running on a host to perform the translation of a symbolic name into an IP address for any host without the need for every host to have a complete database of all other users.

To be able to control and manage the growing Internet, a hierarchical structure was applied, splitting up the Internet in independent units, which were called autonomous systems (see Figure 13.1). An autonomous system can be defined as a set of routers and networks under the same administration [2]. For routing within an autonomous system, interior routing protocols are used (RIP and OSPF); for routing between autonomous systems, exterior routing protocols are used (Border Gateway Protocol, or BGP).

The gateways gather information about the nodes within the autonomous system they service and report that information to the gateways in the other autonomous systems. BGP, the routing protocol between two autonomous systems, has three main functions. First, neighbor acquisition, which enables it to request another gateway to communicate reachability information. Second, a gateway continually tests whether the other gateways are still responding. Finally, the reachability information will be constantly updated by transmitting routing update messages.

Figure 13.1 Communication between two autonomous systems (AS 1 and 2) using BGP between the systems. Within the autonomous system, a different routing protocol can be selected.

13.3 CONGESTION CONTROL

The IP protocol itself has no congestion control mechanism or error control mechanism. This means that when a packet is lost, it will not be retransmitted and that when the network get congested, nothing will be done. The TCP protocol was developed bearing these thoughts in mind. To be able to respond to a congested network (congestion collapse), the TCP standard has incorporated two mechanisms, namely slow start and multiplicative decrease.

Each user transmits its data using a given window size (advertised in the acknowledgments). To control the congestion, TCP has a second window size, the congestion window, which is a limit to the window size. The allowed window is the minimum of the receiver advertisement and the congestion window. In stable operation, the two of them are identical and the user is transmitting at maximum bit rate. When a situation of congestion occurs, TCP will reduce the window size to reduce the amount of traffic injected into the connection. To determine the size of the congestion window, TCP assumes that the datagram loss is caused by congestion (and not, for example, a lossy link). This strategy is called multiplicative decrease congestion avoidance, and it has two main components: In case of congestion the window is reduced by half, up to a minimum of one, and for the remaining segments, the retransmission timer is set back exponentially [3]. The fact that the window is reduced by half when a loss occurs means that the window decrease will be exponential when the congestion persists. The result is a fast reduction of traffic in case of congestion, allowing for the queues to be cleared quickly.

After the congestion has been removed, the system enters in the slow start mode, meaning that the window size is increased by one upon every successful transmission of a packet (this is linear). In this way it prevents the system from oscillating between no traffic and congestion and improves the stability of the system.

Note that under normal circumstances slow start is not slow and it takes only $\log_2 N$ round trips before TCP can transmit N segments. The term slow should be considered relative to the exponential decrease of the window size in case of congestion. To prevent a fast increase in the window size, another mechanism has been implemented, congestion avoidance. Once the congestion window reaches half of the size it had before the congestion was detected, TCP goes into congestion avoidance and the increase in window size is slowed down: the window size is only increased by one if all the packets within the window have been acknowledged.

13.4 IP ROUTING PROTOCOLS

The main task of a routing protocol is to deliver packets from a certain source to a given destination. This must be done efficiently to reduce to cost of delivery (e.g., delay, used capacity) and to ensure that the overhead of the routing protocol is as low as possible.

Two main classes of routing protocols can be distinguished: link state protocols [4] and distance vector protocols [2, 5]. The most widely used routing protocol between gateways at this moment in the Internet is the Routing Information Protocol (RIP). RIP belongs to the class of distance vector protocols, which are also referred to as Bellman-Ford algorithms (respectively, the inventor and the first person who applied the algorithm). In a distance vector protocol each node contains a routing table with a list of shortest paths to the other nodes in the network. At the moment of initializing the network, each node only has knowledge of its own address and the ability to transmit on all the links connecting it to its neighboring nodes. It broadcasts a package to all its neighboring nodes (as does every other node in the network). If the distance to the node is shorter than the value already mentioned in the table, the distance table is updated with the new value. The updated distance vector is transmitted again to the neighboring nodes. When the transmission of distance vector no longer causes an update of the tables, the protocol has converged and the topology of the network has been discovered.

The distance vector protocols are simple, but also offer a limited functionality. As an alternative, link state protocols have been developed. In contrast to the distance vector protocols, the nodes do not simply exchange distances, rather, all nodes maintain a "distributed map" of the network, the link state database. The map will be updated quickly after a modification of the network, by means of a flooding message spread throughout the network. This method results in a route with an equal precision to a centralized calculation of the shortest paths. For the Internet, the IETF has developed Open Shortest Path First (OSPF) protocol as a link state protocol. The algorithm used in link state protocols was first developed by E. W. Dijkstra [4] and is called Shortest Path First. It converges faster than the distance vector protocol.

Several reasons exist to prefer a link state protocol over a distance vector protocol, despite its higher complexity [2]:

- *Support of multiple paths to a destination.* A distance vector protocol would select randomly if two equivalent paths between the nodes exist and thus wastes half the bandwidth. A link state protocol contains a full map of the network and can thus intelligently choose between the paths or use both half of the time.
- *Fast, loopless convergency.* The distance vector protocol requires in the worst case the number of nodes minus one calculation step to calculate the algorithm, whereas the link state protocol only requires a flooding of all the information and a local calculation of the optimal route.

- *Support for precise metrics and, if needed, multiple metrics.* It takes time to obtain exact information about a link; this causes the distance vector protocol to converge more slowly, when the metrics becomes more precise and thus take more time to obtain the exact information.
- *Separate representation of external routes.* If no changes are made, a distance vector protocol automatically addresses the closest (default) gateway. If, however, for some reason another gateway needs to be selected (this is more effective outside the autonomous system), this gateway has to be added explicitly to the distance vector. In a link state protocol this could have been added fairly easily.

Routing protocols, usually, are divided into two classes: proactive and reactive protocols, where proactive protocols use periodic transmission to keep track of the set of neighboring nodes. Whenever a node wants to transmit packets to other nodes, the route to be used is already known (with a certain uncertainty) by the network. The first routing protocols were implemented as proactive protocols, because they are fairly straightforward to implement and provide a relatively fast way to establish routes. Furthermore, in proactive protocols it is easy to determine whether a node belongs to the network or not (whether the destination address is valid). The other class of routing protocols is the reactive routing protocols, also known as on-demand routing protocols. Whenever a route is needed, a route discovery procedure is invoked to determine the route to the requested destination. In principle, a route discovery implies a global flooding search for the destination. If, however, an intermediate node recently has exchanged routing information with the destination node, this information can also be used to deliver the requested information concerning the route to the destination. Especially when the topology is changing, reactive protocols have an advantage, because the network is not flooded with information every time a node or a group of nodes are moving. This saves both bandwidth and power. For wireless systems, these two parameters are scarce. On the other hand, whenever a node is contacted, it is more difficult to determine whether it is reachable, because the exact network topology is not known at that time and has to be determined. In a specific situation a protocol (either reactive or proactive) needs to be selected based on its advantages and disadvantages.

13.5 IPV4 VERSUS IPV6

In the early 1990s the Internet Engineering Task Force began an effort to develop a successor to the IPv4 protocol. A prime motivation for this effort was the realization that the 32-bit IP address space was beginning to become scarce, with new networks and IP devices being attached to the Internet (and hence being allocated a unique IP address) at a high rate. Given that a larger IP address space was needed, and hence a new IP protocol would be developed and deployed, the opportunity was also taken to modify and enhance other aspects of IPv4, based on the accumulated operational experience.

Compared to IPv4, the header of IPv6 packets has undergone some modifications. The IPv6 packet format is shown in Figure 13.2.

Version	Priority	Flow label		
Payload length		Next header		Hop limit
Source address (128 bits)				
Destination address (128 bits)				
Data				

32 bits

Figure 13.2 IPv6 packet header format.

Compared to the IPv4 header [6], the IPv6 header has a much simpler and more streamlined structure. Within the IPv6 packet, the following fields have been defined: The 4-bit *version* field identifies the IP version number; in an IPv6 packet, it is logically fixed at 6. The *priority* field states values between 0 and 7 and can be used to indicate different levels of priority among traffic for congestion-controlled traffic (namely, the source will reduce the transmission window in case of congestion). The values 8 through 15 are used for noncongestion-controlled traffic, such as constant rate real-time traffic. The priority field can be compared to the *type of service* (TOS) field in IPv4. Using the *flow label* field, a certain kind of traffic can be granted special treatment. A 16-bit value indicates the *payload length* field as an unsigned integer giving the number of packages in the payload. The *next header* field indicates to which protocol the data in the payload will be delivered. The *hop limit* field indicates after how many hops the packet will be discarded. Each router that forwards the packet decreases it. *Source* and *destination address* fields are used to identify the source and destination of the packet and to route the packet through the network. The address has the structure, shown in Figure 13.3.

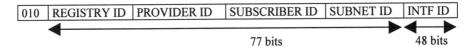

010	REGISTRY ID	PROVIDER ID	SUBSCRIBER ID	SUBNET ID	INTF ID

77 bits   48 bits

Figure 13.3 Structure of an IPv6 address.

The interface ID has 48 bits and the first 3 bits are fixed, leaving 77 bits for the other fields. The payload of the packet contains the *data* to be passed on to the protocol specified in the next-header field, after removing the IP address.

The most important changes to the header in IPv6 compared to IPv4 are an increased addressing capability caused by the increased size of the IP address from 32 bits to 128 bits and the introduction of a new address type, called the *anycast address*. This allows a packet to be delivered to any one of a group of hosts and can, for example, be used to connect to the nearest mirror site of a server. Furthermore, the size of the header is fixed at a 40-byte structure, instead of the size-varying header of IPv4, allowing for faster processing. Finally, IPv6 introduces an exclusive definition of a flow, allowing for a stream of packets to receive a special handling, such as a different quality of service. For example, streaming audio might be treated as a flow, whereas e-mail would still be handled in the conventional manner.

13.6   IP AND QUALITY OF SERVICE

Most traditional Internet applications such as e-mail and ftp are sensitive to packet loss but can tolerate delays. For multimedia applications (voice and video) this is in general the opposite. They can tolerate some packet loss but are sensible to delay and variation of the delay (jitter). This means that without a control mechanism for the available bandwidth, the quality of multimedia services varies depending on the bandwidth available. Therefore, mechanisms guaranteeing a specific quality of service (QoS) for real-time applications are becoming increasingly important in the Internet. QoS can be described as a set of parameters that describes the quality of a specific stream of data.

Two main categories of mechanisms to provide QoS within the Internet exist. They are commonly referred to as IntServ and DiffServ. The first provides a guaranteed end-to-end QoS with a guaranteed capacity, and the latter provides only a priority mechanism to control the packet flow on a hopper mechanism. DiffServ reduces the maintenance overhead compared to IntServ.

In order to support IntServ, an Internet router should be equipped with a traffic control functionality, which performs packet scheduling and admission control and checks for a packet classifier. The packet scheduler forwards the different package based on their QoS class. The scheduler also has the ability to polish the data stream to meet with the desired QoS settings. The packet classifier identifies within a host or router the required/requested QoS level.

Integrated services use the ReSerVation Protocol (RSVP) for signaling of the reservation messages to create specific states in the hosts and the routers between the hosts. RSVP is a simplex protocol, meaning that QoS reservations are only made in one direction. The scalability of RSVP in larger networks raises some concerns because RSVP only offers one type of QoS, the reserved bandwidth, which is probably not required in many cases.

The main target of DiffServ is to provide a means to offer differentiated services for Internet traffic in order to be able to support different types of applications. DiffServ provides predictable performance for a given load at a given time. An important concept when using DiffServ is the service level agreement, an agreement between customer and service provider, which states traffic classification. The service provider should deliver the guaranteed level of service. To distinguish between the different service classes, DiffServ's priority bit can be used.

Unlike the case with IntServ, when using DiffServ, the QoS guarantees are static; all traffic that passes a DiffServ capable router receives the required QoS. The only thing that needs to be done is to mark the packet's TOS or priority bit, which will be interpreted by the network routers. Within DiffServ the specification of how a packet should be handled (forwarding treatment) is called per-hop behavior (PHB).

DiffServ requires a mechanism for queuing and scheduling the received packages according to the desired QoS. The commonly used first-in-first-out (FIFO) queue does not offer service differentiation.

13.7 IP ENHANCEMENTS FOR WIRELESS/MOBILE

When a mobile node is connected to the Internet, the access point will change. The faster the node's speed, the more often this happens. To handle this efficiently, various protocols are currently under development.

Besides mobility management issues [7] within the IP network, the mechanism for congestion control, as described in Section 13.2, also introduces problems, because when developing those mechanisms it was assumed that the only cause for packet loss was congestion, resulting in buffer overflow and hence packet loss. In a wireless system, however, packet loss can also be caused by a low-quality link. In this case it is not desirable to decrease the window size; a retransmission of the packet would be a far better solution. In order to do this, enhancements to IP have been developed. Two basic groups can be identified: split connection approaches and TCP-aware link layer protocols such as Snoop IP [8], which shield the wireless link from the TCP source, with the philosophy that a wireless link is a local problem and should be solved locally. Protocols that hide the wireless link from the TCP sender include link layer protocols.

The first group makes the sender aware of the wireless and prevent the invocation of congestion control messages within TCP. Examples of this class are E2E-NEWRENO, E2E-IETF-SACK, and E2E-SMART. The first remains in fat recovery mode if the first ACK to arrive is partial, i.e., less than the value of the last byte transmitted when fast retransmission was done. This allows the source to recover faster from loss, instead of stalling [8]. The second two allow the user to handle multiple losses within a window based on selective ACKs. In case of losses, however, it is still assumed that congestion is the cause.

In the following the masking protocols are discussed in more detail. At link level, two main mechanisms exist: error correction using techniques, such as forward error correction (FEC), and retransmission of lost packages in response to automatic repeat request (ARQ) messages. Link layer mechanisms used in digital cellular systems nowadays mostly rely on ARQ techniques [9].

It is also possible to effectively split the TCP connection into two TCP sessions, one from the TCP source to the wireless access point and one from the access point to the mobile station over the wireless link. A more elegant solution is the so-called Snoop IP, where a snoop agent is located close to the wireless link. In case of packet loss because of a lossy link, the Snoop agent intercepts the NACK message and forwards the packet from a cache. This masks the radio interface from the TCP source, preventing the multiplicative decrease mechanism from decreasing the window size [10].

Besides problems with congestion control, mobility also introduces problems for mobility management. The mobility management (MM) enhancements to IP can be classified at three levels, namely, macro, micro, and pico. Macro mobility means mobility between different networks and can be compared with roaming in mobile networks. Micro mobility is related to a node moving within a network and can thus be compared with the mobility mechanisms within GSM/GRPS. The lowest level of mobility is the pico mobility used for short-range communication systems; it can also be referred to as ad-hoc networking. A Bluetooth network is an example of the last category. The interrelation between these different kinds of mobility is depicted in Figure 13.4.

Figure 13.4 Relation between different levels of mobility (macro, micro, and pico).

Two levels of security are required: When connecting to the system, a high level of security is needed, and after a handover, authentication is required again. The requirements here, however, are less strict because the user is already connected to the system. At the moment radius servers are available to provide authentication, authorization, and accounting (AAA) mechanisms. The wireless industry is considering the deployment of mobile IP in cellular data systems such as GPRS, UMTS, and cdma2000.

QoS in a wireless environment introduces new problems, because in general the system cannot predict how busy the neighboring cells are, and therefore, it is difficult to offer a guaranteed QoS.

Mobile IP provides mobility between different networks. Two entities are introduced, the home agent and the foreign agent. The home agent resides in the network, where the node usually is connected. If the user moves to another network, the home agent is informed and forwards the packets towards a foreign agent residing in the visited network. The forwarding is done by means of a tunnel[1] of IP packets. The tunnel ends at the care of address (CoA), where the packets are delivered. The use of mobile IP is depicted in Figure 13.5 [11].

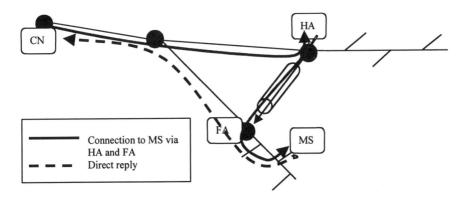

Figure 13.5 Use of mobile IP (CN: correspondent node; MS: mobile station; HA: home agent; FA: foreign agent).

In the home network of a mobile user, a mobile agent is present. The home agent determines whether the user is present in the home network and if he is, the packet is sent to the user directly. If the user has moved to a visited network, the home agent keeps track of the user using the foreign agent in the visited network. A packet transmitted to a user that is not present in its home network is forwarded to the foreign agent in the visited network. A tunnel is set up between the two agents to transmit the packets. In the visited network the foreign agent then further transmits the package. The end of the tunnel is called CoA. The user in the visited network can reply directly to the correspondent node because it knows its address from the received packet. This form of routing is called triangular routing and is not optimal. One optimization method is to have the foreign user detect that a tunnel is being used and transmit a binding update to the correspondent node directly. This way, the tunneling can be reduced to the first packets only. Mobile IP has been implemented in this manner in IPv4. In IPv6 it is done slightly differently: The foreign agent has been removed, and its functionality has been placed in the user equipment itself that is called the colocated CoA [12, 13].

In general, mobile IP has been developed for low mobility applications, like moving with your laptop from one room to another. When used for micro mobility applications, this results in high overhead because of frequent location updates and high latency and disruption during handoff. Furthermore, if QoS mechanisms are in use, obtaining a new CoA could trigger a new QoS negotiation, which may differ from the previous. To be able to handle more advanced applications, mobile IP has to be applied

---

[1] A tunnel encapsulates the packets and adds a new header, with a new destination. Tunneling is done at the same network level.

hierarchically, or rather another extension to IP can be used. Here we discuss two proposed systems: cellular IP and HAWAII.

Cellular systems, however, do not only provide roaming or macro mobility. They also offer micro mobility with fast and optimized handovers. A cellular Internet should leverage the mobility management and handover techniques known from these systems. A scalable protocol that provides this is the proposed cellular IP [13] protocol (see Figure 13.6). It is designed based on the IP paradigm but also incorporates cellular properties such as mobility management, handover, and passive connectivity. The central component in a cellular IP system is the base station, which is, besides an access point, also an IP router, and contains control functionality from BSC and MSC. The cellular IP network is connected to the Internet by means of a gateway router. Mobility between these gateways could be handled by mobile IP, for instance. In cellular IP mobility management and handover are integrated with routing.

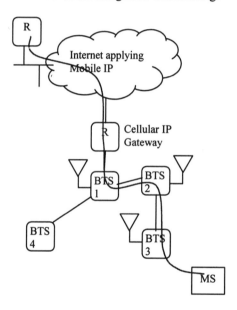

Figure 13.6 Cellular IP basic architecture. Below the cellular IP gateway, special cellular IP routing is used [14].

Uplink packets are transmitted to the gateway hop by hop, and their path is cached by the base stations. To transmit packets to a mobile station (MS), the recently cached uplink path is reversed. When the mobile station does not send data, it regularly sends an empty packet to remain connected to the network and to update the location cached in the base stations. If the MS does not transmit such a packet, the cache is emptied and the MS goes into idle state. Now to reach it, a paging message should be sent. A handover is always MS initiated. The mobile station listens to advertisements transmitted by the base stations. To initiate the handover, the MS tunes to the new BTS and transmits a packet, which will be forwarded to the gateway, updating the caches it encounters on its way. If, before the gateway is reached, the packet reaches a BTS containing correct caching information the handover is completed, otherwise it will take until the gateway is reached. Before the gateway is reached, the MS may be in "semisoft handover state" [14], meaning that packets are still transmitted using the old route, which increases the efficiency of the protocol, especially if the old route can be maintained until the new route is established. When the mobile goes into idle mode it still transmits periodically a paging-update message to the gateway. The packet is sent to the base station with the best signal quality. The result is that idle MSs have a mapping in the paging caches but not in the routing cache. The paging cache has a longer lifetime. If a user is paged in idle mode, first the MS is only paged in areas where it has an entry in the paging cache. Without paging cache entry, the paging package is broadcast throughout the whole access network. Thus, the paging cache reduces the load of the system. An idle terminal that receives a paging package turns into active mode [13].

The Handoff Aware Wireless Access Internet Infrastructure (HAWAII) protocol is another candidate that solves the micro mobility issues [14]. As in cellular IP, host-based forwarding entries are installed in the routers, although here, however, only in specific routers to support intradomain mobility. In HAWAII, the mobile host maintains its IP address, which simplifies the support of QoS mechanisms. There are three main components to HAWAII:

- A two-level hierarchy with separate mobility mechanisms for intradomain and interdomain mobility;
- Special paths assigned between the host and the mobile station;
- Soft-state mechanisms are used, namely, after a certain period of not being used forwarding state reservations time-out.

Figure 13.7 depicts the basic network architecture: The gateway to each domain is called the domain root router, and each mobile station has a home domain and an IP address going with this. While moving in the home domain, the MS keeps its IP address. The root router forwards packets via dynamically established paths, allowing the domains to have sizes of up to hundreds of base stations and thus reducing the interdomain mobility, which is handled using mobile IP in this system.

The path between the root router and the BTS is reserved using the soft-state mechanism, meaning that the MS should periodically update these reservations. For this purpose two messages exist: update and refresh, to establish and maintain a path, respectively.

When the MS is activated, a registration packet is sent to the nearest BTS and forwarded to the root router. Each router between the BTS and the root router adds a forwarding entry to the MS, and finally the root router sends an ACK to the BTS. Now the MS is reachable under the root router.

Figure 13.7 HAWAII basic architecture.

Two mechanisms exist to reestablish the path after a handover: the forwarding path setup scheme and the nonforwarding path setup scheme. In the first mechanism the packets from the new router are first forwarded to the old one before they are diverted at the crossover router. In the nonforwarding case, the crossover router is updated when the update packages move from the new to the old base station. The forwarding method introduces more delay, but the MS remains connected at all times. Note that only the routers up to the crossover router are involved. The higher network layers are unaware of the change of location.

At the level of pico mobility, short-range communication between terminal networks can be set up on an ad-hoc basis. Typically, these networks can have fast-changing topologies, because nodes can leave and enter the networks at will. In the IETF a working area is currently investigating routing protocols for ad-hoc networking. In these mobile ad-hoc networks (MANET), each terminal acts as a router and can forward packets for other users. MANETs have some salient characteristics that put conditions to the development of routing protocols. They are listed as follows:

- Dynamic topologies;
- Bandwidth-constrained operation with varying capacity available on the links;
- Energy-constrained operation. Mobile terminals have a limited power source (a battery) and hence, the forwarding of data streams for other users while not using the terminal to transmit its own data should be taken into consideration in

the system design;

- Limited physical security. A wireless system is more sensitive to eavesdropping than a fixed and an ad-hoc network, especially, because it has a relatively open structure, where many users can join in.

The MANETS need to be scalable, because they are envisioned to be deployed with a size of 10 to 100 nodes and may, for example, be used in disaster areas, where no infrastructure is available, in military applications, or in small indoor networks (e.g., Bluetooth-based) [15].

13.8   TELEPHONY OVER THE INTERNET

There is growing interest in multimedia content over the Internet. The traditional mechanisms and their best-effort performance are in this case no longer sufficient. At this moment two developments are attempting to solve this. First, the ITU started the development of H.323, and second, the IETF is currently developing the Session Initiation Protocol (SIP) to support multimedia calls.

13.8.1   H.323

H.323 is a so-called umbrella recommendation from the ITU to set a standard for multimedia communications. It was initially developed for use over local area networks. The H.323 specification was approved in 1996 by study group 16 of the ITU; version 2 was approved in 1998. H.323 enables applications such as Internet telephony, video conferencing, and e-commerce. Figure 13.8 depicts the basic elements in the H.323 system.

Figure 13.8   H.323 architecture to provide multimedia services over IP.

The gatekeeper is the central component within an H.323 network, and it provides call control to all registered terminal points within the network. Two main functions of the gatekeeper are address translation of the aliases from gateways and terminals into IP addresses, and bandwidth management (e.g., setting a maximum to the number of simultaneous conferences). The gatekeeper can also be used to route calls, allowing for a better control of the information flows (e.g., by offering authorization) and enabling a service provider to bill on the basis of an H.323 call. An H.323 network can be implemented without a gatekeeper, but if one is present, it must be used.

Terminals are the end-points on the network, which provide real-time, bi-directional communication. All the terminals will support voice communication; video and data, however, are optional. The system control will be handled using signaling described in H.245. To communicate with the gatekeeper, the registration/admission/status (RAS) component is being used.

H.323 can use a gateway in a connection to provide extra services such as transformations between different communication procedures and transmission formats. It can also be used for call setup, both in the LAN and circuit-switched networks (PSTN, ISDN).

Although H.323 is able to facilitate multipoint connections, a multipoint control unit (MCU) to host multipoint conferences in a more advanced manner has been developed.

13.8.2   SIP

The IETF is currently developing the Session Initiation Protocol (SIP). SIP is an application layer protocol for creating, modifying, and terminating sessions with one or more participants. These sessions include Internet multimedia conferences, Internet telephone calls, and multimedia distribution. It is possible to set up multicast and unicast calls, or combinations of these, resulting in conference calls [16]. SIP supports five facets to handle multimedia communication:

- User location: determining the location of the end system;
- User capabilities: determining the media and its parameters in use;
- User availability: determining whether the called party is willing to participate in the communication;
- Call setup/signaling: establishing the call, bearer setup;
- Call handling: transfering and terminating calls (advanced functionality).

Figure 13.9 depicts how SIP can be used in combination with a proxy server, showing the (text) messages to be exchanged in order to set up a call.

Both H.323 and SIP have advantages and disadvantages. SIP, for instance, is less complex. H.323, however, has better support for the codecs and call control models developed by the ITU. At this moment H.323 is the more mature product, but in the long run its complexity could cause scalability problems.

To provide circuit-switched services over the Internet, a QoS mechanism and a call control model are required, and it is also necessary to transfer a stream of packets into a circuit-switched connection or vice versa. This transformation of streams is done by a media gateway. Different gateways are still under development to be able to transform both the signaling and data information between the PS and CS domains. Within the IETF media gateways are being developed in the transport area, working group MEGACO (MEdia GAteway COntroller) [17].

Figure 13.9 SIP operation in proxy mode.

Different media gateways exist:

- Trunking gateways are an interface between a telephone network and a VoIP network.
- Access gateways provide analog, primary-rate (PRI) line interfaces over an IP network.
- Network-access modems can attach a "modem" to a telephone circuit and provide data access to the Internet.

13.9 SECURITY IN TCP/IP

The IP protocol suite provides a mechanism, called IP security (IPsec) to secure connections. Besides IPsec, many other protocols exist: Secure Sockets Layer (SSL) at the transport level, some protocols at the application level (proxy, SSH, AAA servers), and security protocols for the data link layer. Here we limit our discussion to IPsec, which operates at the IP level.

IPsec works on the basis of so-called security associations (SA), not specific for IPsec but for security protocols in general. Two modes have been defined, transport and tunnel. The first refers to the mode where the original packet is taken and then modified. In the tunneling mode, the tunneling concept is applied and a new packet is constructed. Because the SAs are unidirectional, two SAs should be defined in case of communication. An SA consists of three parameters: the destination address, the cryptography protocol, and an identifier to separate multiple SAs with the same destination host but using the same cryptography protocol.

IPsec uses three mechanisms to provide traffic security: authentication header (AH), encapsulating security payload (ESP), and Internet key exchange (IKE). We will only consider the first two mechanisms because they closely cooperate and are both within the IP layer. IPsec is independent from existing cryptographic mechanisms of today and it works both with IP version 4 and 6.

The authentication header provides authentication and integrity to IP packages. Encapsulating security provides integrity and confidentiality to IP packets.

The authentication mechanism works on a packet per packet basis (connectionless) and provides both an authentication and integrity check; optional reply protection can be used. The AH mechanism authenticates as much of the packet as possible; however, some parameters change during the connection and can therefore not be authenticated (these include TOS, flags, TTL, and the header checksum). Figure 13.10 illustrates AH header location in an IP packet.

Header	AH header	payload

Figure 13.10 AH header location in an IP packet.

ESP can be used to provide the host with a mechanism to check the integrity, to authenticate, and to encrypt IP datagrams. ESP, like AS, is connectionless. The modifications made to the IP packets are more complicated in the case of ESP than in the case of AH. Besides the insertion of an extra header, a trailer and some authentication data are also included, resulting in the structure shown in Figure 13.11.

Header	ESP header	payload	ESP trailer	ESP trailer

Figure 13.11 ESP header and trailer location in an IP packet.

The trailer includes some padding zeros to complete the payload to the next 32 bits and the number of included zeros. If encryption is used, both the payload and the trailer are encrypted. Again, the two modes are defined: either transport mode, where the ESP modifications are made to the original packet, or tunnel mode, where a new packet is created with the old packet as payload and the modifications made to this new packet.

These protocols may be applied alone or in combination with each other to provide a desired set of security services in IPv4 and IPv6. Each protocol supports two modes of use: transport mode and tunnel mode. In transport mode the protocols provide protection primarily for upper layer protocols; in tunnel mode, the protocols are applied to tunneled IP packets.

13.10 IETF STANDARDIZATION

The IETF is the body concerned with the smooth operation of the Internet, as well as the continuous development and improvement of the protocols and architecture in use [18]. The IETF is an open and international body consisting of vendors, operators, researchers, and designers. The IETF is divided into working groups organized in several areas, such as routing, transport, and security. The area directors, who are members of the Internet Engineering Steering Group (IESG), manage the groups. The Internet Architecture Board (IAB) provides architectural oversight. The IAB also adjudicates appeals when someone complains that the IESG has failed. The IAB and IESG are chartered by the Internet Society (ISOC) for these purposes. The general area director also serves as the chair of the IESG and of the IETF and is an ex-officio member of the IAB [19].

The Internet standards process is as follows [19]. The official publication channel for Internet standards documents is the Request for Comments (RFC) document, published under the responsibility of the RFC editor. RFCs also cover a wide range of topics in addition to Internet standards.

During the development of a specification, draft versions of the document can be published as "Internet Drafts" (I-Ds) in order to solicit comments and review. This ensures rapid availability to a wide audience. An I-D is not a specification; however, once an I-D is posted, the RFC editor can be requested to change the status of the I-D to an informal or experimental RFC to initiate a specification. An I-D that is not changed into an RFC or modified in 6 months will be discarded, in contrast to all RFCs, which are archived.

In the Internet standardization process, two categories can be distinguished: Technical Specifications (TS) and Applicability Statements (AS). A TS is any description of a protocol, service, procedure, convention, or format. An AS specifies how and under what circumstances one or more TSs may be applied to support a particular Internet capability.

Internet specifications undergo different stages of development, called "maturity levels" depending on the stage of development, testing, and acceptance. A standard starts its life as a proposed standard, after approval of the IESG. It is, in general, stable and has a significant support in the Internet community. The proposed standard has resolved known design choices and is believed to be well understood. Neither implementation nor operational experience is required. To be elevated to the level of draft standard, at least two independent and interoperable implementations with sufficient operational experience have to be developed. Elevation to draft standard is a major advance in status, indicating a strong belief that the specification is mature and will be useful. If significant implementations and operational experience have been obtained, a draft standard becomes an Internet standard. The specification receives a STD number and retains its RFC number [20].

13.11 CONCLUSIONS

IPv4 is now commonly used within the Internet. The address space, however, is being stretched to the limit and better QoS mechanisms are required. IPv6 brings relief to both issues, with a four times larger address space and more advanced and efficient mechanisms to provide QoS.

Although IP was not designed so that it could be used for mobile/wireless communications, in the last few years many ongoing developments have taken place to allow for smooth operation of the IP protocol in a tetherless environment. Here we have addressed modifications to IP to prevent the congestion control mechanisms from reacting to packet loss because of the wireless link (e.g., Snoop TCP and advanced protocols to fully handle mobility management within IP, using techniques like mobile IP and cellular IP).

Voice-over-IP in combination with call control models under development, like H.323 and SIP, will enable smooth operation of telephony services over packet (IP) networks. These developments, combined with media gateways that interact with the traditional circuit switched systems like PSTN and GSM, allow for the deployment of IP technology in public communication networks.

To prevent unwanted eavesdropping, security mechanisms for the Internet receive a lot of attention. At the IP transport level, IPsec can be used to secure both header and payload, providing the end user with a secure and reliable link.

In principle, all the building blocks necessary to construct communication networks for both voice and data based on IP are present. What is now needed is the development of a scalable integration of these components to exploit IP to its fullest potential in the future all-IP-based networks discussed in Chapter 14.

REFERENCES

[1]   Stevens, W. R., *TCP/IP Illustrated, Volume 1: The Protocols,* Reading, Massachusetts: Addison-Wesley, 1994.

[2]   Huitema, C., *Routing in the Internet,* Engelwood Cliffs, New Jersey: Prentice Hall, 1995.

[3]   Comer, D. E., *Internetworking with TCP/IP, Volume I: Principles, Protocols, and Architecture,* 2nd ed., Englewood Cliffs, New Jersey: Prentice Hall, 1991.

[4]   Dijkstra, E. W., "A Note on Two Problems in Connection with Graphs," *Numerische Mathematic,* Vol. 1, 1959, pp. 269–271.

[5]   Bellman, R. E., *Dynamix Programming*, Princeton, New Jersey: Princeton University Press, 1957.

[6]   Postel, J., (ed.), "Internet Protocol," RFC 791, 1981.

[7]   Mohan, S., and R. Jain, "Two User Strategies for Personal Communication Services," *IEEE Personal Communications*, Vol. 1, No. 1, 1994, pp. 42–50.

[8]   Díaz, P., D. Airlie, R. Casadevall, and J. Nelson, "Protocols to Enhance TCP Performance on Mobile Systems," *ACTS Mobile Summit 1999*, Sorrento, Italy.

[9]   GSM 03.02, "Digital cellular telecommunications system (Phase 2+); Network architecture," ETSI, 1998.

[10] Balakrishnan, H., et al., "A Comparison of Mechanisms for Improving TCP Performance over Wireless Links," *IEEE/ACM Transactions on Networking*, Vol. 5, No. 6, December 1997, pp. 756–769.

[11] Perkins, C. E., "Mobile Networking Through Mobile IP," *IEEE Internet Computing*, January/February 1998, pp. 58–69.

[12] Perkins, C. E., *Mobile IP: Design Principles and Practices*, Reading, Massachusetts: Addison-Wesley, 1997.

[13] Wijting, C. S., "Use of Mobile IP in the Future UMTS System," *WPMC'99*, Amsterdam, The Netherlands, 1999, pp. 495–500.

[14] Campbell, A. T., J. Gomez, and A. G. Valkó, "An Overview of Cellular IP," *IEEE WCNC'99*, 1999, pp. 606–610.

[15] Ramjee, R., et al., "HAWAII: A Domain-Based Approach for Supporting Mobility in Wide-Area Wireless Networks," *ICNP'99*, 1999, pp. 283–292.

[16] Itawa, A., et al., "Scalable Routing Strategies for AdHoc Wireless Networks," *IEEE Journal on Selected Areas in Communications*, Vol. 17, No. 8, August 1999, pp. 1369–1379.

[17] Handley, M., H. Schulzrinne, E. Schooler, and J. Rosenberg, "SIP: Session Initiation Protocol," *IETF Work in Progress*, RFC 2543, March 1999.

[18] Arango, M., "Media Gateway Control Protocol (MGCP), Version 1.0," Internet RFC 2705, October 1999.

[19] www.ietf.org

[20] Bradner, S., "The Internet Standards Process," RFC 2026, 1996.

Chapter 14

NETWORK ARCHITECTURES

14.1 INTRODUCTION

In this chapter, we discuss the network aspects related to the development of third-generation systems. Because many of the network-related aspects are independent of the radio interface, wideband CDMA itself is not discussed in depth in this section.

The third-generation partnership project (3GPP) has already specified the release '99 standards, which are focused on asynchronous transfer mode (ATM) as the backbone network. The recent developments are focusing on an all-IP-based network to be standardized for the release 2000 (now split into two releases termed release 4 and 5) of the 3GPP. The all-IP network is evolving from the packet switched mobile core network of release '99. 3GPP2 is focusing on the cdma2000 standard and is also developing its own all-IP–based solution evolving from the current IS-41-based architecture. It should be noted, however, that 3GPP architecture has also been presented as a next evolution step for cdma2000 networks. This will be further discussed in Section 14.5.4 and Chapter 16.

Section 14.2 describes the basic technologies of third-generation networks: ATM, signaling system 7 (SS7), and mobile application part (MAP). Section 14.3 describes the GSM/GPRS network architecture, interfaces, and protocols. The 3GPP specifications related to network architecture and interfaces are discussed in Section 14.4. In Section 14.5, the IS-41, IS-95, and cdma2000 network architectures, interfaces, protocols, and their evolution are discussed. Section 14.6 describes Internet Protocol (IP)-based networks.

14.2 BASIC TECHNOLOGIES

Key technologies in building third-generation systems are ATM, SS7, and MAP, which are described in the following sections.

14.2.1 Asynchronous Transfer Mode

Today's transmission protocols of many telecommunication networks are based on pulse code modulation (PCM), and mobile radio systems are no exception. The switching is also based on the switching of 64- or 56-Kbps PCM connections. ATM technology has received a lot of attention during recent years as the next major transport technology. ATM has also been proposed for wireless applications (i.e., ATM cells are transmitted over the air interface). Here, however, we assume that it is only used as a transmission protocol in the infrastructure side.

ATM provides not only the multiplexing gains of packet switching, but also the guaranteed delay characteristics of circuit switching. The fundamental strategy behind ATM is to split the information into small fixed-size units that are easy to handle. The fixed size of the cell allows efficient switching. ATM networks are high-speed switching systems offering large bit pipes, which allow *statistical multiplexing* (i.e., multiplexing of many connections with variable rate characteristics), which altogether reduces the overall bandwidth requirements. Because ATM is based on the transmission of fixed size cells, it can be easily evolved for future services.

The basic unit in an ATM is a cell of 53 bytes, illustrated in Figure 14.1. It consists of header (5 bytes) and information payload fields (48 bytes). The header consists of following fields:

- Generic flow control (GFC);
- Virtual path identified (VPI);
- Virtual channel identifier (VCI).

The VPI and VCI are used to identify the virtual path and virtual channels identified with that path to route the ATM cells from the source node to the destination node.

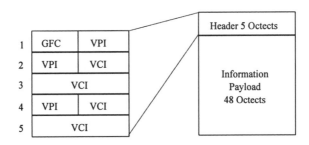

Figure 14.1   ATM cell format.

The ATM and the ATM adaptation layer (AAL) form the data link layer. AAL converts the arbitrarily formatted information supplied by the user into ATM cells. Various forms of AAL protocols are necessary to handle the different types of traffic.

- AAL0 provides direct access to the ATM layer.

- AAL1 assumes constant bit rate traffic, which is intolerant of missequenced information and variation in delay. It offers the following functions: segmentation and reassembly (SAR), handling of delay variation handling, handling of lost and misinserted cells, source clock recovery, monitoring for bit errors, and handling those errors.
- AAL2 is used, for example, for voice and video. It assumes that traffic is bursty and intolerant of missequencing and that a time stamp is needed for packet reassembling. It offers the following functions: multiplexing, SAR, handing delay variation, handling cell lost/error, and source clock recovery. AAL2 will be used for compressing data in third-generation mobile radio systems in the network infrastructure.
- AAL3/4 and AAL5 are geared to traffic that has bursty characteristics with variable frame length. Furthermore, delay is not critical and packets can be resequenced based on sequence numbers. AAL5 is expected to supersede AAL3/4 because it has lower overhead and TCP/IP acknowledgments fit into one cell in AAL5 instead of two cells in AAL3/4.

14.2.2 Signaling System 7

Both IS-41 and GSM make use of protocols specified for SS7, which is a signaling system designed for the transfer of control information between network elements. The layered structure of SS7 is shown in Figure 14.2. SS7 layers are called parts. The ISDN user part (ISUP) of SS7 contains messages carried from ISDN standard devices. For example, call-related signaling of GSM makes use of ISUP when connecting to external networks. The GSM and IS-41 MAP protocols use the transaction capabilities part (TCAP) for network control. Furthermore, the signaling connection control part (SCCP) and message transfer part (MTP) are used in the A-interface of GSM and in the IS-634 interface, which specifies the corresponding A-interface for the North American digital standards, especially for IS-95/IS-41. A tutorial of SS7 can be found in [1], and use of SS7 in IS-41-based systems is discussed in [2].

14.2.3 Mobile Application Part

A unique feature of mobile systems is the support of roaming (i.e., seamless provision of services for users who are subscribed to one operator's network while they are within the coverage of some other operator's network). The functions required for roaming are authentication of the subscriber, transferring subscriber data to the visited network, and a mechanism for routing connections towards the subscribers. The functions to implement roaming are part of the mobility management. For this purpose, both GSM and IS-41 systems have defined mobile application part. MAP defines the application protocols between switches and databases (e.g., MSC, VLR, HLR) for supporting call management, supplementary service management, short message transfer, location management, security management, radio resource management, and mobile equipment management. In principle, the MAP of IS-41 and GSM perform similar functions. Their implementations, however, and the way that protocols are specified are different.

Figure 14.2   SS7 protocols.

14.3   GSM/GPRS NETWORK

Because the 3GPP release '99 is based on evolved GSM/GPRS specifications, in this section, the existing architecture, interfaces, and protocols of the GSM and GPRS networks are described.

14.3.1 Reference Architectures

For GSM, we need to consider two reference architectures, one for the basic GSM and one for GPRS to identify the interfaces subject to evolution. Figure 14.3 depicts the reference architecture of basic GSM.

In the reference architecture the following network elements can be identified:

Mobile station (MS). MS consists of the physical equipment used by a subscriber. The mobile station is usually divided into mobile termination (MT) and terminal equipment (TE) parts. MT terminates the radio path and contains functions such as modulation, error correction, and mobility management. TE is a terminal (e.g., data terminal) connected to an MT possibly via a terminal adaptation function.

Base station system (BSS). The BSS terminates the radio path in the network side. It communicates with the mobile station and connects to the mobile switching center through an interface termed the A-interface. Often, a so-called Abis-interface is implemented in such a way that the BSS consists of one base station controller (BSC) and one or more base transceiver stations (BTS). One BSC controls one or more BTSs. A BTS is a network component, which serves one cell.

Figure 14.3   Basic GSM reference architecture.

Mobile-services switching center (MSC). The MSC is an exchange that performs all the switching and signaling functions for mobile stations located in a geographical area designated as the MSC area. The main difference between a MSC and an exchange in a fixed network is that the MSC has to take into account the mobile nature of the subscribers. Therefore, it has to perform procedures required for the location registration.

Home location register. This functional entity is a database in charge of the management of mobile subscribers. It maintains subscriber information (e.g., international mobile station identity, user profile). Furthermore, it contains some location information that enables the charging and routing of calls towards the MSC where the MS is located (e.g., the MS roaming number, the VLR address, the MSC address, and the local MS identity). One home location register (HLR) can be associated with one or more MSCs.

Visitor location register (VLR). The VLR is the location and management database for the mobile subscribers roaming in the area controlled by the associated MSC(s). Whenever the MSC needs data related to a given mobile station currently located in its area, it interrogates the VLR. When a roaming mobile station enters a new service area covered by an MSC, the MSC forwards a registration request (or a location update request) to the associated VLR, which informs/interrogates the HLR. Usually, one VLR is associated and colocated with one MSC.

Authentication center (AC or AuC). The AC is associated with an HLR, and it manages the authentication and encryption parameter for each individual subscriber.

Equipment identity register (EIR). The EIR is used for keeping track of mobile stations and their identities. Its purpose is to be able to record and prevent the use of possibly stolen or otherwise misused terminals.

Subscriber identity module (SIM). This is part of the mobile station and is used to store all subscriber related information. It is either a smart card (the size of a credit card) or a so-called "plug-in" SIM, which was designed to facilitate an easier implementation of small terminals.

Group call register (GCR). The GCR is the management database for the voice group or broadcast calls in the area controlled by the associated MSC(s). Whenever the MSC needs data related to a requested voice group or broadcast call, it interrogates the GCR to obtain the respective voice group or broadcast call attributes. More information is provided in TS GSM 03.68 and 03.69.

Shared interworking function (SIWF). The SIWF is a network function that provides interworking for circuit data/fax calls. SIWF consists of a SIWF controller (SIWFC) functionality located in MSCs and SIWF server(s) (SIWFS) located in the PLMN. A SIWFS can be accessed by several other network nodes (e.g., any MSC in the same PLMN). More information is provided in TS GSM 03.54.

Gateway MSC (GMSC). If a network delivering a call to the PLMN cannot interrogate the HLR, the call is routed to an MSC. This MSC will interrogate the appropriate HLR and then route the call to the MSC where the mobile station is located. The MSC that performs the routing function to the actual location of the MS is called the gateway MSC (GMSC). For more information, refer to TS GSM 03.04.

SMS gateway MSC (SMS-GMSC). The SMS-GMSC acts as an interface between a short message service center and the PLMN to allow short messages to be delivered from the service center (SC) to mobile stations.

SMS interworking MSC. The SMS interworking MSC acts as an interface between the PLMN and a short message service center to allow short messages to be submitted from mobile stations to the SC.

Interworking function (IWF). The IWF is a functional entity associated with the MSC. The IWF provides the functions necessary to allow interworking between a PLMN and the fixed networks (ISDN, PSTN, and PDN). The functions of the IWF depend on the services and the type of fixed network. The IWF is required to convert the protocols used in the PLMN to those used in the appropriate fixed network. The IWF may have no functionality where the service implementation in the PLMN is directly compatible with that at the fixed network. The interworking functions are described in GSM Technical Specifications GSM 09.04 – 09.

Interfaces. In the basic GSM configuration presented in Figure 12.10, all the functions are considered implemented in different equipments. Therefore, all interfaces within PLMN are external. Interfaces A and Abis are defined in the GSM 08-series of technical specifications. Interfaces B, C, D, E, F, and G need the support of the SS7 MAP to exchange the data necessary to provide the mobile service. No protocols for the H-interface and for the I-interface are standardized. We now describe the main interfaces defined between the different GSM network elements:

- The interface between BSC and the MSC (A-interface) is specified in the following GSM specifications: GSM 08.01, 08.02, 08.04, 08.06, 08.08, and 08.20.
- The interface between the BTS and the BSC (the Abis-interface) is specified in the following GSM specifications: GSM 08.51, 08.52, 08.54, 08.56, 08.58, 08.59, and 08.60.
- The interface between the MSC and the HLR (C-interface): signaling in this interface uses the MAP protocols specified in GSM 09.02.
- The interface between the MS and the BTS (air interface, Um-interface) is specified in the GSM 04 and 05 specification series. The GSM 04 series covers the general principles and higher protocol layers, and the GSM 05 series contains the physical layer specifications.
- The interface between the MSC and its associated VLR (B-interface) is internal to the MSC/VLR, and signaling on it is not standardized.
- The interface between the HLR and the MSC (C-interface): the GMSC must interrogate the HLR of the required subscriber to obtain routing information for a call or a short message directed to that subscriber. Signaling in this interface uses the MAP protocols specified in GSM 09.02.
- The interface between the HLR and the VLR (D-interface) is used to exchange data related to the location of the mobile station and to the management of the subscriber. Signaling in this interface uses the MAP protocols specified in GSM 09.02.
- The interface between MSCs (E-interface) is used for the exchange of data for handover between two MSCs. This interface is also used to forward short messages. Signaling on this interface uses the MAP specified in GSM 09.02.
- The interface between VLRs (G-interface): when an MS initiates a location updating using TMSI, the VLR can fetch the IMSI and authentication set from the previous VLR. Signaling on this interface uses the MAP specified in GSM 09.02.
- The interface between the HLR and the gsmSCF (J-interface) is used by the gsmSCF to request information from the HLR. The support of the gsmSCF-HLR interface is a network operator option. As a network operator option, the HLR may refuse to provide the information requested by the gsmSCF.

GPRS architecture. GPRS is logically implemented on the GSM structure through the addition of two network nodes, the serving GPRS support node (SGSN) and the gateway GPRS support node (GGSN) [3]. Therefore, several new interfaces have been defined. Figure 14.4 illustrates these, together with the GSM interfaces explained previously.

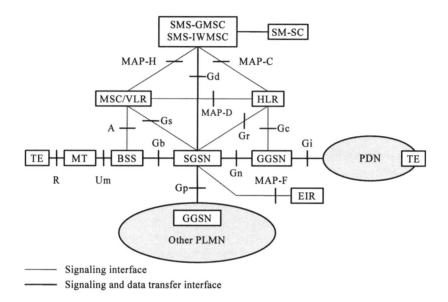

Figure 14.4 GSM/GPRS reference architecture. (*Source*: [3], reproduced with permission from ETSI.)

- The interface between the SGSN and the BSC (Gb-interface);
- The interface between GGSN and HLR (Gc-interface) is optional;
- The interface between SMS-GMSC and SGSN, and between SMS-IWMSC and SGSN (Gd-interface) enables GPRS MSs to send and receive short messages over GPRS radio channels;
- The interface between two GSNs within the same PLMN (Gn-interface);
- The interface between two GSNs in different PLMNs (Gp-interface) has the same functionality as Gn-interface plus a security function required for inter-PLMN communication;
- The interface between SGSN and HLR (Gr-interface);
- The interface between SGSN and MSC (Gs-interface);
- The interface between GPRS and fixed network (Gi-interface) is actually called a reference point since it is not fully specified. This is described in GSM 09.61.

For more details on GPRS, refer to [4–6].

14.3.2 Protocol Stacks

Figure 14.5 illustrates the GSM protocol stack. The layer-3 protocols—radio resource management (RRM), mobility management (MM), and call control (CC)—are specified in GSM 04.08. LAPDm and LAPD are link layer protocols. SCCP' is the GSM specific implementation of the original SS7 SCCP protocol [7]. The A-interface is used for messages between BSC and MSC as well as for messages to and from the mobile station. Therefore, the BSS application part (BSSAP) specified in GSM08.08 is split into two subapplication parts: the BSS management application part (BSSMAP) and the direct transfer application part (DTAP). DTAP is used to transfer the signaling for call and mobility management (CC and MM protocols) transparently to the mobile station. The BSSMAP supports all the procedures between the MSC and the BSS that require interpretation and processing of information related to single calls, and resource management. Both these protocols make use of the SCCP and MTP protocols, which are part of the SS7 protocols. The transaction capabilities application part (TCAP) is used by the MAP protocol for network control. Not shown in Figure 14.5 is the MAP protocol. GSM 09.02 specifies the application protocols between exchanges and data bases (MSCs, GMSCs, VLRs, HLRs, EIRs, and SGSNs) for supporting call management, supplementary service management, short message transfer, location management, security management, radio resource management, and mobile equipment management. It also specifies the applicability of SCCP and TCAP protocols to support these exchanges.

Figure 14.5   GSM protocol stack.

14.3.2.1   GPRS Protocol Stacks

The GPRS protocol architecture makes a distinction between the transmission and signaling planes. The transmission plane consists of a layered protocol structure providing user information transfer, along with associated information transfer control procedures (e.g., flow control, error detection, error correction, and error recovery). Figure 14.6 shows the protocol stack for the GPRS transmission/user plane, while Figure 14.7 presents the GPRS signaling/control plane.

Figure 14.6   GPRS transmission/user plane. (*Source*: [3], reprinted with permission of ETSI.)

Figure 14.7   GPRS signaling/control plane between MS and SGSN. (*Source*: [3], reprinted with permission of ETSI.)

The GPRS tunneling protocol (GTP) transfers data between GPRS support nodes. It encapsulates the protocol data units (PDUs) of packet data protocols such as X.25 or IP. Furthermore, GTP provides a mechanism for flow control between GSNs, if required. The GTP protocol is defined in GSM 09.60. Below the GTP protocol are TCP and UDP protocols. TCP carries GTP PDUs for protocols that need a reliable data link (e.g., X.25) and UDP for protocols that do not need a reliable data link (e.g., IP). IPv4 is used as a GPRS backbone network-layer protocol providing routing functions for user data and control signaling. The IPv6 will replace version 4 in the future.

Between the SGSN and the MS, packet data protocol (PDP) PDUs are transferred with the subnetwork dependent convergence protocol (SNDCP) that maps network-level characteristics into the characteristics of the underlying logical link, and provides multiplexing of multiple layer 3 messages into a single virtual logical link connection. In addition, ciphering, segmentation, and compression are also performed by the SNDCP protocol. Logical link control (LLC) provides reliable logical link. The LLC relay function relays LLC PDUs between Um- and Gp-interfaces. The base station system GPRS protocol (BSSGP) conveys routing and QoS-related information between BSS and SGSN. It is specified in GSM 08.64.

In the radio interface GSM RF, the GSM physical layer specified in GSM 05 series performs modulation, demodulation, encoding, and decoding of data.

The data link layer has two sublayers: radio link control (RLC) and medium access control (MAC). RLC provides a radio solution-dependent reliable link. The MAC function controls the access signaling (request and grant) procedures for the radio channel, as well as the mapping of LLC frames into the GSM physical channel. The RLC and MAC functions are described in GSM03.64.

The signaling plane consists of protocols for control and support of the transmission plane functions such as attaching to and detaching from the GPRS network, activation of a PDP (e.g., IP or X.25) address, controlling the routing path to support user mobility, controlling the assignment of network resources, and providing supplementary services.

The previously described physical layer, RLC/MAC and LLC protocols are also used for signaling transmission between the MS, BSS, and SGSN. The layer 3 mobility management (L3MM) protocol supports mobility management functionality such as GPRS attach, GPRS detach, security, routing update, location update, PDP context activation, and PDP context deactivation between MS and SGSN.

Signaling between the SGSN and HLR is performed using the same protocols, TCAP, SCCP, and MTP, as for the non-GPRS GSM PLMNs. The MAP protocol supports signaling exchanges between SGSN and HLR with enhancements for GPRS.

14.4   3GPP NETWORK

In this section 3GPP-based networks are introduced. The relevant specifications are listed and system architecture, protocols, and interfaces are described. The description of the following is based on the 3GPP specifications listed in Table 14.1.

Table 14.1
3GPP Network and Interface Specifications

Specification number	Name	Scope
UMTS 23.101	General UMTS Architectures	Defines the basic physical and functional separation of UMTS that are common to all UMTS networks independent of their origin. It identifies and names the reference points and functional groupings appearing at this level.
UMTS 23.107	QoS Concept and Architecture	Provides the framework for quality of service in UMTS.
UMTS 23.121	Architectural Requirements for Release 1999	Covers issues related to the evolution of the GSM platform towards UMTS with the overall goal of fulfilling the UMTS service requirements, the support of the UMTS role model, roaming, new functionality, signaling systems, and interfaces.
TR 23.821	Architecture Principles for Release 2000	Lists architectural requirements, features, functions, and solutions of UMTS inside the scope of UMTS release 2000.
TR 23.920	Evolution of the GSM platform towards UMTS	Covers issues related to the evolution of the GSM platform towards UMTS.
TR 23.923	Combined GSM and Mobile IP Mobility Handling in UMTS IP CN	A feasibility study on using mobile IP+ as a tunneling and mobility management protocol in combination with GSM/UMTS mobility management in the packet domain of UMTS CN.
TR 23.930	Iu Principles	Identifies the requirements on the Iu and studies relevant principles to guide further standardization of the related interfaces.
UMTS 25.410	UTRAN Iu Interface: General Aspects and Principles	Introduces UMTS 25.41x series of technical specifications that define the Iu interface.
UMTS 25.411	UTRAN Iu Interface: Layer 1	Specifies the range of physical layer technologies that may be used to support the Iu interface.
UMTS 25.412	UTRAN Iu Interface: Signaling Transport	Specifies the signaling bearers for the RANAP and transport network control plane protocols for both Iu-PS and Iu-CS.
UMTS 25.413	UTRAN Iu Interface: RANAP Specification	Specifies the RANAP protocol for radio network control plane signaling over the Iu interface.
UMTS 25.414	UTRAN Iu Interface: Data Transport and Transport Signaling	Specifies the transport bearers for the user plane of the Iu interface. It also specifies the protocol used to control these transport bearers.
UMTS 25.415	UTRAN Iu Interface: CN-UTRAN User Plane Protocol	Specifies the user plane frame handling protocol for the Iu interface.
UMTS 25.419	UTRAN Iu Interface: Service Area Broadcast Protocol (SABP)	Specifies the communication requirements over the Iu interface towards the BC domain.
UMTS 25.430	UTRAN Iub Interface: General Aspects and Principles	Contains introduction to the TSG RAN TS 25.43x series of UMTS technical specifications that define the Iub interface.
UMTS 25.431	UTRAN Iub Interface: Layer 1	Specifies the standards for the implementation of Layer 1 (physical layer) on the I_{ub} interface.

Table 14.1 (cont.)
3GPP Network and Interface Specifications

UMTS 25.432	UTRAN Iub Interface: Signaling Transport	Specifies the signaling transport related to NBAP signaling to be used across the Iub interface.
UMTS 25.433	NBAP Specification	Specifies the standards for NBAP specification to be used over Iub interface.
UMTS 25.434	UTRAN Iub Interface: Data Transport & Transport Signaling for Common Transport Channel Data Streams	Provides a specification of the UTRAN RNC-node B (Iub) interface data transport and transport signaling for common transport channel data streams.
UMTS 25.435	UTRAN Iub Interface: User Plane Protocols for Common Transport Channel Data Streams	Provides a specification of the UTRAN RNC-node B (Iub) interface user plane protocols for common transport channel data streams.
UMTS 25.426	UTRAN Iur/Iub Interface: Data Transport & Transport Signaling for DCH Data Streams	Specifies the transport bearers for the DCH data streams on UTRAN Iur and Iub interfaces. The corresponding transport network control plane is also specified.
UMTS 25.427	UTRAN Iur/Iub Interface: User Plane Protocol for DCH Data Streams	Provides a specification of the UTRAN Iur and Iub interfaces user plane protocols for dedicated transport channel data streams.

14.4.1 System Architecture

The UMTS network architecture includes the core network (CN), the UMTS terrestrial radio access network (UTRAN), and the user equipment (UE). The UTRAN architecture consists of several radio network subsystems (RNS). The RNS is further divided into the radio network controller (RNC) and several base stations (BTS, referred to as node B in the 3GPP specifications).

The Iu interface connects the CN to UTRAN. The Iur interface enables the exchange of signaling information between two RNCs. The signaling information across the Iur interface is called radio network subsystem application part (RNSAP). The RNSAP is terminated at both ends of the Iur interface by an RNC. S-RNC relocation is a procedure after handover between RNCs is covered within the access network. The signaling and traffic are led from CN to the S-RNC (old RNC) to D-RNC (Target RNC) and then towards the UE. The Iub interface connects an RNC and a node B. The Iub interface allows the RNC and the node B to negotiate about radio resources, for example, to add and delete cells controlled by the node B to support communication of the dedicated connection between UE and S-RNC, information used to control the broadcast and paging channels, and information to be transported on the broadcast and paging channels. One node B can serve one or multiple cells. The UE is connected to the node B through Uu radio interface.

14.4.2 Interfaces and Protocols

The main interfaces of UTRAN are Iu, Iur, and Iub. In the following, the interfaces and their protocol structure are described. The protocol model consists of horizontal layers and vertical planes, which are designed to be independent of each other to allow flexibility to develop the standard further. The protocol structures are designed based on the same generic principles. There are two main horizontal layers: radio network layer and transport network layer, and three vertical planes: control, user, and transport network control plane.

14.4.2.1 Iu Interface

Iu is the interface between UTRAN and core network. The Iu interface is open and has been specified to support multivendor operation. Figure 14.8 shows the logical division of the Iu interface. There are actually three types of interfaces in the Iu interface. The Iu interface towards the packet-switched (PS) domain of the core network is called Iu-ps, and the Iu interface towards the circuit-switched (CS) domain is called Iu-cs. The Iu interface to the broadcast domain is called Iu-bc.

The Iu interface has a hierarchical architecture where one higher layer entity controls several lower layer entities. The hierarchy for the CN-UTRAN signaling connection end-points is as follows:

- Each CN access point may be connected to one or more UTRAN access points.
- For the PS and CS domains, each UTRAN access point shall not be connected to more than one CN access point per CN domain.
- For the BC domain, each UTRAN access point may be connected to one or more CN access points.

Iu interface supports:

- Procedures to establish, maintain, and release radio access bearers;
- Procedures to perform intrasystem handover, intersystem handover, and SRNS relocation;
- Procedures to support the cell broadcast service;
- A set of general procedures, not related to a specific UE;
- The separation of each UE on the protocol level for user specific signaling management;
- The transfer of NAS signaling messages between UE and CN;
- Location services by transferring requests from the CN to UTRAN, and location information from UTRAN to CN. The location information may comprise a geographical area identifier or global coordinates with uncertainty parameters;
- Simultaneous access to multiple CN domains for a single UE;
- Mechanisms for resource reservation for packet data streams.

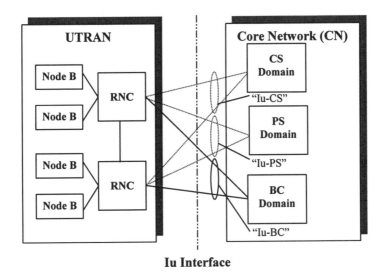

Iu Interface

Figure 14.8 Iu interface architecture. (*Source*: [10], reprinted with permission of ETSI.)

The radio network signaling over Iu consists of the radio access network application part (RANAP). The RANAP protocol consists of mechanisms to handle all procedures between the CN and UTRAN. It is also capable of conveying messages transparently between the CN and the UE without interpretation or processing by the UTRAN. The RANAP protocol is used to:

- Facilitate a set of general UTRAN procedures from the core network such as paging–notification;
- Separate each UE on the protocol level for mobile specific signaling management;
- Transfer of transparent non-access signaling;
- Request of various types of UTRAN radio access bearers;
- Perform the SRNS relocation function.

14.4.2.2 Iu-cs Interface Protocol Structure

Figure 14.9 shows the overall protocol structure for Iu-cs. It consists of three planes: control, transport network control, and user plane.

Figure 14.9 Iu-interface protocol structure towards CS domain. (*Source*: [10], reprinted with permission of ETSI.)

14.4.2.3 Iu-ps Protocol Structure

Figure 14.10 shows the protocol structure for Iu-ps.

14.4.2.4 Iur Interface

Iur interfaces radio network controllers and is especially required to support inter-RNC soft handover. As shown in Figure 14.11, the Iur interface protocol architecture consists of two functional layers:

- The radio network layer defines the procedures related to the interaction of two RNCs within a PLMN. The radio network layer consists of a radio network control plane and a radio network user plane.
- The transport layer defines procedures for establishing physical connections between two RNCs within a PLMN.

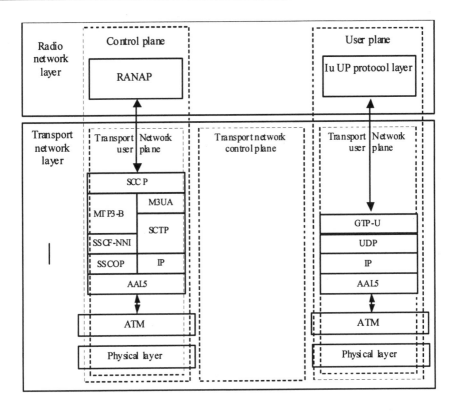

Figure 14.10   Iu-interface protocol structure towards PS domain. (*Source*: [10], reprinted with permission of ETSI.)

Key objectives in the design of the Iur interface have been:

- Interconnection of RNCs supplied by different manufacturers;
- Support of continuation between RNSs of the UTRAN services offered via the Iu interface;
- Separation of Iur interface radio network functionality and transport network functionality to facilitate introduction of future technology.

14.4.2.5   Iub Interface

The Iub is a logical interface that connects base station (referred in 3GPP specifications as node B) to radio network controller. It is an open interface and thus supports multivendor operation. As a general principle for Iub specifications, transmission sharing between the GSM/GPRS Abis interface and the Iub interface shall not be precluded, and the functional division between RNC and node B shall have as few options as possible.

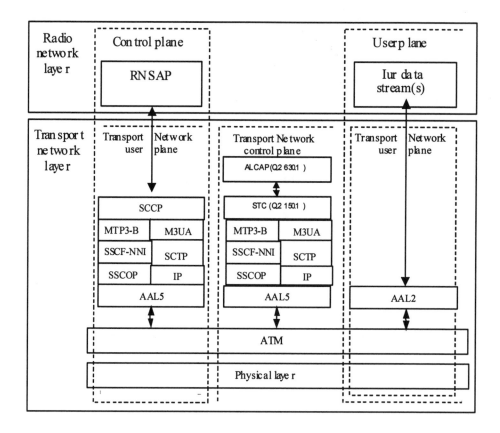

Figure 14.11 Iur interface protocol structure. (*Source*: [10], reprinted with permission of ETSI.)

Figure 14.12 depicts the Iub interface protocol architecture that consists of two functional layers:

- The radio network layer defines procedures related to the operation of node B. The radio network layer consists of a radio network control plane and a radio network user plane.
- The transport layer defines procedures for establishing physical connections between node B and the RNC.

Radio network
control plane

Transport
network
control plane

User plane

Figure 14.12 Iub interface protocol structure. (*Source*: [11], reprinted with permission of ETSI.)

14.5 OVERVIEW OF IS-41/IS-95 NETWORK ARCHITECTURE

In this section, we analyze the existing architecture, interfaces, and protocols of the IS-41/IS-95 networks. Based on this, we introduce a possible evolution scenario and suggest some of the required enhancements to allow the evolution of IS-95/IS-41 towards the third generation. An overview of the different standards required in the implementation of IS-41/IS-95-based systems is given in the next section. For a detailed treatment of the IS-41 standard, refer to [8].

14.5.1 IS-41/IS-95 Standards

When considering a complete IS-95-based network implementation, we need to consult several standards. The main standards are:

- EIA/TIA/IS-41-C, Cellular Radiotelecommunications Intersystem Operations;

- TIA/EIA/IS-707, Data Service Options for Wideband Spread Spectrum Systems;
- EIA/TIA/IS-95B, Mobile Station-Base Station Compatibility Standard for Dual-Mode Wideband Spread Spectrum Cellular System;
- TIA/EIA/IS-658, Data Services Interworking Function Interface for Wideband Spread Spectrum Systems;
- TIA/EIA/IS-634, MSC-BS Interface, Rev A.

14.5.2 Reference Architecture

Figure 14.13 shows a partial view of the reference architecture for the IS-41 core network. The message center (MC) is an entity that stores and forwards short messages. It may also provide supplementary services for SMS.

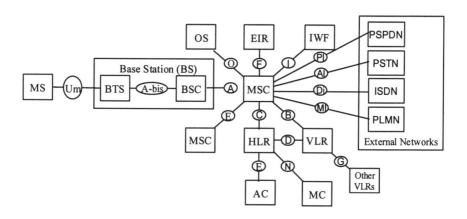

Figure 14.13   Partial view of IS-41 reference architecture.

The following main interfaces are defined between the network elements:

- The interface between the BSC and the MSC (A-interface) is specified in the IS-634 standard.
- The interface between the BTS and the BSC (the Abis-interface) is not specified.
- The interface between the MSC and the HLR (C-interface): signaling is specified in the IS-41 standard.
- The interfaces between the MS and the BTS (air interface, Um-interface) are specified in the IS-95, IS-97, and IS-98 standards.
- The interface between the MSC and its associated VLR (B-interface) is not specified.
- The interface between the HLR and the MSC (C-interface): signaling is specified in the IS-41 standard.

- The interface between the HLR and the VLR (D-interface) is used to exchange data related to the location of the mobile station and to the management of the subscriber. Signaling in this interface is specified in the IS-41 standard.
- The interface between the HLR and the VLR (D-interface): Signaling is specified in the IS-41 standard.
- The interface between IWF and MSC (L-interface) is specified in the TIA/EIA/IS-658 standard.

14.5.2.1 Detailed A-Interface Architecture

The first version (Rev 0) of the IS-634 standard defining the A-interface was published in 1995. The first implementations of IS-95-based systems were thus actually proprietary, and consequently, the standard was heavily influenced by the existing implementations. In the new revision of IS-634 Rev A, published in 1998, two distinct architectures have been defined: architecture A, with the selection/distribution unit (SDU) located inside of the base station, and architecture B, with the SDU located outside of the base station. Architecture A has a direct inter-BS connection option and an indirect (through MSC) BS-to-BS connection option. Architecture B has only a direct inter-BS connection.

There are six "parts" in the IS-634 Rev A. As compared to the earlier version of IS-634, a major restructuring has been performed:

- Part 0 (Base Part): high level function overview;
- Part 1 (Common Protocol): common protocol for architecture A and B;
- Part 2 (Architecture A): specific protocol for architecture A only;
- Part 3 (Architecture B): specific protocol for architecture B only;
- Part 4 (Interworking): interworking protocol between architecture A and B;
- Part 5 (Protocol Details): defines messages, information elements, timers.

The detailed A-interfaces for architectures A and B are shown in Figures 14.14 and 14.15, respectively. The A-interface consists of A1- through A7-interfaces. A3- and A7-interfaces are both TCP/IP connections over ATM and are used for direct inter-BS connections for faster and more efficient soft handover. The A3-interface is divided into A3s (for signaling) and A3t (for traffic). Direct inter-BS connections (A3t, A3s, and A7) support faster soft handover and can add/drop multiple cells in same message.

The A6-interface is used in the indirect (through MSC) BS-to-BS connection option of architecture A (not shown). It is a 16-Kbps subrate circuit carrying user data/speech between the BS and the MSC during soft handover.

SDU is a new functional unit in IS-634. SDU contains functional entities such as traffic handler, signaling layer 2, multiplex sublayer, frame selection/distribution for soft handover, and power control. It also contains the transcoder functionality, which can be located either inside or outside of the base station.

Figure 14.14 IS-634 architecture A.

Figure 14.15 IS-634 architecture B.

14.5.3 Protocol Stacks

14.5.3.1   Air Interface

Figure 14.16 shows the IS-95 air interface protocol structure for the mobile station and the base station. Layer 1 is the physical layer of the digital radio channel, including those functions associated with the transmission of bits, such as modulation, coding, framing, and channelization via radio waves. Between layer 1 and layer 2 is a multiplex sublayer containing the multiplexing functions that allow sharing of the digital radio channel for user data and signaling processes. The multiplex sublayer provides several multiplex options. By *multiplex option* it is meant the ability of the multiplex sublayer and lower layers to be tailored to provide special capabilities. The multiplex option defines such characteristics as the frame format and the rate decision rules. The physical layer and multiplex sublayer are specified in Chapters 6 and 7 of the IS-95 specifications. The IS-95 combines the operation of the network and data link layers and treats them as one layer.

Primary traffic is the main traffic stream carried between the mobile station and the base station on the traffic channel. Secondary traffic is an additional traffic stream that can be carried between the mobile station and the base station on the traffic channel. Signaling traffic means control messages that are carried between the mobile station and the base station on the traffic channel.

For user data, protocol layering above the multiplex sublayer is service option dependent and is described in standards for the service options. Service option means a service capability of the system. A *service option* may be an application such as voice, data, or facsimile.

For the signaling protocol, two higher layers are defined. Signaling protocol layer 2 is the protocol associated with the reliable delivery of layer 3 signaling messages between the base station and the mobile station, such as message retransmission and duplicate detection. Signaling layer 3 is the protocol associated with call processing, radio channel control, and mobile station control, including call setup, handover, power control, and mobile station lockout. Layer 3 signaling is mainly passed transparently to the MSC through the BS.

14.5.3.2   A-Interface

The IS-634 standard defines the MSC-BS messages, message sequencing, and mandatory timers at the base station and the mobile switching center. Figure 14.17 depicts IS-634 functions: call processing and supplementary services, radio resource management, mobility management, and transmission facilities management.

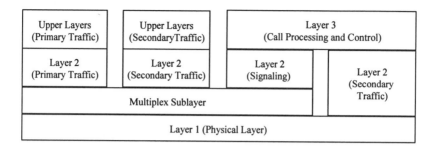

Figure 14.16 Mobile station and base station layers. (*Source*: [9], reproduced under written permission of the copyright holder Telecommunications Industry Association. Note: At the time of the publication, the standard that contains this figure was not finalized, please check with TIA for the correct version.)

Figure 14.17 A-interface functions. (*Source*: IS-634 Rev. A, reproduced under written permission of the copyright holder Telecommunications Industry Association.)

Figure 14.18 shows the A1-interface signaling protocol reference model. Similar to GSM, IS-634 has a base station application part (BSAP), which contains BSMAP and DTAP. The BSAP corresponds to BSSAP in GSM and BSMAP to BSSMAP. Thus, DTAP is used to transfer the mobility management and call control related signaling between the MSC and the mobile station. BSMAP transfers, for example, radio resource management related signaling between BS and MSC. The transport protocols for signaling and user traffic data are shown in Figures 14.19 and 14.20, respectively. The physical layer is based on the use of T1 digital transmission system interface (1.544 Mbps providing 24×56 or 64-Kbps channels), E1 digital transmission interface (consisting of 30×56 or 64 Kbps channels), OC1 digital transmission interface (51.84-Mbps), and OC3 digital transmission interface (155.52 Mbps).

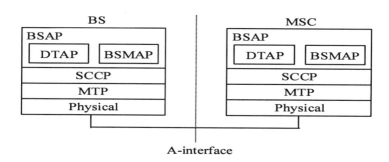

Figure 14.18   A1-interface signaling protocol reference model.

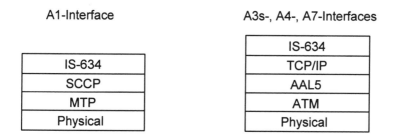

Figure 14.19   Transport protocol options for signaling connections.

Figure 14.20   Transport protocol options for user traffic connection.

14.5.3.3 Data Service Protocols

The data service protocols for IS-95 are specified in TIA/EIA/IS-707 "Data Service Options for Wideband Spread Spectrum Systems." The IS-707.4 specifies asynchronous fax and data services, and IS-707.5 specifies packet data services. The L-interface is specified in TIA/EIA/IS-658.

Figure 14.21 shows an overview of data service protocols based on IS-658. The physical layer in the network side is based on B-channels or H-channels with unrestricted digital information as defined in ANSI T1.607-1990, or in ANSI/IEEE 802.3 LAN. FR SVC is frame relay switched virtual circuit, which conforms to ANSI T1.618. The adaptation layer accepts data octets from the link layer and the relay function, assembles them into blocks of data octets for transmission in the information field of FR SVC frames, or vice versa, accepts blocks of data octets, and disassembles them. Radio link protocol (RLP) reduces the error rate by applying retransmission protocol for erroneous packets. RLP is specified in 707.2.

The link layer is based on the point-to-point protocol (PPP), defined in RFC 1661. PPP provides a multiplexed method to carry higher layer protocols over serial links. Above PPP, asynchronous data and fax data applications use link control protocol (LCP) and Internet protocol control protocols (IPCP). LCP provides a mechanism for the mobile station and the IWF to negotiate various options provided by PPP. IPCP allows the mobile station to request a temporary IP address from the IWF.

Figure 14.21 Mobile data protocol overview. (*Source*: TIA/EIA/IS-658, reproduced under written permission of the copyright holder Telecommunications Industry Association.)

IS-707.5 specifies two protocol options for packet services: relay layer R_m interface protocol option and network layer R_m interface protocol option. The network layer R_m interface protocol option supports terminal equipment (TE) applications where the mobile termination is responsible for aspects of packet mobility management and network address management. The relay layer R_m interface protocol option supports TE applications where the TE is responsible for aspects of packet mobility management and network address management.

14.5.4 IS-41/95 Evolution Towards Third Generation

Different network-related standards of IS-41 and IS-95 systems have been upgraded to support the new functionalities and capabilities of cdma2000. The reader is referred to the appropriate standards to review the changes. Here we describe a more revolutionary evolution scenario based on the so-called hooks and extension concept agreed upon by Operators Harmonization Group (OHG). This evolution is especially suitable for IS-95 operators that deploy WCDMA in the third-generation frequency bands and cdma2000 1x in the existing IS-95 frequency bands.

The modular 3G harmonized system can be implemented with the hooks and extension concept as shown in Figure 14.22. This protocol structure will enable connection of WCDMA and cdma2000 1x radio access into 3G Mobile Packet Core (based on the evolved GSM MAP network).

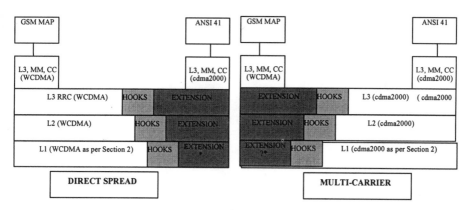

Figure 14.22   Hooks and extension protocol structure.

Currently, only circuit-switched data services are supported in the IS-95 networks. Heavy growth is expected because of the introduction of WAP-based data services. Circuit-switched data services can only support a limited number of simultaneous connections, and when active WAP user penetration reaches maximum capacity, the service quality level will decrease dramatically. Therefore, there is a clear need for the mobile packet core (MPC) network to offload non-real-time services from the circuit-switched network. Figure 14.23 illustrates a network architecture where more radio network capacity has been gained by introducing the IS-2000 radio network as well as by removing the circuit-switched capacity constraint by introducing mobile packet core.

The connection between the cdma2000 RAN and mobile packet core network has been done via wireless gateway (WGW). The WGW is introduced to shorten the time to market with this solution. Eventually, the WGW will not be needed once all the necessary protocols needed to support direct connection to IS-2000 RAN are implemented. Figure 14.24 illustrates the network architecture when WCDMA RAN is introduced to the network.

Figure 14.23   cdma2000 with 3GPP mobile packet core network.

Figure 14.24   Network architecture after WCDMA RAN was introduced to a IS-95 network.

All existing investments are preserved and only WCDMA RAN and 3G MSC have been introduced into the network. This provides true packet radio access as well as a 3G MSC for roaming subscribers and those using release '99-based terminals.

The next step is to introduce all-IP architecture into the network. The all-IP network is implemented by migrating 3G MSC to the call processing server (CPS) and upgrading existing mobile packet core network to support all-IP. The exiting ANSI-41 network is used to serve IS-2000 and IS-95 A/B network users.

14.6 ALL-IP NETWORKS

The tremendous success of the Internet and the need for scalability has raised interest to use IP as the basis for further development of cellular networks. For example, 3GPP has started standardizing IP-based wideband CDMA and EDGE networks for release 2000. These are commonly referred as all-IP networks. In this section, the drivers for developing all-IP networks are discussed; then key targets for all-IP solutions are covered. Thereafter, all-IP architecture and radio issues related to all-IP networks are described.

14.6.1 Drivers for All-IP

The key drivers for all-IP networks are cost and flexibility. An architecture that would reduce the operator cost, including equipment, network deployment, and operational/maintenance cost, is required. This will be realized by off-the-shelf components, migrating the intelligence to the edge of the network, and autoconfiguration. Flexibility can be divided into two categories, flexibility in service creation and in the future evolution of the network. IP-based networks, such as the Internet, provide an open platform for service creation. Flexibility is enabled by separating the applications from the underlying networks, allowing free service creation that is independent of the radio access system. Future evolution is enabled by layered architecture and open interfaces facilitating the evolution of different protocols and subsystems of the network independently. All-IP networks also offer a common platform for all types of services unifying today's circuit-switched (voice) and packet-switched networks into one packet-switched network offering all types of services.

14.6.2 Objectives of All-IP Network Development and Standardization

The 3GPP standardization body has set several objectives for the standardization of all-IP networks. Several other bodies have contributed to the definition of these objectives including 3G.IP, the Mobile Wireless Internet Forum (MWIF), and the UMTS Forum. The objectives include:

- Establishment of flexible service creation environment enabling quick service/application creation;
- End-to-end IP (IP connection terminated at the mobile station);
- Real time including multimedia services;
- Modular and scalable structure;

- Independence of access type: seamless services, common service development, commonality between private and public networks, potential for fixed mobile convergence;
- Separation of services, control, and transport when beneficial;
- Integrated operations and maintenance functionality;
- Leverage IP technology cost factor without compromising quality or performance;
- Open interfaces to ensure a multivendor environment;
- At least the same level of security as in the state-of-the-art mobile networks;
- At least the same level of QoS as in the state-of-the-art mobile networks.

14.6.3 All-IP Architectures

The all-IP architecture is based upon a layered architecture with independent control and transport domains. The basic all-IP model contains four layers. The applications and services layer consists of applications such as e-mail, calendar and WWW. Service control maintains a subscriber service profile, location information, billing of services, and personalization of services. Security and mobility are handled in the network control layer, and the connection layer handles the transport of the user and signaling information.

The main debate over different all-IP approaches has focused on mobility management. Two primary approaches have been proposed: the evolution of current mobility management schemes for mobile networks (GSM MAP and GPRS) and the IETF-based protocol approach using mobile IPv6. Backwards compatibility would favor the former approach and pure IP thinking the latter one. However, the mobile IP protocol still lacks many functionalities already solved in mobile networks such as fast and seamless handover (location update). Most likely the final solution will be comprised of both approaches.

14.6.3.1 3GPP All-IP Reference Architecture

Figure 14.25 shows the reference architecture of the 3GPP all-IP network. The call-state control function (CSCF) performs call-control functions like call set-up/termination and control. Furthermore, it provides location management functions and address handling. The home subscriber server (HSS) contains the HLR, authentication authorization and accounting (AAA), and the user mobility server (UMS). HLR is for mobility-related subscriber data for serving SGSN and GGSN.

Roaming signaling gateway (R-SGW) enables signaling interworking with legacy mobile networks (e.g., GSM MAP, ANSI41). T-SWG stands for transport signaling gateway function. Media gateway control function (MGCF) is the PSTN/PLMN termination point for a defined network. It performs protocol conversion between the legacy and the release 2000 all-IP network call control protocols. Media gateway function (MGW) is the PSTN/PLMN transport termination point for a defined network and is used for interfacing the radio-access network with the all-IP core network. Multimedia resource function (MRF) performs multiparty call and multi-media conferencing functions.

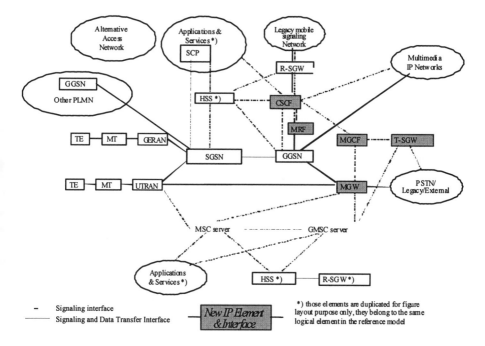

Figure 14.25   3GPP all-IP reference model.

14.6.4 Radio Issues for All-IP Networks

All-IP networks feature low bandwidth and require high spectrum efficiency. Transportation of IP headers over the air results in extra overhead which might be considerable for some services such as low-rate speech.

To reduce the overhead, IP headers can be compressed or totally removed. Header compression method compresses the headers added by the transport and network layers. Header stripping is a header compression mechanism dependent on how the data is transported in the lower layers (link and physical). Typically, all the transport and network layer headers are removed from the packets carrying voice samples. In the case of header compression or header stripping, the network and higher layer protocols terminate at the mobile station. In header stripping the headers are regenerated at the mobile station. However, this scheme does not result in fully transparent end-to-end IP connection as header compression does.

In header removal the transport and network layers are terminated at the network side, and only the payload is transported to the mobile station. The IP headers are not regenerated at the mobile station side.

REFERENCES

[1] Modarressi, A. R., and R. A. Skoog, "Signaling System No. 7: A Tutorial," *IEEE Communications Magazine*, July 1990, pp. 19–35.

[2] Lin, Y.-B., and S. K. DeVries, "PCS Network Signaling Using SS7," *IEEE Personal Communications*, Vol. 2, No. 3, June 1995, pp. 44–55.

[3] GSM 03.60, "General Packet Radio Service (GPRS); Service Description; Stage 2 ETSI," GSM Technical Specification 03.60, Version 5.0.0, June 1997.[1]

[4] Cai, J., and D. J. Goodman, "General Packet Radio Service in GSM," *IEEE Communications Magazine*, October 1997, pp. 122–131.

[5] Hämäläinen, J., *Design of GSM High Speed Data Services*, Ph.D. Dissertation, Tampere University of Technology, Finland, 1996.

[6] Jung, P., Z. Zvonar, and K. Kammerlander (eds.), *GSM: Evolution Towards 3rd Generation,* Boston: Kluwer Academic Publishers, 1998.

[7] Mouly, M., and M.-B. Pautet, *The GSM System for Mobile Communications*, published by the authors, 1992.

[8] Gallagher, M. D., and R. A. Snyder, *Mobile Telecommunications Networking with IS41*, New York: McGraw-Hill, 1997.

[9] TIA/EIA/SP-3693, "Mobile Station-Base Station Compatibility Standard for Dual-Mode Wideband Spread Spectrum Cellular Systems," to be published as ANSI/TIA/EIA-95B.[2]

[10] TS 25.410, "UTRAN Iu Interface: General Aspects and Principles," 3GPP Technical Specification, Version 3.2.0, March 2000.

[11] TS 25.430, "UTRAN Iub Interface: General Aspects and Principles," 3GPP Technical Specification, Version 3.2.0, March 2000.

---

[1] The present technical specifications are the property of ETSI. No reproduction shall be made without ETSI authorization. The original version of these ETSI technical specifications can be obtained from the publication office of ETSI.

[2] To purchase the complete version of any TIA document, call Global Engineering Documents at 1-800-854-7179 or send a facsimile to 303-397-2740.

Chapter 15

STANDARDIZATION

15.1 INTRODUCTION

This chapter addresses the third-generation standardization environment. Standards bodies and industrial interest groups related to the third-generation standardization in different regions are described [1–23]. In addition, regulatory developments such as spectrum licensing are discussed. For IETF standardization, refer to Chapter 13.

15.2 STANDARDIZATION BODIES AND INDUSTRY GROUPS

This section introduces the different standardization bodies for the third-generation systems, their responsibilities, and their organization. We start by describing the third-generation partnership project (3GPP) standardization organization. Next, the standards bodies and the third-generation industry interest groups in Europe, the United States, Japan, Korea, and China are presented. Furthermore, each subsection starts with a description of the general approach each region is taking to the development of mobile radio systems. Industry groups related to all-IP standardization and the ITU IMT-2000 standardization organization are described.

15.2.1 3GPP

In 1998, as WCDMA gained more and more support in different standardization bodies, it was decided to establish a common standardization body to gather global expertise. Therefore, the third-generation partnership project (3GPP) was founded in December

1998 with a statement: "Standards organizations and other related bodies have agreed to co-operate for the production of a complete set of globally applicable Technical Specifications for a 3rd Generation Mobile System based on the evolved GSM core networks and the radio access technologies supported by 3GPP partners (i.e., UTRA both FDD and TDD modes)."

The 3GPP organizational partners are ARIB, CWTS (the China Wireless Telecommunications Standards Group), ETSI, TTA, TTC, and T1P1. In addition to the organizational partners, there are market representation partners: GSM Association, UMTS Forum, Global Mobile Suppliers Association, IPv6 Forum, and Universal Wireless Communications Consortium (UWCC).

The structure of the 3GPP organization is shown in Figure 15.1. The project coordination group has the overall coordination responsibility. The technical specifications are developed within technical specifications groups (TSG). The working groups of each TSG are listed in Table 15.1.

Although 3GPP was originally established to address only the WCDMA air interface, responsibility of the further development of the EDGE air interface was to be transferred from ETSI to 3GPP in 2000. For this purpose, a new TSG, TSG GERAN, has been established.

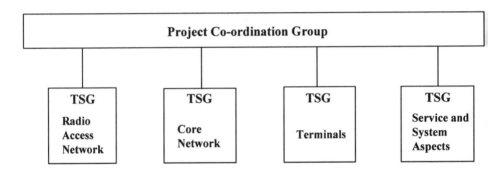

Figure 15.1   3GPP structure.

TSG Core Network (TSG-N) will address the following areas [1]:

- Mobility management, call connection control, and session management signaling between the user equipment and the core network;
- Core network signaling between the core network nodes;
- Interworking with second-generation networks;
- Definition of interworking functions between the core network and external networks;
- Packet related matters such as mapping of QoS, for example, transparency for IP domain applications, general for bearer types, special for optimized applications such as voice-over-IP (VoIP);

- Core network aspects of the Iu interface;
- Core network O&M requirements.

Table 15.1
3GPP Technical Specification Groups and Their Working Groups

3GPP Technical Specification Group Core Network
• 3GPP Working Group CN1 : MM/CC/SM (Iu)
• 3GPP Working Group CN2 : CAMEL3GPP Working Group CN3 : Interworking with external networks
• 3GPP Working Group CN4 : MAP/GTP/BCH/SS
• 3GPP Working Group CN5 : OSA
3GPP Technical Specification Group Radio Access Network
• 3GPP Working Group RAN1: Radio layer 1 specification
• 3GPP Working Group RAN2 : Radio layer 2 specification and radio layer 3 RR specification
• 3GPP Working Group RAN3 : Iub specification, Iur specification, Iu specification, and UTRAN O&M requirements
• 3GPP Working Group RAN4 : Radio performance and protocol aspects from a system point of view–RF parameters and BS conformance
• 3GPP Ad-hoc group on ITU (internal) coordination
3GPP Technical Specification Group Services & System Aspects
• 3GPP Working Group SA1 : Services (including value added services)
• 3GPP Working Group SA2 : Architecture
• 3GPP Working Group SA3 : Security
• 3GPP Working Group SA4 : Codec
• 3GPP Working Group SA5 : Telecom management
3GPP Technical Specification Group Terminals
• 3GPP Working Group T1 : Mobile terminal conformance testing
• 3GPP Working Group T2 : Mobile terminal services & capabilities
• 3GPP Working Group T3 : Universal subscriber identity module (USIM)

The TSG Radio Access Network (TSG-R) is responsible for the radio access part including radio aspects of terminal equipment and UTRAN functions (FDD & TDD), requirements, and interfaces. TSG-R will address the following areas of work [1]:

- Radio layer 1 specification;
- Radio layer 2 specification;
- Radio layer 3 RR specification;
- Iub specification (including logical O&M);
- Iur specification;
- Iu specification;
- UTRAN O&M requirements;

- Transport of implementation specific O&M between the management system and node B;
- Conformance test specifications for testing of all aspects of base stations;
- Specifications for radio performance and RF system aspects.

The TSG Service and System Aspects (TSG-SA) is responsible for the overall architecture and service capabilities of systems based on 3GPP as well as cross TSG coordination. Its specific responsibilities include [1]:

- Definition, evolution, and maintenance of the overall system architecture including the assignment of functions to particular subsystems (UTRAN, CN, terminal, SIM), identification of key information flows, and definition of required bearers and services offered by these different subsystems;
- Development of a framework for services, service capabilities, service architecture, charging and consideration of need for "default" services and/or applications;
- Definition of a security framework and review of security aspects of overall system;
- Management of work items including assignment of tasks to other TSGs and monitoring of progress.

The TSG Terminals (TSG-T) is responsible for specifying the terminal equipment including:

- Terminal equipment performance specifications;
- USIM and its interface specifications.

15.2.2 3GPP2

Following the establishment of 3GPP, another partnership project named 3GPP2 was founded to develop specifications for cdma2000-based networks. Similar to 3GPP, 3GPP2's technical work is performed in technical specification groups:

- TSG-A (A-interface system);
- TSG-C (cdma2000);
- TSG-P (wireless packet data interworking);
- TSG-R (interface of 3GPP radio access technology to 3G core network evolved from ANSI-41);
- TSG-S (service and system aspects);
- TSG-N (ANSI-41/WIN).

15.2.3 Europe

In Europe, telecommunication standards are developed by ETSI. The general approach in Europe has been to build a consensus with an emphasis on long-term solutions. The European Union is involved in the standards setting process by developing enabling regulatory policies. The global success of GSM is partly based on this type of standardization approach, and the Europeans are trying to carry this over to the third generation.

As noted earlier, responsibility of further development of GSM/EDGE development was transferred from ETSI to 3GPP in 2000. The last TC SMG (technical committee Special Mobile Group) was held in June 2000. In ETSI, the internal successor bodies of TC SMG are the ETSI project "New SMG9" and TC MSG (Mobile Standards Group).

The structure of old TC SMG is depicted in Figure 15.2. The actual standardization work is carried out in subtechnical committees. SMG2 has been responsible for the UMTS air interface evaluation and will standardize the UMTS radio access network. SMG3 has been responsible for GSM/UMTS core network standardization, and SMG12 has been responsible of the overall UMTS architecture. SMG5, the original subtechnical committee responsible for UMTS, was discontinued in 1997. For more information on the ETSI standardization, see [3].

The UMTS Forum is a nonprofit association under the Swiss law. The UMTS Forum was established in 1996 on the recommendation of the UMTS Task Force [4]. It provides advice and recommendations to the European Commission, European Radiocommunications Office (ERC), ETSI, and national administrations. The UMTS Forum is divided into working groups (e.g., WG1 Regulatory, WG2 Spectrum, and WG3 Market Aspects). For more information on the UMTS Forum, see [5].

The GSM MoU Association is the GSM operators' organization for promoting and evolving the GSM cellular platform worldwide. Regional interest groups have been established to take requirements and special needs from the various regions into account: Asia Pacific Interest Group (APIG), Arab Interest Group (AIG), European Interest Group (EIG), North American Interest Group (NAIG), Central/Southern African Interest Group (CSAIG), East Central Asia Interest Group (ECAIG), and India Interest Group (IIG). In addition to GSM operators, government regulators/administrations can be members of the GSM MoU Association. Within GSM MoU, the Third Generation Interest Group (3GIG) was formed in 1994. The 3GIG is responsible for conveying GSM MoU vision and requirements into the standardization of UMTS. Furthermore, it addresses third-generation market aspects and licensing issues. For more information on the GSM MoU, see [6].

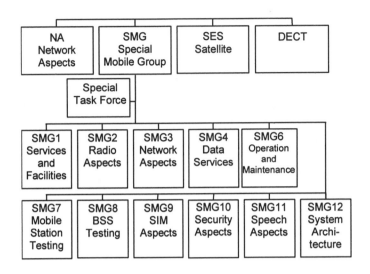

Figure 15.2   Partial view of the ETSI organization relevant to UMTS.

15.2.4 The United States

In the United States, the standards setting process is driven by market inputs and industry interests more than in Europe. On one hand, this leads to faster reaction to changes, but on the other, it leads to a larger number of standards. The selection between standards is then left to the market forces. The market-driven approach is reflected also in the standardization bodies. There are two main standardization bodies for mobile radio systems. The Telecom Industry Association (TIA) is a trade organization that provides its members with numerous services including standardization. The committee T1 on telecommunications was established in 1984 as a consequence of the breakup of the Bell System [7]. Both the TIA and T1 can set American National Standards Institute (ANSI) accredited standards. Within TIA, the Wireless Communications Division is responsible for standardization of wireless technologies. The main committees from a third-generation point of view are TR45 and TR46 (Public Mobile and Personal Communications Standards). Within T1, subcommittee T1P1 is responsible for management activities for personal communications systems (PCS). For more information on the TIA standardization activities, see [8], and on T1P1 activities, see [9].

15.2.4.1   TR45

Figure 15.3 shows the structure of TR45. In consists of six permanent subcommittees and six ad hoc groups not shown in the figure. The responsibilities of the subcommittees are as follows:

- TR45.1 is responsible for the AMPS standardization.
- TR45.2 is responsible of the development of the IS-41 standard.
- TR45.3 is responsible for IS-136, the US TDMA system, and its evolution to UWC-136.
- TR45.4 is responsible for the IS-634 standard (i.e., the A-interface standard between BS and MSC). For the third generation, TR45.4 will most likely develop the interface between the access and core networks.
- TR45.5 is responsible for IS-95 technology and its evolution to cdma2000.
- TR45.6 work items include support for packet data and Internet access. TR45.6 coordinates its activities with TR45.2 to define network architecture for IMT-2000 that complies with packet data requirements.

Figure 15.3   Structure of TIA TR45.

TR45.5 consists of four subcommittees as shown in Figure 15.4. TR45.5.1 is responsible for user needs and services such as multimedia services, speech, feature requirements, and data services. TR45.5.2 covers signaling and protocols including authentication/privacy, intersystem issues, call processing, and protocol tests. TR45.5.3 is responsible for the physical layer and has five working groups: RF parameters, supervision and malfunction detection, handoff and system timing, MS minimum performance, and BS minimum performance. TR45.5.4 is developing alternative technologies and has been responsible for development of the framework for wideband cdmaOne.

15.2.4.2   TR46 and T1P1

TR46 was created to address the TIA standards development for the US PCS. The main subcommittees within TR46 are listed below and represented in Figure 15.5.

- TR46.1 WIMS (wireless ISDN and multimedia service);
- TR46.2 cross-technology;
- TR46.5 PCS1900;
- TR46.6 composite CDMA/TDMA.

Figure 15.4   Structure of TIA TR45.5.

Figure 15.5   Structure of TIA TR46.

Standards committee T1P1—wireless/mobile services and systems belonging to the T1 organization—is closely related to TR46. They have an almost similar structure and jointly manage the above-mentioned TR46 standards. T1P1 has a cooperative arrangement with ETSI SMG regarding PCS1900/GSM technology.

15.2.4.3   Industry Interest Groups

Each of the main PCS standards in the United States has its own industry interest group: the CDMA Development Group (CDG) for IS-95, Universal Wireless Communications Consortia (UWCC) for IS-136, GSM North America (GSM MoU regional interest group), and GSM Alliance. The last two are consortia for North American PCS1900 operators using GSM technology.

CDG is an industry consortium of companies developing IS-95. The CDG is comprised of IS-95 service providers and manufacturers of subscriber and network infrastructure equipment. For more information on the CDG activities, see [10]. The CDG organizational structure consists of the following entities [10]:

- Executive board: a carrier-only board that oversees the CDG process. Participation on the board is limited to chief technical officers, chief operating officers, and chief executive officers.

- Steering committee: a committee consisting of executives from member organizations. The objective of this committee is to review and approve the work of the supporting teams and address resource issues.
- CDG working groups (teams): teams consisting of carrier and manufacturer business and technical resources. Advanced systems group is focusing on third-generation technologies.

The UWCC is a limited liability corporation in Washington state, established to support an association of carriers and vendors developing, building, and deploying products and services based on IS-136 TDMA and IS-41 wireless intelligent network (WIN) standards. The UWCC structure consists of a board of governors and three forums: Global TFMA Forum (GTF), Global Win Forum, and Global Operators Forum. For more information on the UWCC activities, see [11].

15.2.5 Japan

In Japan, the telecommunication operators and manufacturers have traditionally moved as one single entity. NTT has been in the leading role, dominating the wireless standardization scene, and the others have followed NTT in unison. The selection of IS-95 by DDI and IDO, two Japanese wireless operators, for replacing the analog system was an unusual break from the tradition of following NTT [12]. It has been very difficult for international companies to participate in the closed standards setting. One further obstacle has been the language. For third generation, Japan has opened standardization in order to create a global standard. Instead of Japanese, English has been selected as the working language for the documentation of the third-generation standards.

In Japan, the IMT-2000 standardization is divided between two standardization organizations: Association of Radio Industries and Businesses (ARIB) and Telecommunication Technology Committee (TTC). ARIB is responsible for radio standardization and TTC for network standardization. Contributions to ITU-R and ITU-T are done through the Telecommunication Technology Council of the Ministry of Posts and Telecommunications (MPT). MPT has established a Study Group on Next-Generation Mobile Communication, which gives recommendations to TTC and ARIB.

The organization of IMT-2000 Study Committee within ARIB is described in Figure 15.6. The Coordination Group (CG) in the Standards Subcommittee is responsible for international coordination with other standards bodies such as TTA, ETSI, and TIA. The harmonization discussions concerning WCDMA and cdma2000 have been carried out in this group. SWG1 is responsible for system description; SWG2 is responsible for the specification of the air interface layer 1; and SWG3 is responsible for the air interface specification for layers 2 and 3. The requirements and objectives are established in the Application Group. For more information on the ARIB activities, see [13].

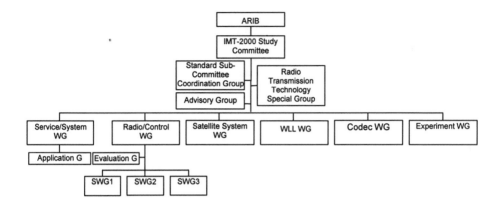

Figure 15.6   ARIB organization for IMT-2000.

Figure 15.7   TTC organization for IMT-2000.

TTC, represented in Figure 15.7, is responsible for the development of system architecture, information flows, requirements for network control, layer 2 in cooperation with ARIB, layer 3, RAN-CN interface, network-to-network interface (NNI), call associated signaling, and interworking with other networks. Because Japan has chosen to base its third-generation network on the GSM core network, close cooperation with ETSI is under way. For more information on the TTC standardization, see [14].

15.2.6 Korea

In Korea, the government sets the policies for the mobile radio system standardization. For example, the adoption of IS-95 over other wireless technologies for Korean PCS was decided by the Korean government. The Korean goals for IMT-2000 are a global standard, global roaming, a low cost terminal, to meet the ITU requirements, and backwards compatibility, if possible. Korea has a parallel approach to standards: international cooperation and development of domestic standards for implementation after 2002. The international cooperation includes relations with ARIB, ETSI, and TIA, as well as participation in the ITU RTT process.

The Telecommunication Technology Association (TTA) was established by the Korean Ministry of Communications (MIC) in 1988 to develop Korean Information Communication Standard (KICS) related to telecommunications. Within TTA, the IMT-2000 subcommittee has been established for the development of third-generation standards. The structure of the subcommittee is depicted in Figure 15.8. For more information on the TTA standardization, see [15].

Figure 15.8   Structure of TTA.

The Electronics and Telecommunications Research Institute (ETRI) is an R&D organization developing technologies considered important for Korea. The Mobile Telecommunications Technology Division of ETRI is developing technology for IMT-2000. For more information on ETRI, see [16].

15.2.7 China

In China, the Ministry of Post and Telecommunications (MPT) is driving IMT-2000 standardization. The Research Institute for Telecom & Transmission (RITT) is responsible for domestic standardization. It has actively cooperated with outside standardization bodies such as ETSI and ARIB. An MPT delegation has participated, for example, in the ETSI/SMG meetings as an associate member. MPT also participates actively in the ITU process.

15.2.8 All-IP Industry Groups

In order to accelerate the development of all-IP, industry interest groups have been established. In 1999, driven by AT&T Wireless Systems, a group of operators and vendors established an organization called 3G.IP to drive initially all-IP architecture based on GPRS for EDGE networks. The objectives of 3G.IP are as follows [www.3gip.org]:

- Define a 3G-network architecture based on packet technologies and IP telephony for simultaneous real and non-real-time services;
- Common core network for EDGE and UTRA–based on evolved GPRS;
- Seamless service support between EDGE and UTRA networks;
- Personal mobility and interoperability between mobile and fixed networks.

The Mobile Wireless Internet Forum (MWIF) was established in February 2000 to drive development of all-IP-based architecture for cdma2000-based networks.

15.2.9 ITU

Figure 15.9 shows the organization of ITU for the IMT-2000 standardization. The telecommunications standardization in the ITU is divided into two sectors: Radiocommunications Sector (ITU-R) and Telecommunication Standardization Sector (ITU-T). The Intersector Coordination Group (ICG) is coordinating the IMT-2000 radio and network standards. The IMT-2000 radio aspects are being standardized in the Task Group (TG) 8/1 in ITU-R. TG8/1 is also responsible for the overall system architecture for IMT-2000 within ITU. In ITU-T, SG8 has been identified as the lead study group for coordination of standardization activities on IMT-2000 network standards. For more information on the ITU standardization, see [1].

After the 3GPP standardization body was founded, the role of the ITU shifted towards setting global frameworks for requirements, spectrum allocations, interworking, and so on, rather than to the detailed drafting of standards.

15.3 REGULATION

Fundamentally, regulatory matters are national issues. In some cases, however, regional and global harmonization of regulations and policy issues is desirable [20]. Areas where common regulations may be required are licensing, services, provisioning, interconnection, infrastructure, frequencies, numbering, security, and policies [20]. With regard to a radio access system, the most important regulatory areas are spectrum coordination and licensing.

The process of obtaining frequencies can be divided into three phases: identification of spectrum, allocating the spectrum for specific purposes, and licensing the spectrum. Because the identified spectrum is usually already used for some other purposes, it needs to be cleaned from existing users before it can be used for the new purpose. Therefore, proactive long-term planning is required to ensure the availability of spectrum when required.

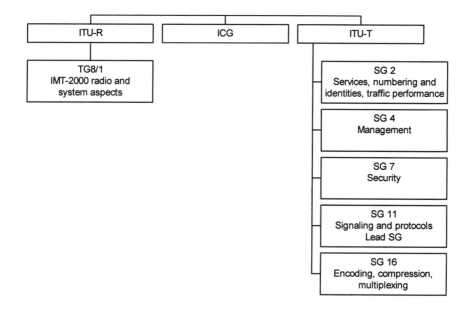

Figure 15.9   IMT-2000 standardization structure in ITU.

ITU plays an important role in spectrum regulation. The Radio Regulations (RR) of ITU are updated in World Radio Conference (WRC). The national regulators are not bound to follow the ITU guidelines for spectrum allocation. The ITU RRs, however, form a tool to encourage them to do that in order to achieve a global harmonization of spectrum [20]. The IMT-2000 spectrum was identified in 1992 by WARC92 as a result of ITU studies on IMT-2000. These studies indicated that the minimum spectrum for IMT-2000 should be 230 MHz.

Spectrum is allocated for a specific purpose. It has to be decided what services can be provided, what technology is allowed to be used, and how the spectrum is allocated between operators. In addition to the IMT-2000 services, it might be desirable to provide some other services such as broadband data for indoor service within the IMT-2000 spectrum [21]. Whether this should be allowed or not must be decided. The spectrum use can be bound to a specific standard, or the operator may be allowed to use any technology to deliver their services. The latter approach was taken in the U.S. PCS spectrum assignments. The spectrum can be allocated on an exclusive basis or by using spectrum sharing. For an exclusive allocation, what is the minimum bandwidth for a single operator and what is the maximum number of operators that can be accommodated within the given frequency allocation?

The last phase in spectrum allocation is licensing. It has to be decided who can operate the networks. This can happen by "beauty contests" (i.e., who is evaluated to be most suitable to run a network), by lotteries, or by auctions. It has been argued whether the spectrum auction is a fair and effective way of assigning licenses for third-generation mobile operators [21]. Furthermore, because spectrum is so scarce, it is a valuable asset. Thus, governments are eager to cash in that asset, as was seen in the United States for the PCS licenses. If it is decided to auction the spectrum, it has to be decided how auctions will be held and who can bid. The are three options for auctions: multiple round open auctions (used in the United States for the PCS frequencies), sealed bid auction, or royalty auction, where the payment takes form of a percentage of qualifying revenue. An important question is whether an existing operator can bid for the third-generation license. This has to be weighed against the competitive advantage they might gain if they already have a nationwide coverage while the new operators have to build the coverage from scratch. The regulation could require national roaming between operators to alleviate the newcomers' position [21]. The regulation can also rule what the conditions are for a license. These include coverage requirements and the length of a license.

In general, Europe follows ITU recommendations for spectrum issues. The Europe-wide harmonization is carried out by the Conference of European Post and Telecommunications (CEPT). The European Commission can issue a directive to create harmonized frequency allocations for specific technologies. Examples of such directives are GSM in 1987 and DECT in 1991.

The commission also published its proposed UMTS decision, which planned to set in place timescales and actions for national licenses and spectrum harmonization by 2000 [22]. The European Radiocommunications Committee (ERC) of CEPT makes decisions which usually form the basis for the harmonized spectrum designations. CEPT has designated most of the IMT-2000 spectrum for UMTS with the adoption of the ERC decision on UMTS, which identifies a total of 155 MHz of spectrum for terrestrial UMTS services, with an additional 60 MHz set aside for UMTS satellite services [23].

In the United States, the Federal Communications Commission (FCC) is managing the spectrum issues. The Inter-American Telecommunication Commission (CITEL) is issuing recommendations on spectrum issues for North and South America. The recommendations of CITEL are nonbinding but carry significant weight. The third-generation spectrum for the most part is already used by PCS systems. The United States, however, is currently developing spectrum positions for the third generation including the identification of new frequency bands.

In Japan, the Ministry of Post and Telecommunication (MPT) is responsible for spectrum regulation. The Japanese spectrum plan follows the ITU recommendation for IMT-2000 frequencies.

15.3.1 UMTS Spectrum Licensing

The UMTS spectrum licensing started in Finland where four UMTS licenses were issues using the "beauty contest" method in March 1999. Spain has also issued four licenses in March 2000. The most attention so far has been paid to the spectrum auction of the United Kingdom where five licenses were auctioned with a total price of £22.5 billion. Because of higher revenues than expected, other governments are also considering spectrum auctions.

REFERENCES

[1] 3GPP Web site http://www.3gpp.org/
[2] ITU Web site, http://www.itu.int/imt/
[3] ETSI Web site, http://www.etsi.fr/
[4] UMTS Task Force Report, February 21, 1996.
[5] The UMTS Forum Web site, http://www.umts-forum.org/
[6] GSM MoU Web site, http://www.gsmworld.com/
[7] Dimolitsas, S., "Standards Setting Bodies," in *The Mobile Communications Handbook*, J. D. Gibson (ed.), Boca Raton: Florida: CRC Press, 1996.
[8] TIA Web site, http://www.tiaonline.org/
[9] T1 Web site, http://www.t1.org/
[10] CDG Web site, http://www.cdg.org/
[11] UWCC Web site, http://www.uwcc.org/
[12] Donegan, P., "Domestic Rivalries Unsettle Japan's Third Generation Push," *Mobile Communications International*, No. 40, April 1997, pp. 51–53.
[13] ARIB Web site, http://www.arib.or.jp/arib/english/index.html
[14] TTC Web site, http://www.ttc.or.jp/
[15] TTA Web site, http://www.tta.or.kr/
[16] ETRI Web site, http://www.etri.re.kr/
[17] RAST Web site, http://www.rast.etsi.fi/
[18] Kohno, R., and M. Nagakawa, "International Cooperative Research and Development of Wireless Personal Communications in Asian-Pacific Countries," *IEEE Personal Communications*, Vol. 4, No. 2, April 1997, pp. 6–12.
[19] ITU Circular Letter 4 April 1997 (8/LCCE/47), "Request for Submission of Candidate Radio/Transmission Technologies (RTTs) for IMT 2000/FPLMTS Radio Interface."
[20] Leite, F., R. Engelman, S. Kodama, H. Mennenga, and S. Towaij, "Regulatory Considerations Relating to IMT-2000," *IEEE Personal Communications*, August 1997, pp. 14–19.
[21] U.K. Department of Trade and Industry, "Mobile Communications on the Move," *A Consultation Document from the Department of Trade and Industry*, July 1997, can be found from Web site http://www.open.gov.uk/radiocom/rahome.htm.
[22] European Commission, "Proposal by the Commission for a Decision of the European Parliament and Council on the co-ordinated introduction of mobile and wireless communications (UMTS) in the Community," 1997.
[23] "European Radiocommunications Committee ERC Decision of 30 June 1997 on the frequency bands for the introduction of the Universal Mobile

Telecommunications System (UMTS)," ERC/DEC/(97)07 can be found on the ERO Web site http://www.ero.dk/.

Chapter 16

FUTURE DIRECTIONS

16.1 INTRODUCTION

In this chapter, we discuss future directions of third-generation systems. In the earlier chapters we described the different third-generation standards and technologies. Third-generation systems, however, will not be deployed in isolation. Several other systems will coexist with third-generation systems and thus impact their development.

Section 16.2 discusses the technology landscape of third-generation systems and introduces different competing and complementing systems. Section 16.3 introduces enabling technologies. In Section 16.4 the third-generation market and the role of second-generation systems in establishing it are discussed. Section 16.5 discusses different evolution scenario. A third-generation service/application scenario is described in Section 16.6. The chapter is concluded in Section 16.7.

16.2 TECHNOLOGY LANDSCAPE

In this section, different complementing and competing technologies for third generation are presented. Figure 16.1 shows how different technologies are positioned with respect to data rate and range. WCDMA FDD provides wide area coverage with a maximum data rate of 2 Mbps, while WCDMA TDD is restricted to a smaller range. EDGE also provides wide area coverage but with smaller data rates as compared to WCDMA. WLANs offer very high data rates but only local area coverage. Bluetooth provides data rates up to 1 Mbps with a very limited coverage.

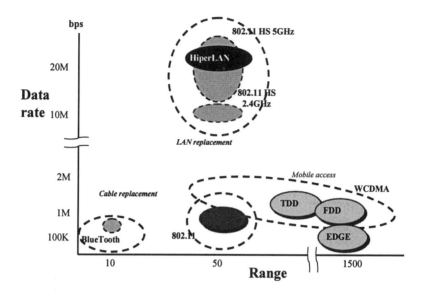

Figure 16.1   Positioning of different wireless technologies with respect to range and data rates.

16.2.1 Cellular Air Interface Evolution

A need for more bandwidth and the cost pressure calling for higher spectrum efficiency drive the development of new air interface technologies. The higher the throughput in the network, the better it is from the end user's perspective. On the other hand, spectrum efficiency determines how efficiently the bandwidth can be delivered to the end user and is of primary concern for the mobile operator.

To increase throughput, existing air interfaces will be stretched as far as is technically and economically possible. Figure 16.2 depicts this development. The common feature-of all of these developments was the evolution of the standard within the originally designed waveform using the change-of-modulation technique. Limiting factors have been receiver complexity and transmitter linearity requirements. As can been seen from Figure 16.2, mobile systems have three standard carrier bandwidths; 200 kHz, 1.25 MHz, and 5 MHz, using time division multiple access (TDMA) and code division multiple access (CDMA) techniques.

The US-TDMA system, also known as the digital advanced mobile phone system (DAMPS), and IS-136 (Interim Standard) first evolved within the 30-kHz carrier using multislot and higher order modulation (8-PSK, phase shift keying) techniques. After the evolution of the 30-kHz channel became too difficult, the US-TDMA community selected the enhanced data rates for global evolution (EDGE) as its future air interface evolution. Global System for Mobile Communications (GSM) also used multislot techniques (high-speed circuit-switched data, HSCSD, and general packet radio service, GPRS) and higher order modulation (8-PSK, EDGE) to evolve the 200-kHz carrier. The next step could be multicarrier EDGE.

The main third-generation standard is wideband CDMA, commonly known as WCDMA. WCDMA high-speed packet access (HSPA), intended to be used in the downlink, uses a higher order modulation to increase the maximum data rate from 2 Mbps up to 10 Mbps. WCDMA time division duplex (TDD) will be used in the so-called unpaired frequency bands allocated for third-generation networks.

The IS-95A standard, also known as cdmaOne, evolved into IS-95B, which uses parallel CDMA codes to create high-speed data services. The next step in the IS-95 evolution is cdma2000. 1X (1X means that the carrier bandwidth is the same as that in the IS-95 standard, namely 1.25 MHz) adds coherent detection into the uplink to increase radio performance and data rates. 1X-EV (evolution) uses, in addition to parallel CDMA codes, higher order modulation techniques to further increase the maximum data rates up to 2.4 (data only) or 5.4 Mbps (voice and data).

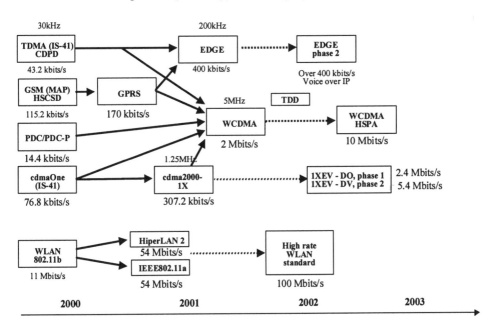

Figure 16.2 Evolution of mobile radio standards.

16.2.2 Wireless LAN (WLAN)

Figure 16.3 depicts the different WLAN standards. The 2-Mbps IEEE WLAN standard, IEEE 802.11 [1], was released in 1997. It operates in ISM band (2.4 GHz) and offers three physical interfaces: direct sequence (DS), frequency hopping (FH), and infrared (IR). In about the same time, the 20-Mbps HIPERLAN/1 standard [2] for the 5-GHz band was released by ETSI [3]. 802.11 has had moderate commercial success, while HIPERLAN/1 has not appeared in the market at all.

Very soon afterwards, a standardization of new high-rate versions of these standards began. It is IEEE 802.11a for the 5-GHz NII band and 802.11b for the 2.4-GHz ISM band. In Europe, it is ETSI BRAN HIPERLAN/2 for 5 GHz. It is worth noting that 802.11a and HIPERLAN/2 (and MMAC in Japan) have harmonized PHY layers, all OFDM with 54 Mbps.

The difference between 802.11a and HIPERLAN/2 lies in the higher layers. 802.11a/b is a wireless Ethernet with carrier sense multiple access/collision avoidance MAC. (Ethernet uses collision detection (CD).) HIPERLAN/2 is centralized, using TDD/TDMA and thus allows for efficient quality of service support over the radio interface.

Technical standardization work on first releases of 802.11a/b and HIPERLAN/2 has been completed.

Figure 16.3   WLAN standards.

16.2.3 Bluetooth

The Bluetooth (BT) standard is owned by the Bluetooth Special Interest Group [4]. BT was originally intended to offer very short range (10m) wireless replacement for serial links between devices, such as a mobile phone and a headset. As the work on the standard progressed, several scenarios emerged: data and voice access point, cable replacement, and personal ad hoc networking.

BT works in the 2.4-GHz IMS band using frequency hopping of 1600 hops/s with modulation of 1 Msymbol/s (GFSK). BT was adopted by IEEE 802.15 (Personal Area Networks) [5] as a starting point for further development. There are three important changes/improvements that 802.15 is probably going to contribute: a backwards compatible (bi-modal) high rate WPAN of 20 Mbps, up-scaleable system to 5.8-GHz band, and IEEE 802-conformant protocol stack.

16.2.4 Satellites

Satellites can provide coverage to remote, rural, and maritime areas where it would be too expensive or impossible to set up terrestrial infrastructure. Satellite-based systems providing mobile services are termed mobile satellite services (MSS) or satellite personal communications networks (S-PCN).

The planned S-PCN systems use nongeostationary orbits (non-GEO). Non-GEOs can be subdivided into low earth orbit (LEO) and medium earth orbit (MEO), also called intermediate circular orbit (ICO), systems. LEO systems have altitudes from 800 to 1600 km and MEO systems around 10,000 km. Examples of non-GEO systems are Globalstar, Iridium, Loral, and Odyssey [6]. It should be noted that these systems are being deployed within the time frame of 1998–2002 (i.e., close to IMT-2000 deployment). The first services are speech and data up to 9.6 Kbps. The European third-generation satellite activities are addressed in [7].

16.3 ENABLING TECHNOLOGIES

The applications and terminals need to be developed to a new level in order to utilize the extensive capabilities of third-generation networks. The improvement in the processing power of DSPs and ASICs makes the radio part smaller and smaller, thus creating room for additional devices such as better displays, digital cameras, and many others. The increased software, higher number of application devices, and higher data call for improvements in battery technologies. Solar power technology might be used as a complementing energy source.

Digital cameras facilitate digitization of images. Although IMT-2000 will bring significant improvements to data rates, multimedia applications such as image transmission will require even higher bandwidths. Thus, efficient compression technologies are vital. Fractal coding is a new technology that could facilitate a breakthrough in this area [8].

Users viewing multimedia information will want to do it through a high-quality, high-resolution display. Mobile users use their terminals in very variable lighting conditions. Consequently, the display has to be capable of presenting the picture clearly regardless of external lighting. Of course, size is of essential importance for portable devices.

In order to capitalize on mass market, user interfaces of wireless devices must be developed far beyond today's standards. Applications have to be easy to use, non-technical, and understandable to a lay person. Voice recognition is one possible technique that can help with building user-friendly applications. Virtual reality is used to create a virtual environment for one user: a mobile user could imitate office conditions, for example, in a hotel room and could see the others in a realistic meeting environment. Interactive virtual reality opens new possibilities for developing more attractive games that can be played against other users over the wireless link.

Context-aware applications are applications that change their behavior according to the user's present context—their location, who they are with, the time of day, and so on [9]. In wireless systems, the context information can be used to either transfer information from the network to the user, or transfer information from the user to the network. One challenge to implement such applications is how to sense the user's context (location, surrounding people, time of the day, etc.). The Global Positioning System (GPS) is one way of locating users, but it may not be accurate enough. Another possibility is to use the location information from the mobile radio system. The environment can be sensed by active badges, attached to people, identifying their carrier [9]. A standard way of creating context-aware applications (i.e., creating an actual program triggered by the context) has been presented in [9]. The methodology is based on a language similar to HTML, which is used to create World Wide Web pages. The language is used to create *notes* that can then be executed based on the context. The context and notes can be passed from the user terminal to the backbone network and vice versa. An example would be when a user enters a certain area, notes for that area are passed from the network and then triggered by the terminal when arriving at specific locations within the given area. GSM's Short Message Service (SMS) has been used to pass context and notes between a GSM terminal and backbone network [9]. Some of the features for context-aware applications can be realized by the multicast feature of wireless networks (i.e., information is transmitted only within certain areas, as for price information in a supermarket). A more advanced environment for context-aware applications, however, can take factors other than location into account, such as surrounding people, and thus, the applications can be tailored for individual users.

Operating systems form a platform for application development. EPOCH is an operating system for wireless information devices such as communicators and smart phones. Windows CE is an operating system developed for palmtop computers. The new version 2.0 adds wireless connectivity to this product. Sun Personal Java 1.0 is a similar type of product for designing application programming interfaces (API). The real value of these products is that they create platforms for application development with which many are already familiar. These developments indicate that wireless Web services are going to be taken seriously. The key to mass market development is to create open interfaces for application developers.

A very significant development for creating wireless connectivity is the wireless application protocol (WAP), which aims for a global wireless network language and protocol stack. This would facilitate a common platform for developing applications for wireless Internet devices. The applications developed on top of WAP can run on future bearers and devices. The WAP Forum was created with the goal of making WAP an industry standard [10].

The new type of content places requirements for new types of charging mechanisms. Minute-based charging is not appropriate for data calls, but, for example, volume-based charging could be used. The importance of security will increase along with the number of users in wireless networks, as computer fraud and wireless fraud are serious problems. Over-the-air security and end-to-end security are important.

16.4 MARKET CONSIDERATIONS

In this section, we discuss the wireless market development. Prediction of the future is always a difficult task, and the wireless communications field is not an exception to this rule. Long-term market projections for mobile radio systems are difficult when companies are struggling even with short-term projections. In this section we present some subscriber predictions for third-generation systems in order to assess potential users at the time of the deployment of third-generation systems. In addition, these predictions performed at the time of first edition of this book are compared against the actual market developments.

There are considerable differences between several market estimates for mobile users. The UMTS Forum has predicted that there would be 1.7 billion mobile users by 2010. This is 20% of the world's user population. The UMTS Forum also estimates that 45% of the mobile users will use high-speed data services by 2010. It is clear, however, that there is no single right figure for the IMT-2000 market. The development of the third-generation market depends on several factors including coverage, tariffing, lifestyle changes, and terminal offerings. Actually, the bold figure of 1.7 billion seems now quite natural since there will be one billion subscribers already by 2002.

Table 16.1 presents world subscriber growth for different compounded annual growth percentages for 1998–2001. In addition, the actual or latest estimate of the growth for these years is presented. According to [11], growth of 40% to 45% is considered most plausible and actually this has been the case during recent years. Between 1996 and 1997, the annual growth worldwide averaged 51.6%, and since 1990, compounded annual growth has averaged 50.9%.

The market for mobile multimedia services will depend on several factors. The market analysis group in the UMTS Forum has conducted a study that analyzed four different market scenarios for the mobile multimedia market [8]. These scenarios give an indication of the factors that will determine market development. A short summary of these scenarios follows.

Table 16.1

Forecast World Subscriber Growth (Millions)

	Assumed compounded annual growth					
Year	25%	30%	35%	40%	45%	Actual/latest est.
1998	258	268	279	289	299	311
1999	322	349	376	404	434	467
2000	403	453	508	566	629	632
2001	504	589	685	793	912	792

Source: [11] and Strategy Analytics.

Scenario 1, slow evolution. Mobile multimedia development is slow because of limited applications and high service and terminal prices. Unsuccessful liberalization has led to low competition. Fragmented standards have resulted in no economies of scale. In addition, fragmented standards hinder global application development. No global standard has been achieved, which has led to high equipment prices. Applications are difficult to reconfigure and to adapt to personal needs, and there have been no breakthrough developments for critical terminal technologies such as displays.

Scenario 2, business centric. Mobile multimedia takes off in the business sector but not in the consumer sector, as there is a lack of innovation in consumer applications. A lack of competition because of unsuccessful liberalization contributes to high premiums to access the service. High access costs, however, can be afforded by business sector users. Terminals are intelligent and configurable to user needs but still difficult to use.

Scenario 3, sophisticated mass market. A mass market for mobile multimedia has emerged. Terminals and applications have a large set of features and can be customized for personal needs. User interfaces, however, are still rather complicated, being suitable for IT-literate users. Liberalization has resulted in high competition and thus low access premiums. Global standards for both systems and application platforms have facilitated low equipment costs and rich application developments. Spectrum was auctioned, and the resulting costs are reflected in initially high prices that quickly drop.

Scenario 4, commoditized mass market. A real mobile multimedia mass market has emerged and comprises both business and consumer users. Liberalization and adoption of global standards have resulted in economies of scale. Spectrum is cheap; simple and cheap terminals are available from the beginning.

Table 16.2 shows the number of mobile users and multimedia users by 2005 for these different scenarios. The essential differences between the first two and the last two scenarios concern the worldwide platform in radio and traffic delivery and in application. The common platform facilitates economies of scale and sets a standardized basis for application development. The main difference between scenarios 1 and 2 is that in the second scenario users can configure their terminals according to their needs because they understand IT. In the first scenario, the user interface is more limited and thus not adequate even for most business users. The main difference between scenarios 3 and 4 is the higher initial cost in scenario 3 because of spectrum pricing and the price and ease of use for terminals. In scenario 3, spectrum is auctioned and terminals are bulky and expensive in the beginning. Furthermore, in scenario 3, tariffing is initially high because of high investment into spectrum and equipment. These factors contribute to the slower development of mass market as with scenario 4.

The potential new users for voice for data services can also be viewed from the perspective of the penetration rate by 2000. The higher the penetration rate of a service, the less potential there is for new business. On the other hand, a very high penetration rate might drive a deployment of a new system to support the required capacity.

Table 16.2
Number of Mobile and Multimedia Users in Europe by 2005

Scenario	Mobile users by 2005 (penetration)	Multimedia users by 2005
Slow evolution	82 M (22%)	7.5 M
Business centric	82 M (22%)	9 M
Sophisticated mass market	123 M (35%)	19 M
Commoditized mass market	140 M (40%)	27 M

Source: [8].

If we look the situation today, it seems that regarding second-generation development we are heading towards commoditized mass market. Despite the high auction prices, the same is expected to happen for the third-generation market. The penetration figures are actually very conservative. In Europe 40% penetration has already been exceeded in 1999, and in 2005 the penetration is expected to be between 80% and 90%.

One of the biggest drivers of the wireless multimedia market will be wireless gaming and entertainment. It has been estimated (Datamonitor) that by 2005, 200 million people will be playing Internet games on mobile phones and that revenue from wireless gaming will stand at $600 million by 2003, rising to $6 billion by 2005.

Second-generation systems are also tools to educate operators and customers on the data market. For example, operators can test the impact of tariffing on service demand, without the heavy investments required for third-generation services. When users are already used to new types of services, adoption of data services will be easier. For many services, third generation will just mean better quality of service (i.e., faster service and higher reliability) when compared to services provided by second-generation systems. Thus, second-generation systems can be viewed as a logical stepping stone towards third-generation systems, and certainly not something we need to discard as quickly as possible.

16.5   DEVELOPMENT SCENARIOS

In this section we discuss possible scenarios of how wireless systems could develop. The most important factor in the development is how wireless data in general will develop (i.e., how the customers will start to use wireless data services). In this respect, the application development and end user terminal developments are critical.

Originally, the cdma2000 and WCDMA air interface proposals were poised to continue the competition between the IS-95 and GSM standards. However, the competitive landscape is now becoming more blurred. In the United States, EDGE has emerged as a viable evolution path not only for GSM operators but also for TDMA operators. In Asia, the IS-95 operators are considering the deployment of either WCDMA or cdma2000 as the third-generation air interface. Korean IS-95 operators have announced their support for WCDMA. This seems to be pointing to WCDMA as the main third-generation air interface choice in the new frequency bands and to EDGE and cdma2000 1X for the evolution of the current frequency bands.

WCDMA TDD technology has been designed to be used in the unpaired frequency bands of the third-generation spectrum. Release 2000 (now called Release 4) will be the first release to contain full TDD specifications. A critical factor in TDD development will be the availability of dual mode terminals. The focus of TDD development seems to be on the indoor radio environment. However, it has also been proposed for outdoor use. The initial third-generation deployments will focus on FDD bands. TDD has been positioned as the capacity extension for FDD.

In the core network evolution, the big question is how fast IP will emerge. How much backwards compatibility will be retained to the existing systems such as GPRS and how many new revolutionary items will be adopted? This will depend on both the success of current systems such as GPRS and the technology development.

The roles of WLANs and Bluetooth are interesting in the indoor environment where they provide high data rates, much higher than wideband CDMA schemes at the moment. However, the WCDMA High Speed Packet Access (HSPA) proposal will increase WCDMA's maximum data rate significantly up to 10 Mbps. Integration of WLANs with cellular systems could provide users with seamless services regardless of the air interface used. Availability of multimode terminals is critical for this kind of system.

16.6   A WIRELESS DAY

Mike wakes up in the morning in Dallas and makes a video call to his wife Maureen, who is traveling and is on the other side of the globe in Tokyo. She answers from her hotel room and high-quality video connects them in a nice discussion. After the call, while getting ready for work, Mike wakes up his children, Amy and Matt, who want to first connect their Tamagotchi virtual pets with their friends to exchange vital information about their pets' dreams. This happens using improved real-time short message service, consuming hardly any bandwidth, and thus being cheap.

During his lunch break, Mike contacts his friend to play an interactive virtual reality video game through the wireless network. Bursts of images are transmitted through the wireless link. On the way back home Mike remembers that he needs to transmit his latest technical report to his boss. He stops on the road, connects to the corporate intranet, and a 10-Mbytes engineering document is transferred instantly through the IMT-2000 wireless connection. After driving a while, his car stops. What now? he wonders. There seems to be something wrong. He connects through video to his car service to get instructions. He focuses the digital camera in his wireless device into the car alarm lights. This view is transmitted to the car service, and based on it, they give him instructions on how to fix the problem.

Finally at home, Mike reads his e-mail. Something from Mom, he notices. A photograph taken in Hawaii, where she is on holiday. She took the photo with a digital camera and said, "Send this to my son." Her IMT-2000 wireless terminal compressed the image, connected to the backbone network, and e-mailed the image to Mike over the wireless link.

16.7 CONCLUDING REMARKS

As we postulated in [11], human nature, characterized by zero-tolerance and a need for instant gratification, is driving the development of wireless communication systems to add more functionality and higher bit rates. In the future, user devices will enable communications using all senses: hearing, vision, and touch. New services, impossible to predict today, will emerge encouraged by the open platform provided by the IP-based network. Air interface development will be based on the evolution of currently known waveform structures. Future network architectures will be centered on two scenarios; hierarchical versus distributed, with the distributed architecture as the dominating one.

Finally, we predict that the most exciting development of the next 20 years will be the emergence of a new technology for communications, not based on currently known radio [12]. This is something that is impossible to imagine today as the current use of mobile phones was 100 years ago.

REFERENCES

[1] LAN Medium Access Control (MAC) and Physical Layer (PHY) Specifications, IEEE Std. 802.11, 1997.

[2] "Radio Equipment and Systems (RES); High Performance Radio Local Area Network (HIPERLAN) Type 1; Functional specification" ETS 300 652.

[3] ETSI Web site, http://www.etsi.org/

[4] Bluetooth SIG Web site, http://www.bluetooth.com/

[5] IEEE 802.15 Working Group for Wireless Personal Area Networks, http://grouper.ieee.org/groups/802/15/

[6] Comparetto, G., and R. Ramirez, "Trends in Mobile Satellite Technology," *IEEE Computer*, February 1997, pp. 44–52.

[7]   Guntsch, A., M. Ibnkahla, G. Losquadro, M. Mazzella, D. Rovidas, and A. Timm, "EU's R&D Activities on Third-Generation Mobile Satellite Systems (S-UMTS)," *IEEE Personal Communications*, Vol. 36, No. 2, February 1998.

[8]   The UMTS Market Aspects Group, *UMTS Market Forecast Study*, 1997.

[9]   Brown, P. J., J. D. Bovey, and X. Chen, "Context-Aware Applications: From the Laboratory to the Marketplace," *IEEE Personal Communications*, October 1997, pp. 58–63.

[10] WAP Forum Web site, http://www.wapforum.org/

[11] Shosteck, H., "The Explosion in World Cellular Growth," *The 2nd Annual UWC Global Summit*, Vancouver, British Columbia, Canada, April 15–17, 1998.

[12] Ojanpera, T., contribution in Webb, W., *Wireless Communications—Next 20 Years*, Norwood, Massachusetts: Artech House, forthcoming in 2001.

ABOUT THE AUTHORS

Tero Ojanperä, Nokia Networks vice president, is responsible for research, standardization, and technology within IP Mobility Networks at Nokia Networks. The IP Mobility Networks business division covers a full portfolio of radio access networks (e.g., GSM, GPRS, EDGE, TDMA/EDGE, WCDMA, and WLAN), broadband wireless systems (point-to-point radio links, point-to-multipoint radio links, and wireless mesh networks), and core networks (GSM, GPRS, and all-IP). He was born in Korsnäs, Finland, on November 12, 1966. He received his M.Sc. degree from the University of Oulu, Finland, in 1991 and his Ph.D. degree from Delft University of Technology, The Netherlands, in 1999.

He joined Nokia Mobile Phones, Oulu, Finland, in 1990 and worked as a research engineer from 1991 to 1992. From 1992 to 1995, he led a radio systems research group concentrating on wideband CDMA, GSM WLL, and US TDMA. From 1994 to 1995, he was also a project manager of the wideband CDMA concept development within Nokia. Later, this concept formed the basis for the FRAMES wideband CDMA.

From 1995 to 1997, Dr. Ojanperä was a research manager in the Nokia Research Center, Helsinki, Finland, heading Nokia's third-generation air interface research program, which consisted of several air interface projects such as wideband CDMA, packet data, radio resource management algorithms, OFDM, and wideband TDMA. He was actively involved in setting up the FRAMES project and during 1995–1996, he was also the leader of the technical area air interface and the multiple access work package in the project that was responsible for the selection of the FRAMES Multiple Access (FMA) scheme. The FRAMES wideband CDMA was the basis for the UMTS WCDMA concept in ETSI. From 1994 to 1997, he was a Nokia representative for the UMTS radio interface issues in the ETSI SMG5 and SMG2 standardization committees.

From August 1997 to August 1998, he worked as a principal engineer at the Nokia Research Center in Irving, Texas. He was involved in U.S. third-generation standards activities for the cdma2000. In addition, he was involved in technical/strategic work for Nokia's proposal for the UWC-136.

In September 1998, Dr. Ojanperä joined Nokia Telecommunications (now renamed Nokia Networks), Finland, as the head of research, radio access systems. In July 1999 he was promoted to vice president, radio access systems research. In 1999 he was also assigned as general manager of Nokia Networks, Korea.

Dr. Ojanperä has authored several conference and magazine papers as well as a chapter in three books: *Wireless Communications TDMA versus CDMA* (Kluwer Academic Publishers, 1997), *GSM: Evolution towards 3rd Generation* (Kluwer Academic Publishers, 1998), and *The Mobile Communications Handbook*, Second Edition (CRC Press, 1999). He has three patents. During 1996–1997, he was a member of the Nokia patent committee and during 1997–1998, he was a member of the patent committee Nokia Mobile Phones.

Dr. Ojanperä is a member of the IEEE and in 1996 he was secretary of the IEEE Finland Section. He was a member of the Technical Program Committee of the Vehicular Technology Conference 1999, and is vice chair for the International Communications Conference (ICC) 2001.

Ramjee Prasad was born in Babhnaur (Gaya), Bihar, India on July 1, 1946. He is now a Dutch citizen. He received his B.Sc. (Eng.) from Bihar Institute of Technology, Sindri, India, and his M.Sc. (Eng.) and Ph.D. degrees from Birla Institute of Technology (BIT), Ranchi, India, in 1968, 1970, and 1979, respectively.

He joined BIT as a senior research fellow in 1970 and became an associate professor in 1980. While he was with BIT, he supervised a number of research projects in the area of Microwave and Plasma Engineering. During 1983–1988 he was with the University of Dar es Salaam (UDSM), Tanzania, where he became professor in telecommunications in the Department of Electrical Engineering in 1986. At UDSM, he was responsible for the collaborative project "Satellite Communications for Rural Zones" with Eindhoven University of Technology, The Netherlands. From February 1988 until May 1999 he was with the Telecommunications and Traffic Control Systems Group at the Delft University of Technology (DUT), The Netherlands, where he was actively involved in the area of wireless personal and multimedia communications (WPMC). He was the founding head and program director of the Center for Wireless and Personal Communications (CEWPC) of International Research Center for Telecommunications–Transmission and Radar (IRCTR). Since June 1999, Dr. Prasad has been with Aalborg University, Denmark, as codirector of the Center for PersonKommunikation (CPK) and holds the chair of wireless information and multimedia communications. He was involved in the European ACTS project FRAMES (Future Radio Wideband Multiple Access Systems) as a DUT project leader. He is a project leader of several international industrial-funded projects. He has published over 300 technical papers, contributed to several books, and authored, coauthored, and edited five books, *CDMA for Wireless Personal Communications*, *Universal Wireless Personal Communications*, *Wideband CDMA for Third Generation Mobile Communications*, *OFDM for Wireless Multimedia Communications*, and *Third Generation Mobile Communication System*s, all published by Artech House. His current research interest lies in wireless networks, packet communications, multiple-access protocols, advanced radio techniques, and multimedia communications.

Dr. Prasad has served as a member of the advisory and program committees of several IEEE international conferences. He has also presented keynote speeches, and delivered papers and tutorials on WPMC at various universities, technical institutions, and IEEE conferences. He was also a member of European cooperation in the scientific and technical research (COST-231) project dealing with "Evolution of Land Mobile Radio (including personal) Communications" as an expert for the Netherlands, and a member of the COST-259 project. He was the founder and chairman of the IEEE Vehicular Technology/Communications Society Joint Chapter, Benelux Section, and is now the honorary chairman. In addition, Dr. Prasad is the founder of the IEEE Symposium on Communications and Vehicular Technology (SCVT) in the Benelux, and he was the symposium chairman of SCVT'93.

In addition, Dr. Prasad is the coordinating editor and editor-in-chief of the Kluwer *International Journal on Wireless Personal Communications* and a member of the editorial board of other international journals including the *IEEE Communications Magazine* and *IEE Electronics Communication Engineering Journal.* He was the technical program chairman of the *PIMRC'94 International Symposium* held in The Hague, The Netherlands during September 19–23, 1994 and also of the Third Communication Theory Mini-Conference in conjunction with GLOBECOM'94 held in San Francisco, California, from November 27–30, 1994. He was the conference chairman of the *50th IEEE Vehicular Technology Conference* and the steering committee chairman of the *2nd International Symposium on Wireless Personal Multimedia Communications (WPMC)* both held in Amsterdam, The Netherlands, from September 19–23, 1999. He is the general chairman of WPMC'01, to be held in Aalborg, Denmark on September 9–12, 2001.

Dr. Prasad is also the founding chairman of the European Center of Excellence in Telecommunications, known as HERMES. He is a fellow of IEE, a fellow of IETE, a senior member IEEE, a member of The Netherlands Electronics and Radio Society (NERG), and a member of IDA (Engineering Society in Denmark).

Index

The Artech House Universal Personal Communications Series

Ramjee Prasad, Series Editor

CDMA for Wireless Personal Communications, Ramjee Prasad

OFDM for Wireless Multimedia Communications, Richard van Nee and Ramjee Prasad

Third Generation Mobile Communication Systems, Ramjee Prasad, Werner Mohr, and Walter Konhäuser, editors

Universal Wireless Personal Communications, Ramjee Prasad

WCDMA: Towards IP Mobility and Mobile Internet, Tero Ojanperä and Ramjee Prasad, editors

Wideband CDMA for Third Generation Mobile Communications, Tero Ojanperä and Ramjee Prasad, editors

For further information on these and other Artech House titles, including previously considered out-of-print books now available through our In-Print-Forever® (IPF®) program, contact:

Artech House
685 Canton Street
Norwood, MA 02062
Phone: 781-769-9750
Fax: 781-769-6334
e-mail: artech@artechhouse.com

Artech House
46 Gillingham Street
London SW1V 1AH UK
Phone: +44 (0)20 7596-8750
Fax: +44 (0)20 7630-0166
e-mail: artech-uk@artechhouse.com

Find us on the World Wide Web at:
www.artechhouse.com